MCAT® General Chemistry

2025–2026 Edition: An Illustrated Guide

Copyright © 2024
On behalf of UWorld, LLC
Dallas, TX
USA

All rights reserved.
Printed in English, in the United States of America.

Reproduction or translation of any part of this work beyond that permitted by Sections 107 and 108 of the United States Copyright Act without the permission of the copyright owner is unlawful.

The Medical College Admission Test (MCAT®) and the United States Medical Licensing Examination (USMLE®) are registered trademarks of the Association of American Medical Colleges (AAMC®). The AAMC® neither sponsors nor endorses this UWorld product.

Facebook® and Instagram® are registered trademarks of Facebook, Inc. which neither sponsors nor endorses this UWorld product.

X is an unregistered mark used by X Corp, which neither sponsors nor endorses this UWorld product.

Acknowledgments for the 2025–2026 Edition

Ensuring that the course materials in this book are accurate and up to date would not have been possible without the multifaceted contributions from our team of content experts, editors, illustrators, software developers, and other amazing support staff. UWorld's passion for education continues to be the driving force behind all our products, along with our focus on quality and dedication to student success.

About the MCAT Exam

Taking the MCAT is a significant milestone on your path to a rewarding career in medicine. Scan the QR codes below to learn crucial information about this exam as you take your next step before medical school.

| Basic MCAT Exam Information | Scores and Percentiles | MCAT Sections | Registration Guide |

Preparing for the MCAT with UWorld

The MCAT is a grueling exam spanning seven subjects that is designed to test your aptitude in areas essential for success in medicine. Preparing for the exam can be intimidating—so much so that in post-MCAT questionnaires conducted by the AAMC®, a majority of students report not feeling confident about their MCAT performance.

In response, UWorld set out to create premier learning tools to teach students the entire MCAT syllabus, both efficiently and effectively. Taking what we learned from helping over 90% of medical students prepare for their medical board exams (USMLE®), we launched the UWorld MCAT Qbank in 2017 and the UWorld MCAT UBooks in 2024. The MCAT UBooks are meticulously written and designed to provide you with the knowledge and strategies you need to meet your MCAT goals with confidence and to secure your future in medical school.

Below, we explain how to use the MCAT UBooks and MCAT Qbank together for a streamlined learning experience. By strategically integrating both resources into your study plan, you will improve your understanding of key MCAT content as well as build critical reasoning skills, giving you the best chance at achieving your target score.

MCAT UBooks: Illustrated and Annotated Guides

The MCAT UBooks include not only the printed editions for each MCAT subject but also provide digital access to interactive versions of the same books. There are eight printed MCAT UBooks in all, six comprehensive review books covering the science subjects and two specialized books for the Critical Analysis and Reasoning Skills (CARS) section of the exam:

- Biology
- Biochemistry
- General Chemistry
- Organic Chemistry
- Physics
- Behavioral Sciences
- CARS (Annotated Practice Book)
- CARS Passage Booklet (Annotated)

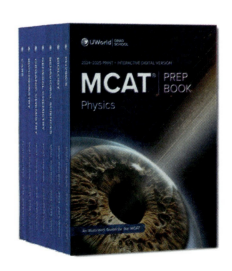

Each UBook is organized into Units, which are divided into Chapters. The Chapters are then split into Lessons, which are further subdivided into Concepts.

MCAT Sciences: Printed UBook Features

The MCAT UBooks bring difficult science concepts to life with thousands of engaging, high-impact visual aids that make topics easier to understand and retain. In addition, the printed UBooks present key terms in blue, indicating clickable illustration hyperlinks in the digital version that will help you learn more about a scientific concept.

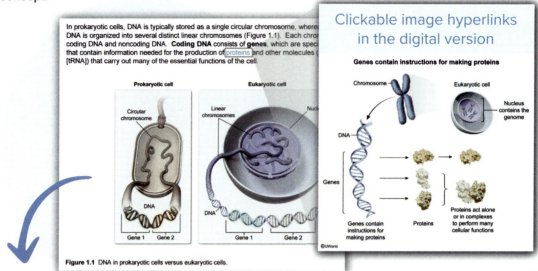

Thousands of educational illustrations in the print book

Test Your Basic Science Knowledge with Concept Check Questions

The printed UBooks also include 450 new questions—never before available in the UWorld Qbank—for Biology, General Chemistry, Organic Chemistry, Biochemistry, and Physics. These new questions, called Concept Checks, are interspersed throughout the entire book to enhance your learning experience. Concept Checks allow you to instantly test yourself on MCAT concepts you just learned from the UBook.

Short answers to the Concept Checks are found in the appendix at the end of each printed UBook. In addition, the digital version of the UBook provides an interactive learning experience by giving more detailed, illustrated, step-by-step explanations of each Concept Check. These enhanced explanations will help reinforce your learning and clarify any areas of uncertainty you may have.

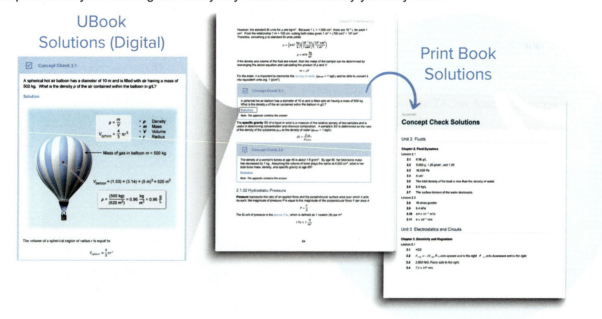

MCAT CARS Printed UBook Features

For CARS, the main book, or Annotated Practice Book, teaches you the specialized CARS skills and strategies you need to master and then follows up with multiple sets of MCAT-level practice questions.

Additionally, the CARS Passage Booklet includes annotated versions of the passages in the CARS Main Book. From these annotations, you will learn how to break down a CARS passage in a step-by-step manner to find the right answer to each CARS question.

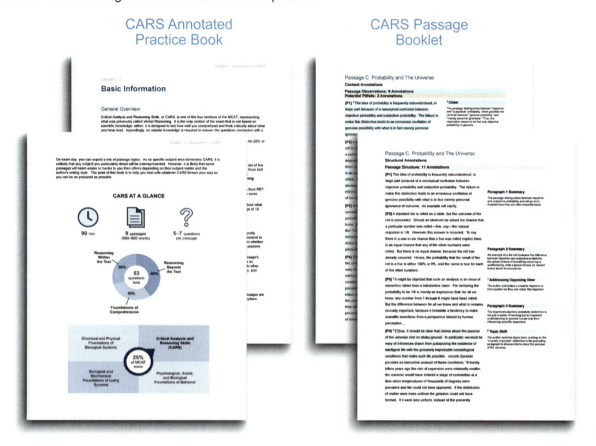

CARS Annotated Practice Book

CARS Passage Booklet

MCAT-Level Exam Practice with the UWorld Qbank

UWorld's MCAT UBooks and Qbank were designed to be used together for a comprehensive review experience. The UWorld Qbank provides an active learning approach to MCAT prep, with thousands of MCAT-level questions that align with each UBook.

The printed UBooks include a prompt at the end of each unit that explains how to access unit practice tests in the MCAT Qbank. In addition, the MCAT UBooks' digital platform enables you to easily create your own unit tests based on each MCAT subject.

To purchase MCAT Qbank access or to begin a free seven-day trial, visit gradschool.uworld.com/mcat.

Boost Your Score with the #1 MCAT Qbank

Scan for free trial

Why use the UWorld Qbank?

- Thousands of high-yield MCAT-level questions
- In-depth, visually engaging answer explanations
- Confidence-building user interface identical to the exam
- Data-driven performance and improvement tracking
- Fully featured mobile app for on-the-go review

Special Features Integrating Digital UBooks and the UWorld Qbank

The digital MCAT UBooks and the MCAT Qbank come with several integrated features that transform ordinary reading into an interactive study session. These time-saving tools enable you to personalize your MCAT test prep, get the most out of our detailed explanations, save valuable time, and know when you are ready for exam day.

My Notebook

My Notebook, a personalized note-taking tool, allows you to easily copy and organize content from the UBooks and the Qbank. Simplify your study routine by efficiently recording the MCAT content you will encounter in the exam, and streamline your review process by seamlessly retrieving high-yield concepts to boost your study performance—in less time.

Digital Flashcards

Our unique flashcard feature makes it easy for students to copy definitions and images from the MCAT UBooks and Qbank into digital flashcards. Each card makes use of spaced repetition, a research-supported learning methodology that improves information retention and recall. Based on how you rate your understanding of flashcard content, our algorithm will display the card more or less frequently.

Fully Featured Mobile App

Study for your MCAT exams anytime, anywhere, with our industry-leading mobile app that provides complete access to your MCAT prep materials and that syncs seamlessly across all devices. With the UWorld MCAT app, you can catch up on reading, flip through flashcards between classes, or take a practice quiz during lunch to make the most of your downtime and keep MCAT material top of mind.

Book and Qbank Progress Tracking

Track your progress while using the MCAT UBooks and Qbank, and review MCAT content at your own pace. Our learning tools are enhanced by advanced performance analytics that allow users to assess their preparedness over time. Hone in on specific subjects, foundations, and skills to iron out any weaknesses, and even compare your results with those of your peers.

Explore the Periodic Table

You will need to use the periodic table to answer questions on the MCAT for specific sections. Introductory general chemistry concepts constitute 30% of the material tested in the Chemical and Physical Foundations of Biological Systems section of the exam. In addition, General Chemistry constitutes 5% of the Biological and Biochemical Foundations of Living Systems section of the MCAT. Using and understanding the periodic table is a crucial skill needed for success in these sections.

1 H 1.0																	2 He 4.0
3 Li 6.9	4 Be 9.0											5 B 10.8	6 C 12.0	7 N 14.0	8 O 16.0	9 F 19.0	10 Ne 20.2
11 Na 23.0	12 Mg 24.3											13 Al 27.0	14 Si 28.1	15 P 31.0	16 S 32.1	17 Cl 35.5	18 Ar 39.9
19 K 39.1	20 Ca 40.1	21 Sc 45.0	22 Ti 47.9	23 V 50.9	24 Cr 52.0	25 Mn 54.9	26 Fe 55.8	27 Co 58.9	28 Ni 58.7	29 Cu 63.5	30 Zn 65.4	31 Ga 69.7	32 Ge 72.6	33 As 74.9	34 Se 79.0	35 Br 79.9	36 Kr 83.8
37 Rb 85.5	38 Sr 87.6	39 Y 88.9	40 Zr 91.2	41 Nb 92.9	42 Mo 95.9	43 Tc (98)	44 Ru 101.1	45 Rh 102.9	46 Pd 106.4	47 Ag 107.9	48 Cd 112.4	49 In 114.8	50 Sn 118.7	51 Sb 121.8	52 Te 127.6	53 I 126.9	54 Xe 131.3
55 Cs 132.9	56 Ba 137.3	57 La* 138.9	72 Hf 178.5	73 Ta 180.9	74 W 183.9	75 Re 186.2	76 Os 190.2	77 Ir 192.2	78 Pt 195.1	79 Au 197.0	80 Hg 200.6	81 Tl 204.4	82 Pb 207.2	83 Bi 209.0	84 Po (209)	85 At (210)	86 Rn (222)
87 Fr (223)	88 Ra (226)	89 Ac+ (227)	104 Rf (261)	105 Db (262)	106 Sg (266)	107 Bh (264)	108 Hs (277)	109 Mt (268)	110 Ds (281)	111 Rg (280)	112 Cn (285)	113 Uut (284)	114 Fl (289)	115 Uup (288)	116 Lv (293)	117 Uus (294)	118 Uuo (294)

	58 Ce 140.1	59 Pr 140.9	60 Nd 144.2	61 Pm (145)	62 Sm 150.4	63 Eu 152.0	64 Gd 157.3	65 Tb 158.9	66 Dy 162.5	67 Ho 164.9	68 Er 167.3	69 Tm 168.9	70 Yb 173.0	71 Lu 175.0
*														
+	90 Th 232.0	91 Pa (231)	92 U 238.0	93 Np (237)	94 Pu (244)	95 Am (243)	96 Cm (247)	97 Bk (247)	98 Cf (251)	99 Es (252)	100 Fm (257)	101 Md (258)	102 No (259)	103 Lr (260)

Table of Contents

UNIT 1 ATOMIC THEORY AND CHEMICAL COMPOSITION

CHAPTER 1 STRUCTURE AND PROPERTIES OF ATOMS 1
- Lesson 1.1 Atomic Structure 3
- Lesson 1.2 Nuclear Decay 9
- Lesson 1.3 Electronic Structure of Atoms 15
- Lesson 1.4 Periodic Table Organization 29
- Lesson 1.5 Atomic Properties and Trends on the Periodic Table 37
- Lesson 1.6 Measurements and Chemical Composition 47

UNIT 2 INTERACTIONS OF CHEMICAL SUBSTANCES

CHAPTER 2 CHEMICAL BONDING, REACTIONS, AND STOICHIOMETRY 59
- Lesson 2.1 Chemical Bonding 61
- Lesson 2.2 Bond Polarity and Molecular Polarity 83
- Lesson 2.3 Intermolecular Forces 99
- Lesson 2.4 Chemical Reactions 111
- Lesson 2.5 Stoichiometry of Chemical Reactions 125

UNIT 3 THERMODYNAMICS, KINETICS, AND GAS LAWS

CHAPTER 3 THERMODYNAMICS 137
- Lesson 3.1 Zeroth Law and Temperature 139
- Lesson 3.2 First Law of Thermodynamics 151
- Lesson 3.3 Second Law of Thermodynamics 155
- Lesson 3.4 Heat Transfer in Physical Changes 161
- Lesson 3.5 Heat Transfer in Chemical Changes 183
- Lesson 3.6 Calorimetry 195
- Lesson 3.7 Gibbs Free Energy 199

CHAPTER 4 KINETICS 205
- Lesson 4.1 Reaction Mechanisms 205
- Lesson 4.2 Reaction Rates 213
- Lesson 4.3 Chemical Catalysis 223

CHAPTER 5 CHEMICAL EQUILIBRIUM 227
- Lesson 5.1 Equilibrium in Reversible Reactions 227
- Lesson 5.2 Law of Mass Action 231
- Lesson 5.3 Reaction Quotient 233
- Lesson 5.4 Le Châtelier's Principle 243

CHAPTER 6 GAS LAWS 251
- Lesson 6.1 Pressure, Volume, and Temperature Relationships 251
- Lesson 6.2 Pressure and Volume Relationships to Molar Amount 259
- Lesson 6.3 Ideal Gas Model 265

UNIT 4 SOLUTIONS AND ELECTROCHEMISTRY

CHAPTER 7 SOLUTIONS 275
- Lesson 7.1 Solutes in Solution 277
- Lesson 7.2 Solubility of Solutes 285

CHAPTER 8 ACID-BASE CHEMISTRY 297
- Lesson 8.1 Definitions of Acids and Bases 297
- Lesson 8.2 Acids and Bases in Aqueous Solution 303
- Lesson 8.3 Strength of Acids and Bases 307

Lesson 8.4	Buffers	317
Lesson 8.5	Hydrolysis of Salts	323
Lesson 8.6	Acid-Base Titrations	333

CHAPTER 9 REDOX REACTIONS AND ELECTROCHEMISTRY 353

Lesson 9.1	Review of Redox Reactions	353
Lesson 9.2	Redox Titrations	355
Lesson 9.3	Electrochemical Cells	359
Lesson 9.4	Types of Electrochemical Cells	369
Lesson 9.5	Applications of Electrochemical Cells	373

APPENDIX

CONCEPT CHECK SOLUTIONS **379**

INDEX **387**

Unit 1 Atomic Theory and Chemical Composition

Chapter 1 Structure and Properties of Atoms

1.1 Atomic Structure

    1.1.01   Atomic Number and Mass Number
    1.1.02   Isotopes
    1.1.03   Atomic Mass

1.2 Nuclear Decay

    1.2.01   Alpha Decay
    1.2.02   Beta Decay, Positron Emission, and Electron Capture
    1.2.03   Gamma Emission
    1.2.04   Half-life and Activity

1.3 Electronic Structure of Atoms

    1.3.01   Bohr Model of the Atom
    1.3.02   Absorption and Emission
    1.3.03   Heisenberg Uncertainty Principle
    1.3.04   Orbitals and Quantum Numbers
    1.3.05   Pauli Exclusion Principle
    1.3.06   Aufbau Principle and Electron Configurations
    1.3.07   Magnetism and Hund's Rule

1.4 Periodic Table Organization

    1.4.01   Periods and Groups
    1.4.02   Element Classifications
    1.4.03   Named Element Groups

1.5 Atomic Properties and Trends on the Periodic Table

    1.5.01   Effective Nuclear Charge
    1.5.02   Size of Atoms (Atomic Radius)
    1.5.03   Size of Atomic Ions (Ionic Radii)
    1.5.04   Electronegativity
    1.5.05   Electron Affinity
    1.5.06   Ionization Energy

1.6 Measurements and Chemical Composition

    1.6.01   Metric Units
    1.6.02   Density
    1.6.03   Molecular Weight
    1.6.04   Mass Percent
    1.6.05   Empirical and Molecular Formulas
    1.6.06   Avogadro's Number and the Mole
    1.6.07   Compound Stoichiometry
    1.6.08   Molarity

Lesson 1.1
Atomic Structure

Introduction

Samples of matter can be broadly classified into one of two categories: pure substances or mixtures.

Pure substances contain only **one chemical species** (ie, either an element or a compound) that has a distinct identity. An **element** is a substance that cannot be separated into simpler substances by chemical interactions, and a **compound** consists of two or more chemically bonded elements that form a new substance with a new identity that can be separated back into its component elements by chemical means.

A summary of the classifications of matter is shown in Figure 1.1.

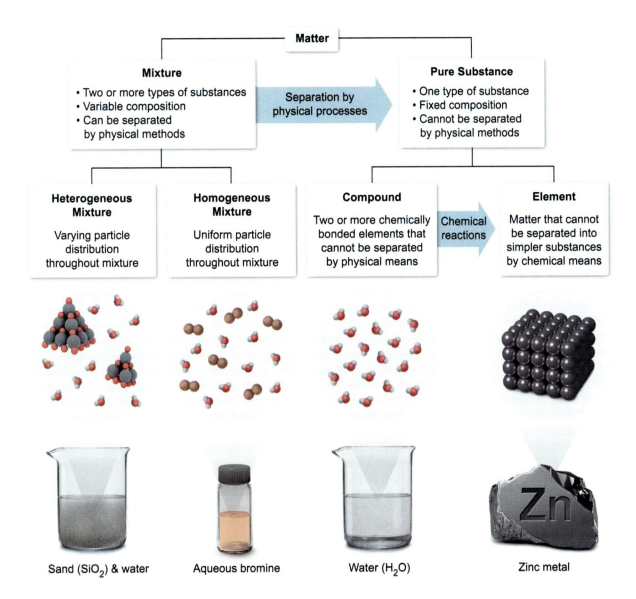

Figure 1.1 Classifications of matter.

Elements are composed of **atoms**, the smallest units into which a sample of matter can be divided without changing the sample's identity or properties. As such, the elemental level is the limit to which matter can be chemically separated. Each element is made up of its own unique type of atom, but all atoms share the same structural characteristics.

All atoms consist of a central **nucleus** containing a mixture of positively charged particles called **protons** and chargeless particles called **neutrons**, and the nucleus is surrounded by negatively charged particles called **electrons**. The various elements differ in the number of protons in their respective nuclei. A model of an atom of the element carbon is shown in Figure 1.2.

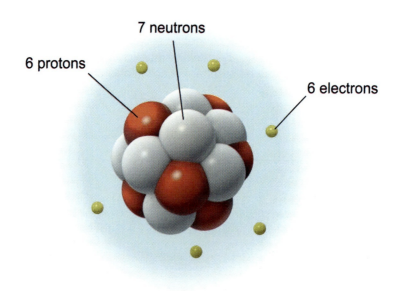

Figure 1.2 Model of a carbon atom.

1.1.01 Atomic Number and Mass Number

The **atomic number** of an atom refers to the number of protons in its nucleus and determines the *elemental identity* of the atom (ie, all atoms of the same element have the same number of protons). Atoms with different numbers of protons (ie, different atomic numbers) are different elements with different properties.

The **mass number** of an atom indicates the sum of the number of protons and neutrons contributing to the total mass of the nucleus. As such, the number of neutrons in an atom can be determined by subtracting the atomic number from the mass number.

$$\text{Number of neutrons} = \text{Mass number} - \text{Atomic number}$$

Any atom in its *elemental state* is electrically neutral (ie, no net charge) because the number of negatively charged electrons (e⁻) is equal to the number of positively charged protons. An example of the number of each type of particle in a carbon atom with a mass number of 13 is shown in Figure 1.3.

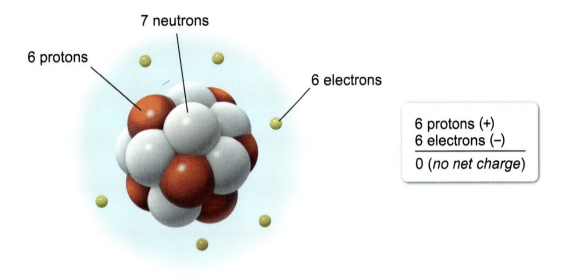

Figure 1.3 Model of a carbon atom in its elemental state with a balance of positive and negative charges.

An atom in its elemental state that gains or loses electrons becomes an **atomic ion** (ie, an atom with a net charge). A neutral atom acquires a net charge of +1 for each electron lost and a net charge of −1 for each electron gained because the numbers of protons and electrons become unequal.

For example, the number of each type of particle in an atomic ion formed from a fluorine atom with a mass number of 19 is shown in Figure 1.4. Ion formation is discussed further in Lesson 2.1.

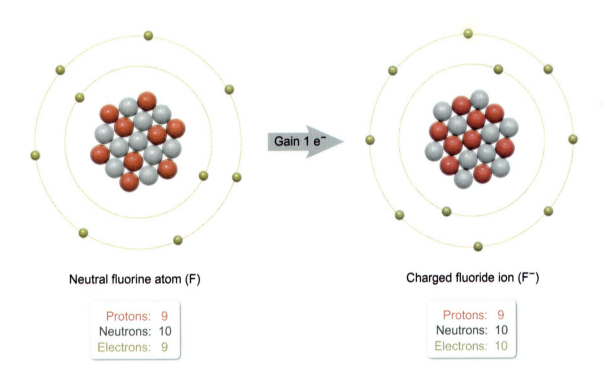

Figure 1.4 Model of a fluorine atom converted to an atomic ion from its elemental state.

> **☑ Concept Check 1.1**
>
> Atoms of calcium (Ca) ionize during reactions to form Ca^{2+} ions. How many neutrons and electrons are in a single Ca^{2+} ion that has a mass number of 42?
>
> **Solution**
>
> Note: The appendix contains the answer.

1.1.02 Isotopes

The number of neutrons contributes to the mass of an atom but does not determine the elemental identity of an atom. Isotopes are atoms of the *same element* (ie, same number of protons) that have different numbers of neutrons in their nuclei. For this reason, isotopes are identified by their mass number (the sum of the number of protons and neutrons), which appears at the end of the elemental name (eg, zinc-65) or placed as a superscript to the upper left of the element symbol (eg, ^{65}Zn).

In nuclear notation, the atomic number is sometimes included as a subscript below the mass number (eg, $^{65}_{30}Zn$); however, the atomic number is often omitted because all atoms of a given element have the same number of protons (eg, all zinc atoms have 30 protons). By knowing the name (or symbol) of an element, the number of protons can be determined by referencing the atomic number on the periodic table.

Most naturally occurring elements exist as a mixture of different isotopes. The percentage of each isotope found in an average sample of an element is determined by the **natural abundance** of each isotope (ie, the relative amounts naturally present on Earth). The distribution of isotopes can be found by analyzing a pure elemental sample using mass spectrometry. The **mass spectrum** of a pure sample of elemental zinc is shown in Figure 1.5.

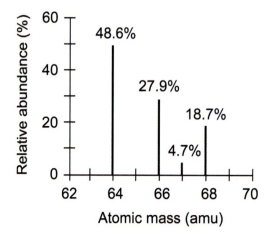

Figure 1.5 Distribution of isotopes in the mass spectrum of an average sample of pure elemental zinc.

The chemical behavior of atoms is determined primarily by their electron configuration, not the number of neutrons. Isotopes of the same element have nearly identical chemical properties (eg, bonding, reactivity) because they have the same electron configuration but differ in their *physical* properties (eg, density, mass). The reasons why electrons largely determine the reactivity of atoms are examined in Lesson 2.1.

1.1.03 Atomic Mass

Because all atoms contain protons and neutrons, the masses of different isotopes can be conveniently measured and compared by the atomic mass unit (amu), which is defined as one twelfth the mass of a carbon-12 atom. However, the atomic mass of a single isotope cannot be used in chemical calculations for samples used in a lab because most samples of substances contain a mixture of isotopes.

Instead, lab measurements use the atomic mass (atomic weight), which is an average of the masses of each isotope proportionally weighted to its natural abundance. These weighted averages are listed below each element symbol on the periodic table. An example of the weighted average for the isotopes in a sample of zinc is shown with its mass spectrum in Figure 1.6.

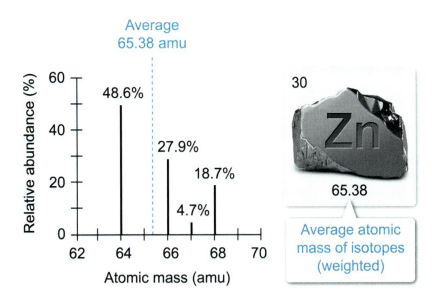

Figure 1.6 Average mass of isotopes in an average sample of pure elemental zinc.

Chapter 1: Structure and Properties of Atoms

✓ Concept Check 1.2

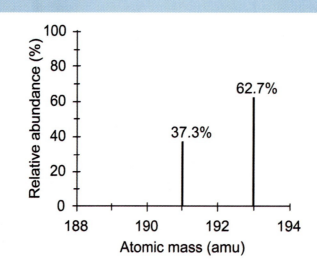

Based on the masses of the isotopes and their natural abundance in a pure sample of iridium shown in the mass spectrum above, calculate the average atomic mass of iridium.

Solution

Note: The appendix contains the answer.

Lesson 1.2

Nuclear Decay

Introduction

The **nucleons** (protons and neutrons) in an atomic nucleus are held together by a close-range attraction called the **strong nuclear force**. However, the positively charged protons also repel each other according to Coulomb's law. As such, the nucleus remains intact by achieving a balance between the competing attractive and repulsive forces between the particles.

Neutrons assist this balance of forces by separating and distributing the protons within the nucleus. As a result, the stability of the nucleus depends, in part, on the number of nucleons and the ratio of neutrons to protons (ie, the n^0/p^+ ratio). A nucleus is most stable when the n^0/p^+ ratio is around 1.

In larger nuclei with a greater number of nucleons, n^0/p^+ ratios that deviate from 1 are less stable and have a less effective balance between the competing nuclear forces. If the repulsive forces are not sufficiently balanced by the attractive forces, a nucleus can eject nuclear particles in a process called **radioactive decay**. The emission of **radiation** (ie, nuclear particles and/or high-energy photons) allows an unstable nucleus to achieve a more stable configuration by releasing energy and fragmenting into smaller nuclei with more favorable n^0/p^+ ratios.

Several types of radioactive decay are known, but the three most common types are **alpha decay**, **beta decay**, and **gamma emission**.

1.2.01 Alpha Decay

During **alpha decay**, an unstable nucleus ejects an **alpha (α) particle** consisting of 2 protons and 2 neutrons (ie, a helium-4 nucleus without its electrons). Accordingly, the resulting nucleus formed following an alpha decay has a mass number that is 4 units less and an atomic number that is 2 units less than the nucleus that underwent the decay. An example of alpha decay is shown in Figure 1.7.

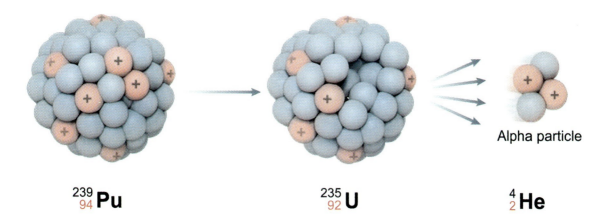

$^{239}_{94}$**Pu** $^{235}_{92}$**U** $^{4}_{2}$**He**

Figure 1.7 Alpha decay of plutonium-239.

Alpha particles are the least energetic and most massive of the common radioactive emissions. As such, an alpha particle can be easily stopped by collisions with other matter. Alpha particles ultimately acquire electrons from the surrounding environment to form helium atoms. As a result, natural underground deposits of helium gas are produced over time by the radioactive decay of elements such as uranium.

> ### ✓ Concept Check 1.3
>
> If actinium-227 is formed by the alpha decay of element X, what is the identity, atomic number, and mass number of element X?
>
> **Solution**
>
> *Note: The appendix contains the answer.*

1.2.02 Beta Decay, Positron Emission, and Electron Capture

Radioactivity involving **beta (β) decay** occurs in one of three forms: β⁻ decay (electron emission), β⁺ decay (positron emission), and electron capture. In all three forms of beta decay, the mass number remains unchanged but the atomic number either increases or decreases by 1 unit, depending on the type of emission.

In **β⁻ decay**, a neutron converts to a nuclear proton and ejects a high-speed electron (represented symbolically as e^-, $_{-1}^{0}e$, or $_{-1}^{0}\beta$) and an antineutrino (represented symbolically as $\bar{\nu}_e$ or $_{0}^{0}\bar{\nu}_e$), which increases the atomic number by 1 unit. For example, the β⁻ decay of potassium-40 yields the results shown in Figure 1.8.

$$_{19}^{40}K \rightarrow {}_{20}^{40}Ca + {}_{-1}^{0}e + {}_{0}^{0}\bar{\nu}_e$$

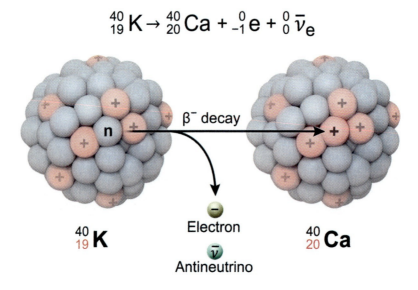

Figure 1.8 β⁻ decay of potassium-40 showing the emission of a beta particle (electron) and an antineutrino.

In **β⁺ decay**, which is commonly called **positron emission**, a nuclear proton converts to a neutron (the opposite of β⁻ decay) and ejects a positron (represented symbolically as e^+, $_{+1}^{0}e$, or $_{+1}^{0}\beta$) and a neutrino (represented symbolically as ν_e or $_{0}^{0}\nu_e$), which decreases the atomic number by 1 unit. For example, the β⁺ decay of potassium-40 yields the results shown in Figure 1.9.

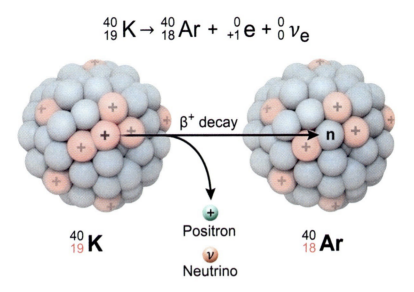

Figure 1.9 β⁺ decay of potassium-40 showing the emission of a positron (antielectron) and a neutrino.

In **electron capture**, a nuclear proton captures an electron orbiting near the nucleus and the combined particles convert to a neutron without emitting a particle. Like β⁺ decay, this process decreases the atomic number by 1 unit. For example, the nuclear decay of potassium-40 by electron capture yields the results shown in Figure 1.10.

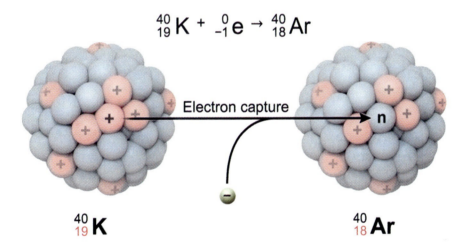

Figure 1.10 Radioactive decay of potassium-40 by electron capture.

Although neutrinos and antineutrinos are present in β⁺ decay and β⁻ decay, respectively, these essentially massless particles are not always explicitly stated when discussing beta decay in a chemistry context. Some sources choose to omit neutrinos and antineutrinos from nuclear equations to give abbreviated statements of the nuclear conversions that focus on the particles that are most relevant to chemistry.

> ## ✓ Concept Check 1.4
>
> Magnesium-23 has been determined to undergo beta decay. Which of the two isotopes (^{23}Na or ^{23}Al) detected in the magnesium sample would provide evidence of β⁻ decay?
>
> ### Solution
>
> *Note: The appendix contains the answer.*

1.2.03 Gamma Emission

During **gamma (γ) decay**, an unstable nucleus in an excited state (ie, a higher-energy state) releases excess energy by emitting a **gamma ray** (represented symbolically as γ, $^{0}_{0}\gamma$, or $h\nu$), which is a **high-energy photon** (ie, an electromagnetic wave packet). Because only energy is released in a gamma emission, the number of protons and neutrons in the nucleus remains the same, and the isotopic and elemental identity of the nucleus is unchanged. An example of gamma emission is shown in Figure 1.11.

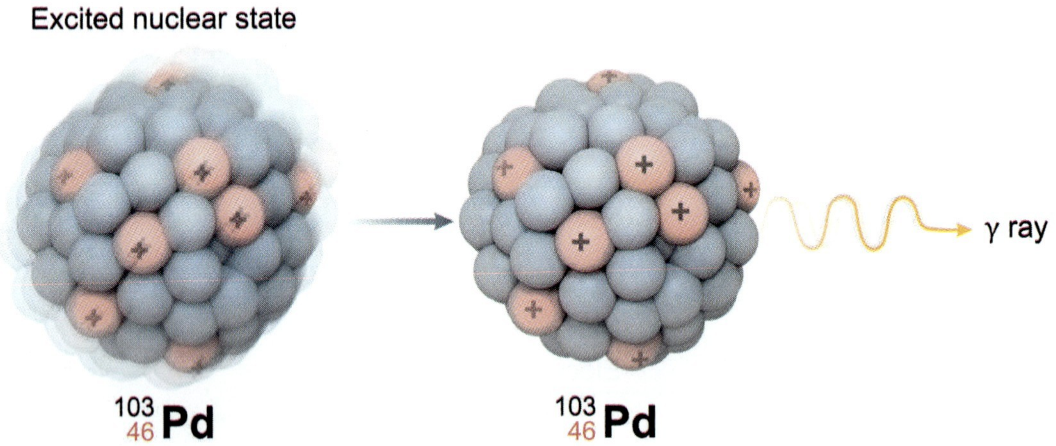

Figure 1.11 Gamma emission by an atom of palladium-103.

1.2.04 Half-life and Activity

The rates at which different radioisotopes decay vary significantly because some nuclei are much more stable than others. The relative stability of an isotope can be assessed from its **half-life**, which is defined as the time required for half a sample of an isotope to decay. A very short half-life indicates an unstable isotope that decays very rapidly whereas a longer half-life indicates a more stable isotope that decays more slowly. The half-life of an isotope is an intrinsic property that does not change over time.

The amount of a radioisotope in a sample decreases by half with each half-life that passes, as illustrated by the diagram in Figure 1.12. This decay causes the fraction of the initial amount of the radioisotope remaining in the sample to become exponentially smaller, and eventually approach zero.

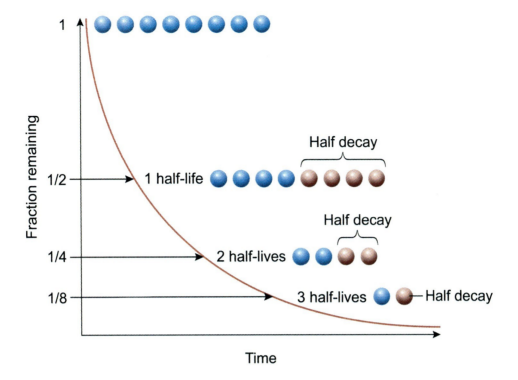

Figure 1.12 Radioactive decay plot and half-life diagram.

The fraction of radioisotope remaining after *n* half-lives can be calculated as:

$$\text{Fraction remaining} = \frac{1}{2^n}$$

After one half-life ($n = 1$), $\frac{1}{2}$ or 50% of the original substance remains; after two half-lives ($n = 2$), $\frac{1}{4}$ or 25% remains; after three half-lives ($n = 3$), $\frac{1}{8}$ or 12.5% remains; and so on. Accordingly, the activity of the sample decreases as time passes.

The **activity** of a radioisotope refers to the number of decay events per unit of time. An older unit of activity called the **curie (Ci)** was initially developed based on the number of decays per second detected from 1 g of the radium-226 isotope:

$$1 \text{ curie (Ci)} = 3.7 \times 10^{10} \text{ decays/second}$$

However, the preferred SI unit of activity is the **becquerel (Bq)**:

$$1 \text{ becquerel (Bq)} = 1 \text{ decay/second}$$

The activity of a radioisotope is proportional to both the amount of decaying radioisotope (ie, more atoms yield a higher activity) and to the half-life of the radioisotope (ie, a shorter half-life gives more activity). Activity will decrease by 50% with each successive half-life that passes.

$$\text{Activity} = (\text{Fraction remaining}) \times (\text{Initial activity})$$

$$\text{Activity} = \frac{1}{2^n} \times (\text{Initial activity})$$

When interpreting radioactive decay data, the half-life of a sample can be read directly from a radioactive decay plot corresponding to the time at which 50% of the initial amount has decayed. If the decay data is presented in a table, the half-life can be roughly estimated by finding a time interval between two data points that gives an approximately 50% reduction in the amount of the sample or its activity.

 Concept Check 1.5

The radioactive decay of a sample of rhodium-106 was found to have the activity shown in the table below.

Time (s)	Activity (Bq)
15	5,303
30	3,750
45	2,652
60	1,875
75	1,326
90	938

What is the approximate half-life of ^{106}Rh, and what will be the activity of the sample after three half-lives?

Solution

Note: The appendix contains the answer.

Lesson 1.3

Electronic Structure of Atoms

Introduction

As discussed in Lesson 1.1, all atoms consist of a central **nucleus** made of positively charged **protons** and chargeless **neutrons**, which is surrounded by negatively charged **electrons**. However, the electrons are not arranged randomly around the nucleus but are organized within specific shells and subshells according to their energy. One key feature of electron energy is that it is **quantized** (ie, electrons can only occupy specific energy levels and cannot exist between these discrete levels), as illustrated by the staircase and ramp analogy in Figure 1.13.

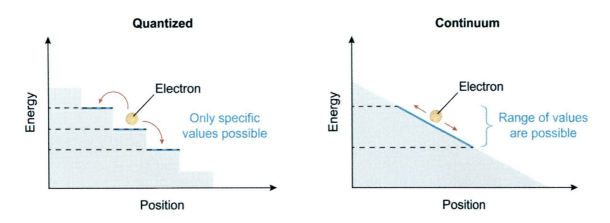

Figure 1.13 Illustration of quantized energy (discrete values) compared to an energy continuum (a continuous range of values).

The quantized nature of the electron's energy fundamentally determines how the electrons are organized around the nucleus of an atom. This lesson examines the arrangement of electrons around the nucleus and the quantum mechanical principles governing the arrangement.

1.3.01 Bohr Model of the Atom

The first atomic model to incorporate the quantized nature of electrons was the model of the hydrogen atom proposed by Danish physicist Niels Bohr. In the **Bohr model** of the atom:

- Electrons move around the nucleus in fixed circular orbits (shells), which correspond to specific energy levels that are allowed only at specific distances from the nucleus.
- Electrons can exist only in these specific orbits around the nucleus (not between these orbits) because the electron energy is quantized.
- Electrons in orbits farther from the nucleus have higher energies than electrons in orbits closer to the nucleus.
- Electrons move from a lower orbit to a higher orbit by absorbing energy (eg, heat, light), and electrons return to a lower orbit from a higher orbit by emitting energy (as a photon).
- The energy absorbed or emitted by electrons must be equal to the energy difference between two orbits.

A diagram of the Bohr model of the atom is shown in Figure 1.14.

Chapter 1: Structure and Properties of Atoms

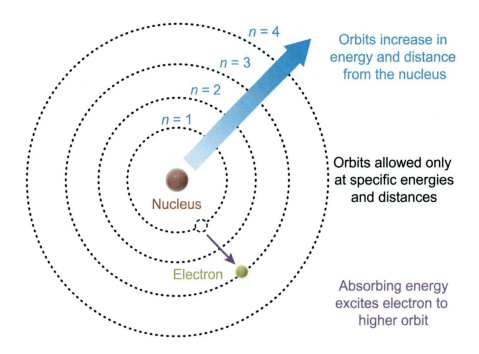

Figure 1.14 Bohr model of the atom with quantized electron orbits.

1.3.02 Absorption and Emission

Electromagnetic radiation consists of **photons**, which are discrete packets of electromagnetic energy that have properties of both particles and waves. The energy E of a photon on the **electromagnetic spectrum** (Figure 1.15) is directly proportional to the photon's frequency v and inversely proportional to its wavelength λ, as given by the equation:

$$E = hv = \frac{hc}{\lambda}$$

where c is the speed of light (3.0×10^8 m/s) and h is Planck's constant (6.626×10^{-34} J s). This equation shows that λ is also inversely proportional to v (ie, longer wavelengths correspond to lower frequencies).

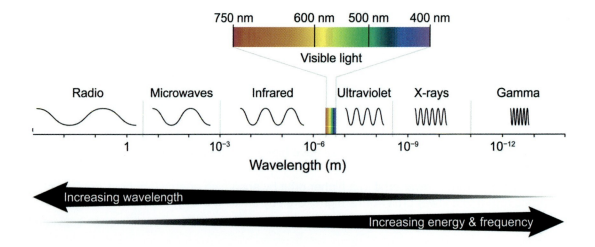

Figure 1.15 The electromagnetic spectrum.

Interestingly, when photons from the visible region of the electromagnetic spectrum interact with electrons around atoms, the atoms do not absorb photons at every wavelength. Instead, only photons with certain amounts of energy (ie, photons at specific wavelengths) are absorbed or emitted by the atom. This produces a **line spectrum** with **absorption or emission lines** at specific wavelength**s** instead of a continuous spectrum with an absorption or emission band across all wavelengths.

The **emission line spectrum** and **absorption line spectrum** for hydrogen are shown in Figure 1.16.

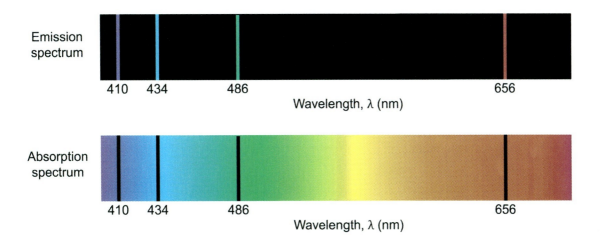

Figure 1.16 Emission line spectrum and absorption line spectrum for a sample of hydrogen.

These line spectra are explained by the Bohr model of the atom, which states that electrons can transition from one allowed energy level to another only by absorbing or emitting an amount of energy equal to the difference between the two energy levels. As such, photons with less energy are not absorbed because they are insufficient to enable the electron to make the transition between the two levels, and emitted photons always have an energy equal to the difference between the associated levels. A diagram of an electronic transition involving photon absorption and emission is shown in Figure 1.17.

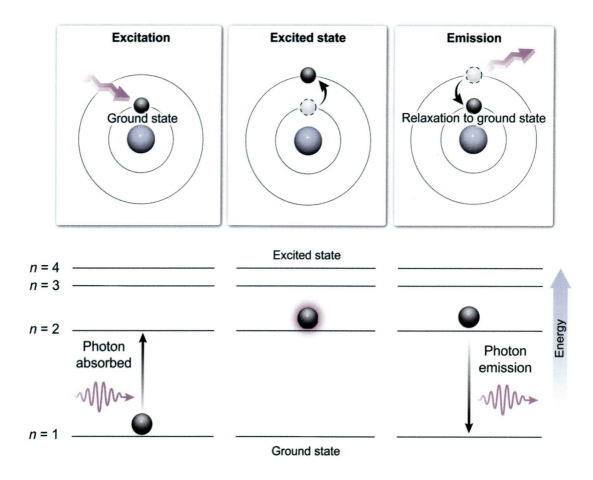

Figure 1.17 Electronic transitions between energy levels involving photon absorption and emission.

As shown in the diagram, the electrons around an atom ordinarily exist in the **ground state** (ie, the lowest allowed energy level or configuration). An electron may transition to an **excited state** (ie, occupying a higher energy level than the ground state) by absorbing a photon. Conversely, an electron in an excited state can return to the ground state by emitting a photon. The energies of the absorbed photon and the emitted photon are the same if the energy levels involved in the transitions are the same.

Applying the Bohr model to the electronic transitions possible in the hydrogen atom explains its line emission spectrum (Figure 1.16). Accordingly, the electronic transitions from four possible excited states to a lower energy state are correlated to the line emission spectrum of hydrogen as shown in Figure 1.18.

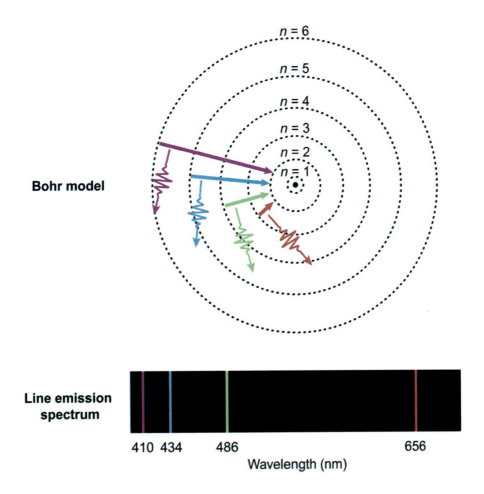

Figure 1.18 Bohr model of the atom showing the electronic transitions corresponding to the line emission spectrum of hydrogen.

1.3.03 Heisenberg Uncertainty Principle

Although the Bohr model of the atom successfully predicted the electronic transitions (ie, the absorption and emission line spectra) in atoms and ions with one electron (eg, H, He⁺), it was unsuccessful in fully describing the more complex behavior of larger atoms with more than one electron.

The Bohr model was unsuccessful, in part, because it did not consider the repulsion between electrons. The model was also unsuccessful for larger atoms because the model assumed a defined path (ie, a circular orbit) for the electron; however, determining the exact path of an electron around an atom ultimately proved impossible due to electrons having properties of both particles and waves.

As such, the **Heisenberg uncertainty principle** states that it is not possible to accurately know both the position x (ie, the path) of an electron and its momentum p (ie, its motion) around the atom at the same time. This uncertainty is quantified in relation to the reduced Planck constant $\hbar$ by the equation:

$$\Delta x \Delta p \geq \frac{\hbar}{2}$$

This relationship is represented visually in Figure 1.19.

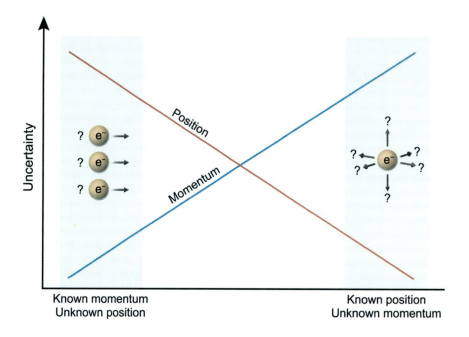

Figure 1.19 Illustration of the Heisenberg uncertainty principle.

As illustrated, the uncertainties associated with the position and momentum of an electron are inversely proportional: The more accurately an object's position is known, the less accurately its momentum can be known, and vice versa. The Heisenberg uncertainty principle most strongly impacts the study of small particles such as electrons.

1.3.04 Orbitals and Quantum Numbers

Because the position and momentum of an electron cannot simultaneously be known with complete certainty (ie, Heisenberg uncertainty principle), electrons around atoms are better modeled using probability distributions rather than discrete paths. These probability distributions, called **orbitals**, represent regions of space around the atom in which electrons are most likely to be found at a given point in time.

To illustrate the concept of an orbital, consider the diagrams in Figure 1.20 for an electron with a given energy and average distance from the nucleus.

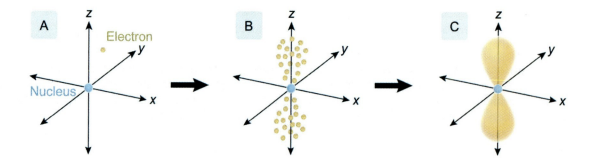

Figure 1.20 Illustration of the concept of an orbital.

In Part A, the position of an electron is plotted in space at one point in time. This process is repeated at numerous points in time as the electron changes position, resulting in the plot shown in Part B. The path by which the electron arrived at the points in Part B cannot be known; however, over time, a three-dimensional shape (ie, an orbital) begins to emerge, showing where the electron is most likely to be found (Part C).

Orbitals have different shapes and orientations based on the quantum numbers and specific probability distribution of the associated electrons, as shown in Figure 1.21.

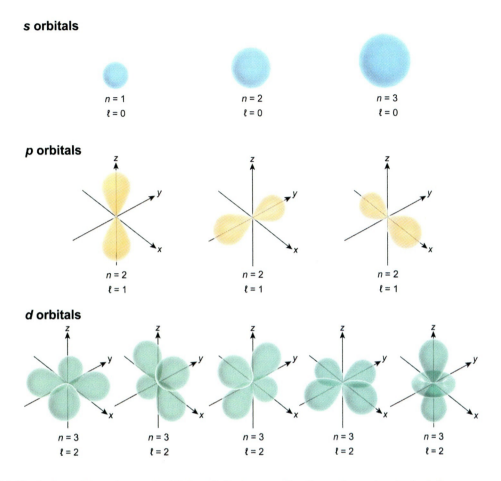

Figure 1.21 Illustration of three types of orbitals with their respective three-dimensional orientations.

All electrons around an atom can be described using four **quantum numbers** (n, ℓ, m_ℓ, m_s):

- The **principal quantum number** n describes the main energy level (shell) of the electron and its most probable distance from the nucleus, where $n = 1, 2, 3, 4....$
- The **angular momentum quantum number** ℓ corresponds to the subshell type (s, p, d, f) and its corresponding orbital shape, where $\ell = 0, 1, 2, 3, 4..., (n-1)$.
- The **magnetic quantum number** m_ℓ specifies the three-dimensional orientation of an orbital in a subshell. A given subshell contains orbitals with values of m_ℓ that include all integers ranging from $-\ell$ to $+\ell$. Accordingly, each s subshell contains one s orbital, each p subshell contains three p orbitals, each d subshell contains five d orbitals, and each f subshell contains seven f orbitals.
- The **electron spin quantum number** m_s describes the *intrinsic* angular momentum of an electron, which is a vector quantity with a value of either $+\frac{1}{2}$ (spin up) or $-\frac{1}{2}$ (spin down).

The maximum number of electrons in a shell is equal to $2n^2$, and within the shell the electrons are disbursed into subshells and their orbitals. Each orbital in a subshell can hold a maximum of two electrons. Therefore, s, p, d, and f subshells hold a maximum of 2, 6, 10, and 14 electrons, respectively.

The four quantum numbers of each electron are a unique set of parameters for that electron around the atom. A summary of the significance of each quantum number is given in Table 1.1.

Table 1.1 Summary of the quantum numbers and their corresponding parameters.

Quantum number	Allowed values	Corresponds to
Principal, n	1, 2, 3, 4…	Shell (energy and distance from nucleus)
Angular momentum, ℓ	0, 1, 2, 3, 4…, $(n-1)$	Subshell (shape of orbitals: s, p, d, f)
Magnetic, m_ℓ	$-\ell$ to $+\ell$	Orbital orientations
Electron spin, m_s	$-½$ or $+½$	Spin of electron

1.3.05 Pauli Exclusion Principle

All electrons around an atom can be described using four quantum numbers, and each set of four quantum numbers acts as a unique set of parameters for a particular electron around an atom. As such, the **Pauli exclusion principle** states that no two electrons in an atom can have the same set of four quantum numbers.

For electrons within the same orbital, three of the four quantum numbers (n, ℓ, m_ℓ) are the same. Consequently, the Pauli exclusion principle dictates that each orbital within a subshell can hold a maximum of two electrons, and those electrons must have opposite spins (ie, their fourth quantum number for electron spin, m_s, must be opposite in sign).

In schematic diagrams, orbitals are often represented by blanks or boxes, and the electrons are represented by arrows. The spin of each electron is indicated by the orientation of the arrow (up or down), as shown in Figure 1.22.

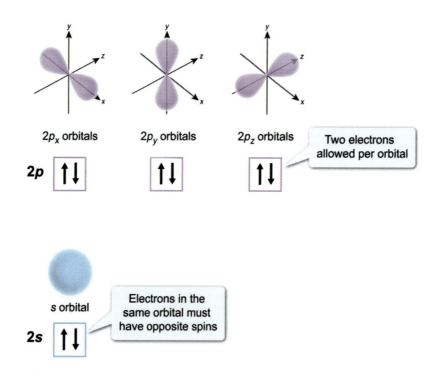

Figure 1.22 Orbital diagram following the Pauli exclusion principle.

1.3.06 Aufbau Principle and Electron Configurations

An **electron configuration** is a sequence that accounts for the placement of all electrons within the shells and subshells of an atom or ion. Shells are indicated using the principal quantum number *n*, which relates to the shell's distance from the nucleus. Subshells are labeled by type as *s, p, d,* or *f*, which can hold a maximum of 2, 6, 10, or 14 electrons, respectively, and the number of electrons in each subshell orbital is denoted with a superscript. For example, the electron configuration of helium (He) is:

$$\text{He: } 1s^2$$

For atoms with multiple electrons, the **Aufbau principle** (or "building-up" principle) states that electrons fill the lower energy levels first (ie, a lower energy subshell must fill completely before electrons can occupy the next higher subshell). As a result, electron configurations are listed in order of increasing energy. The typical order in which subshells are filled is shown in Figure 1.23.

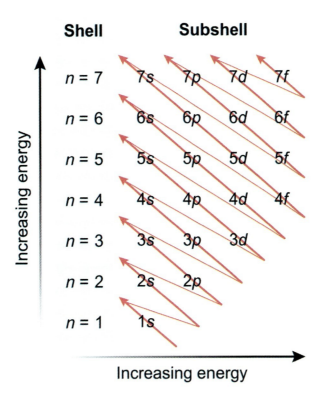

Figure 1.23 Typical filling order of electron subshells in an atom.

For example, the atomic number of selenium (Se) indicates that a Se atom in its elemental state has 34 protons and 34 electrons. Recalling the maximum capacities of the *s, p, d,* and *f* subshells (ie, 2, 6, 10, and 14 electrons, respectively) and applying the order shown in the diagram in Figure 1.23 for these 34 electrons, the electron configuration of selenium is:

$$\text{Se: } 1s^2 2s^2 2p^6 3s^2 3p^6 4s^2 3d^{10} 4p^4$$

Alternatively, the electron configuration can be determined more simply by counting through the 34 electrons using subshell blocks on the periodic table. By this method, one electron is added to the appropriate orbital in each block for each square that is crossed (moving from left to right) until the element is reached. A visual representation of this method is shown in Figure 1.24.

Chapter 1: Structure and Properties of Atoms

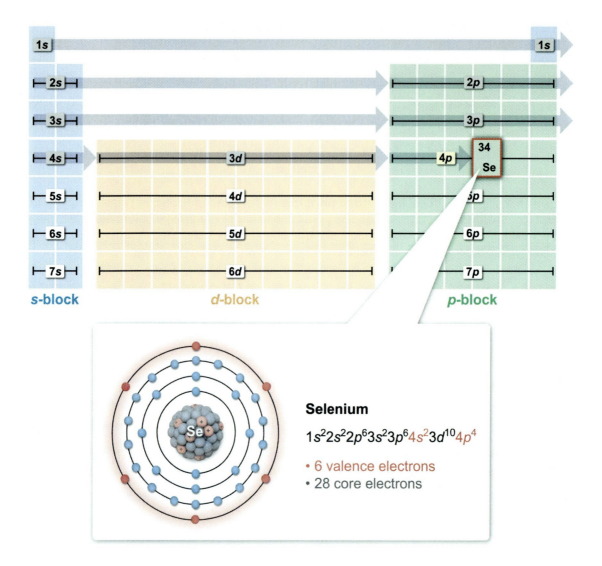

Figure 1.24 Electron configuration of an element determined using subshell blocks on the periodic table.

In an electron configuration, the outermost shell (ie, the shell with the highest value of the principal quantum number n) contains the valence electrons, and the lower shells contain the core electrons. Elements in the same column of the periodic table have the same number of valence electrons and, as examined in Chapter 2, it is the valence electrons that primarily determine the chemical properties and reactions of an element.

Exceptions to the Aufbau Principle

Although the subshell blocks on the periodic table and the ordering diagram of Figure 1.23 are useful for predicting the electron configuration of most elements, there are several exceptions found in the d-block and f-block metals. These exceptions occur because half-filled and completely filled orbitals are more stable than those that are unevenly filled. In the d-block exceptions, an electron from a valence s orbital can be shifted to a d orbital, as seen in the electron configurations of chromium (Cr) and copper (Cu):

Element	Predicted by Aufbau		Actual configuration	
Cr	$1s^22s^22p^63s^23p^6\mathbf{4s^23d^4}$	→	$1s^22s^22p^63s^23p^6\mathbf{4s^13d^5}$	(half-filled 3d subshell)
Cu	$1s^22s^22p^63s^23p^6\mathbf{4s^23d^9}$	→	$1s^22s^22p^63s^23p^6\mathbf{4s^13d^{10}}$	(filled 3d subshell)

Similarly, in most of the f-block exceptions, an electron in an *f* orbital is shifted to a *d* orbital.

Noble Gas Abbreviation

For elements with many electrons, a **noble gas abbreviation** can be used to express the electron configuration more concisely. This abbreviation uses the bracketed symbol for a preceding noble gas on the periodic table to represent the portion of the electron configuration corresponding to the lower-level electrons, followed by the configuration for the remaining higher-level electrons of the element.

For example, of the 34 electrons in selenium, the first 18 electrons have a configuration identical to that of preceding noble gas argon (Ar); therefore, [Ar] can be used as a symbolic abbreviation to denote this configuration.

$$\text{Se (complete): } \mathbf{1}s^22s^22p^63s^23p^64s^23d^{10}\mathbf{4}p^4$$

$$\text{Ar (complete): } \mathbf{1}s^22s^22p^63s^23p^6$$

$$\text{Se (abbreviated): } [\text{Ar}]4s^23d^{10}4p^4$$

Electron Configuration of Ions

In their elemental state, atoms contain an equal number of protons and electrons (ie, no net charge). If an atom gains or loses electrons, an atomic ion is formed, and the charge of the ion is determined by how many electrons are gained or lost. Electron configurations of atomic ions are expressed by adjusting the number of electrons in the configuration of the neutral atom accordingly.

If electrons are gained by an atom, the additional electrons are placed in an orbital in the lowest unfilled shell. For example, selenium forms an anion by gaining two electrons.

$$\text{Se} + 2\ e^- \rightarrow \text{Se}^{2-}$$

The electron configuration of selenium shows that the $4p$ orbital is only partially filled (ie, $4p^4$). As such, this is the lowest unfilled orbital, and the two electrons gained fill this vacancy first. Consequently, the electron configuration of the Se^{2-} anion is:

$$\text{Se}^{2-}\text{: } 1s^22s^22p^63s^23p^64s^23d^{10}\mathbf{4}p^6$$

Similarly, if electrons are lost, the electrons in the highest energy orbitals of the valence shell (ie, the shell with the highest value of the principal quantum number *n*) are removed first because they are farther from the nucleus and less tightly bound. For example, the formation and electron configuration of a Ga^{3+} ion is shown in Figure 1.25.

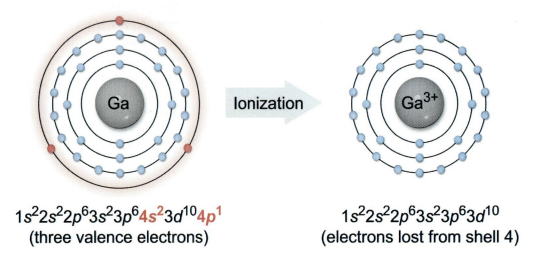

$1s^2 2s^2 2p^6 3s^2 3p^6 4s^2 3d^{10} 4p^1$
(three valence electrons)

$1s^2 2s^2 2p^6 3s^2 3p^6 3d^{10}$
(electrons lost from shell 4)

Figure 1.25 Formation of a Ga^{3+} ion and its electron configuration.

✓ Concept Check 1.6

What is the electron configuration of a Bi^{3+} cation?

Solution

Note: The print book appendix contains the short-form answer.

1.3.07 Magnetism and Hund's Rule

The atoms of many elements have electron configurations that end with partially filled sublevels. For example, the electron configuration of an atom of nitrogen (N) is $1s^2 2s^2 2p^3$. The 2p sublevel is only partially filled because p sublevels contain three p orbitals, which can each hold two electrons (ie, a maximum capacity of six electrons). This raises the question of how electrons are arranged in a partially filled sublevel.

A diagram of three possibilities for the electron configuration of nitrogen is shown in Figure 1.26.

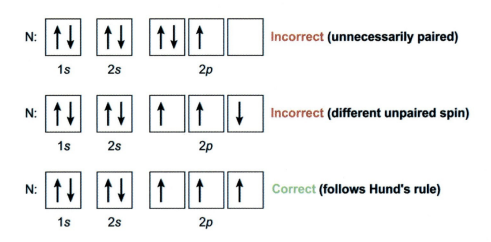

Figure 1.26 Hund's rule applied to the electron configuration of nitrogen.

According to **Hund's rule**, atoms achieve a lower energy state by filling orbitals with electrons in such a way as to maximize the number of unpaired electrons within a sublevel. As a sublevel is filled, each orbital in the sublevel receives one unpaired electron with the same spin before any electrons are paired. Based on Hund's rule, only the third diagram in Figure 1.26 gives the correct electron arrangement.

An evaluation of the electron pairing in an atom or ion's electron configuration is important because the magnetic properties depend on the number of unpaired electrons present in its orbitals. If all electrons are paired, then the atom is **diamagnetic** and the paired electrons are repelled by an external magnetic field. If unpaired electrons are present, then the atom is **paramagnetic** and the unpaired electrons are attracted to an external magnetic field. Diamagnetic and paramagnetic characteristics are summarized in Figure 1.27.

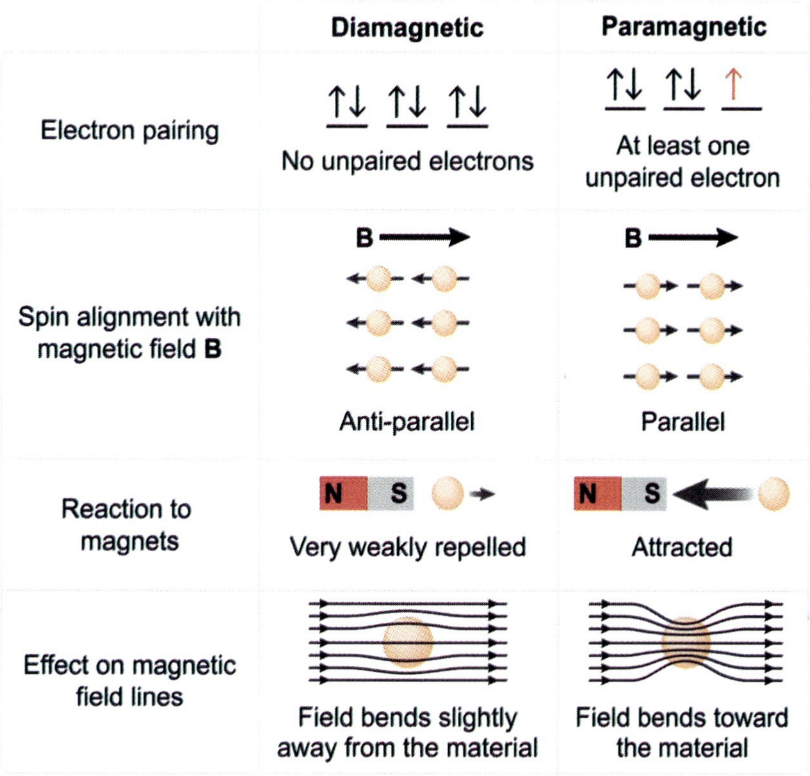

	Diamagnetic	Paramagnetic
Electron pairing	↑↓ ↑↓ ↑↓ No unpaired electrons	↑↓ ↑↓ ↑ At least one unpaired electron
Spin alignment with magnetic field **B**	Anti-parallel	Parallel
Reaction to magnets	Very weakly repelled	Attracted
Effect on magnetic field lines	Field bends slightly away from the material	Field bends toward the material

Figure 1.27 Characteristics of diamagnetic and paramagnetic atoms.

Lesson 1.4
Periodic Table Organization

Introduction

The incremental nature of the atomic structure of atoms and the repeating patterns in the electronic configurations of atoms allow the various elements to be organized in a systematic arrangement called the **periodic table**.

The recurring (ie, periodic) patterns within the periodic table are useful to compare the atomic structures and properties of the various elements. This lesson highlights the key features of the periodic table and examines how its organization arranges the elements into groups that share physical or chemical similarities.

1.4.01 Periods and Groups

The elements are listed on the periodic table in order of increasing atomic number (ie, number of protons) in horizontal rows called **periods**, as shown in Figure 1.28. Within each period, the elements are further organized into vertical columns called **groups** based on their valence electron configuration (ie, elements within a group have the same number of valence electrons). Groups are typically numbered 1 through 18 from left to right across the table.

An alternative numbering system for the element groups involves a number (often expressed as a Roman numeral) along with a letter (A or B). Although this older numbering system is now considered obsolete, it is sometimes still used for Groups 1–2 (IA–IIA) and 13–18 (IIIA–VIIIA) because the numerical value of the Roman numeral correlates to the number of valence electrons in each group (except for helium). For example, elements in Group IIA (2A) each have two valence electrons.

1 IA																	18 VIIIA
1 **H** 1.0	2 IIA				Group number Period Number							13 IIIA	14 IVA	15 VA	16 VIA	17 VIIA	2 **He** 4.0
3 **Li** 6.9	4 **Be** 9.0											5 **B** 10.8	6 **C** 12.0	7 **N** 14.0	8 **O** 16.0	9 **F** 19.0	10 **Ne** 20.2
11 **Na** 23.0	12 **Mg** 24.3	3 IIIB	4 IVB	5 VB	6 VIB	7 VIIB	8	9 VIIIB	10	11 IB	12 IIB	13 **Al** 27.0	14 **Si** 28.1	15 **P** 30.1	16 **S** 32.1	17 **Cl** 35.5	18 **Ar** 39.9
19 **K** 39.1	20 **Ca** 40.1	21 **Sc** 45.0	22 **Ti** 47.9	23 **V** 50.9	24 **Cr** 52.0	25 **Mn** 54.9	26 **Fe** 55.8	27 **Co** 58.9	28 **Ni** 58.7	29 **Cu** 63.5	30 **Zn** 65.4	31 **Ga** 69.7	32 **Ge** 72.6	33 **As** 74.9	34 **Se** 79.0	35 **Br** 79.9	36 **Kr** 83.8
37 **Rb** 85.5	38 **Sr** 87.6	39 **Y** 88.9	40 **Zr** 91.2	41 **Nb** 92.9	42 **Mo** 95.9	43 **Tc** (98)	44 **Ru** 101.1	45 **Rh** 102.9	46 **Pd** 106.4	47 **Ag** 107.9	48 **Cd** 112.4	49 **In** 114.8	50 **Sn** 118.7	51 **Sb** 121.8	52 **Te** 127.6	53 **I** 126.9	54 **Xe** 131.3
55 **Cs** 132.9	56 **Ba** 137.3	57 **La*** 138.9	72 **Hf** 178.5	73 **Ta** 180.9	74 **W** 183.9	75 **Re** 186.2	76 **Os** 190.2	77 **Ir** 192.2	78 **Pt** 195.1	79 **Au** 197.0	80 **Hg** 200.6	81 **Tl** 204.4	82 **Pb** 207.2	83 **Bi** 209.0	84 **Po** (209)	85 **At** (210)	86 **Rn** (222)
87 **Fr** (223)	88 **Ra** (226)	89 **Ac+** (227)	104 **Rf** (261)	105 **Db** (262)	106 **Sg** (266)	107 **Bh** (264)	108 **Hs** (277)	109 **Mt** (268)	110 **Ds** (281)	111 **Rg** (280)	112 **Cn** (285)	113 **Uut** (284)	114 **Fl** (289)	115 **Uup** (288)	116 **Lv** (293)	117 **Uus** (294)	118 **UUo** (294)

*	58 **Ce** 140.1	59 **Pr** 140.9	60 **Nd** 144.2	61 **Pm** (145)	62 **Sm** 150.4	63 **Eu** 152.0	64 **Gd** 157.3	65 **Tb** 158.9	66 **Dy** 162.5	67 **Ho** 164.9	68 **Er** 167.3	69 **Tm** 168.9	70 **Yb** 173.0	71 **Lu** 175.0
+	90 **Th** 232.0	91 **Pa** (231)	92 **U** 238.0	93 **Np** (237)	94 **Pu** (244)	95 **Am** (243)	96 **Cm** (247)	97 **Bk** (247)	98 **Cf** (251)	99 **Es** (252)	100 **Fm** (257)	101 **Md** (258)	102 **No** (259)	103 **Lr** (260)

Figure 1.28 Periods and groups of the periodic table.

Because they have the same number of valence electrons, elements in the same group have *similar* chemical properties (eg, bonding, reactivity). For example, lithium and sodium are both in Group 1 (IA) and have one valence electron each. As such, sodium can be expected to have chemical properties more similar to those of lithium than to those of other groups' elements. Accordingly, when going across a period, the elements have different numbers of valence electrons and display *different* chemical properties. The similarity of the chemical properties of elements in the same groups is examined in Chapter 2.

1.4.02 Element Classifications

Within the organization of the periodic table, the elements can be broadly classified as **metals**, **metalloids**, and **nonmetals**, as shown in Figure 1.29. The heavy zigzag line shown on some periodic tables separates metals from nonmetals, with metals located to the left of the line and nonmetals to the right. The metalloids are located along the line (excluding Al and Po).

Chapter 1: Structure and Properties of Atoms

1																	18
1 H 1.0	2			Metals								13	14	15	16	17	2 He 4.0
3 Li 6.9	4 Be 9.0			Metalloids‡								5 B 10.8	6 C 12.0	7 N 14.0	8 O 16.0	9 F 19.0	10 Ne 20.2
11 Na 23.0	12 Mg 24.3	3	4	5	6	7	8	9	10	11	12	13 Al 27.0	14 Si 28.1	15 P 30.1	16 S 32.1	17 Cl 35.5	18 Ar 39.9
19 K 39.1	20 Ca 40.1	21 Sc 45.0	22 Ti 47.9	23 V 50.9	24 Cr 52.0	25 Mn 54.9	26 Fe 55.8	27 Co 58.9	28 Ni 58.7	29 Cu 63.5	30 Zn 65.4	31 Ga 69.7	32 Ge 72.6	33 As 74.9	34 Se 79.0	35 Br 79.9	36 Kr 83.8
37 Rb 85.5	38 Sr 87.6	39 Y 88.9	40 Zr 91.2	41 Nb 92.9	42 Mo 95.9	43 Tc (98)	44 Ru 101.1	45 Rh 102.9	46 Pd 106.4	47 Ag 107.9	48 Cd 112.4	49 In 114.8	50 Sn 118.7	51 Sb 121.8	52 Te 127.6	53 I 126.9	54 Xe 131.3
55 Cs 132.9	56 Ba 137.3	57 La* 138.9	72 Hf 178.5	73 Ta 180.9	74 W 183.9	75 Re 186.2	76 Os 190.2	77 Ir 192.2	78 Pt 195.1	79 Au 197.0	80 Hg 200.6	81 Tl 204.4	82 Pb 207.2	83 Bi 209.0	84 Po (209)	85 At (210)	86 Rn (222)
87 Fr (223)	88 Ra (226)	89 Ac+ (227)	104 Rf (261)	105 Db (262)	106 Sg (266)	107 Bh (264)	108 Hs (277)	109 Mt (268)	110 Ds (281)	111 Rg (280)	112 Cn (285)	113 Uut (284)	114 Fl (289)	115 Uup (288)	116 Lv (293)	117 Uus (294)	118 UUo (294)

*Lanthanoid series	58 Ce 140.1	59 Pr 140.9	60 Nd 144.2	61 Pm (145)	62 Sm 150.4	63 Eu 152.0	64 Gd 157.3	65 Tb 158.9	66 Dy 162.5	67 Ho 164.9	68 Er 167.3	69 Tm 168.9	70 Yb 173.0	71 Lu 175.0
+Actinoid series	90 Th 232.0	91 Pa (231)	92 U 238.0	93 Np (237)	94 Pu (244)	95 Am (243)	96 Cm (247)	97 Bk (247)	98 Cf (251)	99 Es (252)	100 Fm (257)	101 Md (258)	102 No (259)	103 Lr (260)

Figure 1.29 Element categorization on the periodic table.

Metals tend to be shiny, ductile, and good conductors of heat and electricity whereas nonmetals tend to be dull, brittle, and poor electrical conductors (making them good electrical insulators). Metalloids share some characteristics of metals and some characteristics of nonmetals; as a result, metalloids tend to be better electrical conductors than nonmetals but not as conductive as metals, making them useful as semiconductors in some instances and as insulators in others.

To illustrate the differences between these types of elements, a comparison of selected properties for a metal, metalloid, and nonmetal is shown in Table 1.2.

Table 1.2 Property comparison between calcium, silicon, and iodine.

Calcium (Ca)	Silicon (Si)	Iodine (I)
Metal	Metalloid	Nonmetal
Silvery	Shiny, blue-gray	Lustrous, purple-black
Ductile, malleable	Brittle	Brittle
Can hammer into sheets	Shatters when hammered	Shatters when hammered
Conductor: 2.9×10^7 S/m	Semiconductor: 1×10^3 S/m	Nonconductive: 1×10^{-7} S/m
Density: 1.55 g/cm³	Density: 2.33 g/cm³	Density: 4.93 g/cm³
Melting point: 842°C	Melting point: 1414°C	Melting point: 114°C

1.4.03 Named Element Groups

In addition to classification by type (ie, metal, nonmetal, or metalloid), the elements on the periodic table can be categorized as part of either the representative elements (also called the main group elements) or as part of the transition metals. The **representative elements** include Groups 1–2 and Groups 13–18 (ie, the A group elements) and have electronic configurations that place them in the s-block and the p-block of the periodic table. The elements in Groups 3–12 (ie, the B group elements) are the **transition metals** and occupy the d-block of the periodic table.

Of the common representative elements, the seven elements highlighted in Figure 1.30 exist as diatomic molecules in their standard elemental state. These seven elements achieve a more stable electron configuration by forming atomic pairs (ie, H_2, N_2, O_2, F_2, Cl_2, Br_2, and I_2) that share valence electrons and exist as gases at room temperature, with exceptions of bromine (a liquid) and iodine (a solid).

Chapter 1: Structure and Properties of Atoms

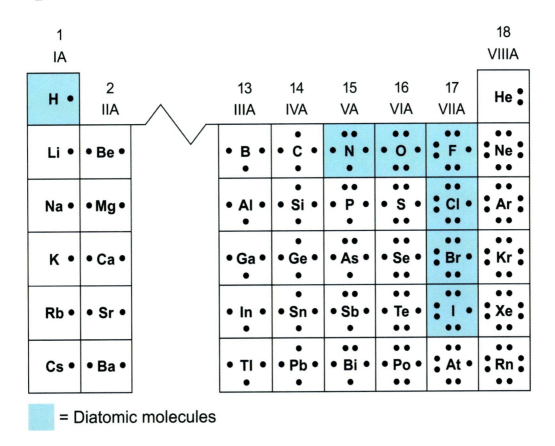

Figure 1.30 Representative elements that exist as diatomic molecules in their elemental form.

Because elements in a given column have the same number of valence electrons, and thus similar chemical properties, the elements of some columns are given a group name, as shown in Figure 1.31.

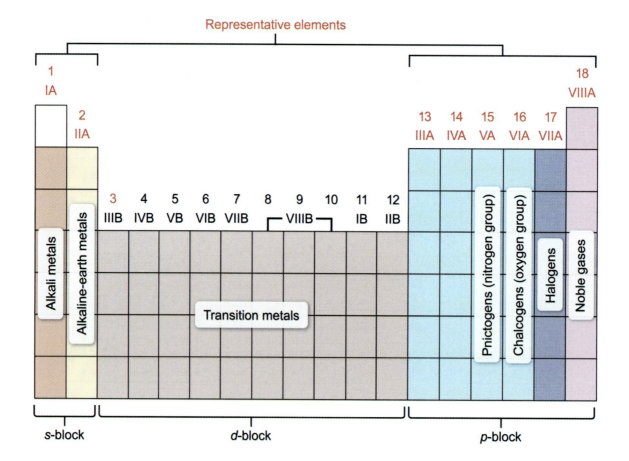

Figure 1.31 Named groups of the periodic table.

The following is a summary of key features of selected named groups.

Alkali Metals

The alkali metals in Group 1 (IA) exhibit general metallic characteristics and are considered highly reactive. These metals have a single electron in their valence shell (ns^1), which can be readily donated to form strong ionic bonds with nonmetals and result in the formation of a cation with a +1 charge (eg, K^+). Although included in Group IA due to its electron configuration, hydrogen behaves as a nonmetal (unless under extreme pressure) and is not considered an alkali metal.

Alkaline-Earth Metals

Similar to alkali metals, alkaline-earth metals are a highly reactive group of elements occupying Group 2 (IIA). Alkaline-earth metals possess two valence electrons in their valence shell (ns^2), which are readily lost to form a cation with a +2 charge (eg, Mg^{2+}).

Transition Metals

The transition metals in Groups 3–12 (the B group elements) are unique in that the number of valence electrons they can lose from their s and d orbitals varies, resulting in the ability to form ions with different charges. The resulting metal cations often result in the formation of compounds that produce colorful solutions. For example, cobalt can form an ion with a +2 charge or an ion with a +3 charge, which can produce red, green, or purple solutions depending on the solution conditions.

Chalcogens

The chalcogens (also known as the oxygen group) are found in Group 16 (VIA). Each element has six electrons in their valence shell (ns^2np^4) and tends to fill the partially filled p orbital by gaining two additional electrons and forming an anion with a −2 charge (eg, O^{2-}). However, the physical properties of each of these elements are very distinctive.

Halogens

The halogens, which occupy Group 17 (VIIA), are a group of highly reactive nonmetals. Because each halogen has seven electrons in their valence shell (ns^2np^5), they have a strong tendency to fill the nearly-filled p orbital by gaining one additional electron, forming an anion with a −1 charge (eg, F^-). Halogens exist as diatomic molecules in their standard elemental state to achieve a more stable electron configuration. Interestingly, the physical properties of halogens vary under standard conditions (ie, F_2 and Cl_2 are gases, but Br_2 is a liquid and I_2 is a solid).

Noble Gases

The noble gases are found in Group 18 (VIIIA) and, as their name implies, they exist in the gas state at room temperature. Apart from helium, which has only a full s orbital (ns^2) in its valence shell, noble gases have full s and p orbitals in their valence shells (ns^2np^6). Because the valence orbitals are filled, accepting another electron at that energy level is not possible and accepting more electrons at higher levels is energetically unfavorable. As a result, noble gases tend to be unreactive (ie, inert).

Lesson 1.5

Atomic Properties and Trends on the Periodic Table

Introduction

Periodic trends are specific, repeating patterns found on the periodic table that relate to the different characteristics and chemical behavior of the elements. These trends have a periodic nature because of the progressive, incremental changes in the atomic structure of the elements across a period (row) and the similarities of the electronic configuration of the outer shell within their respective column (group). In this lesson, the periodic trends of several important properties are examined.

1.5.01 Effective Nuclear Charge

Each proton in the nucleus of an atom contributes one unit of positive charge that exerts an electrostatic attraction on the negatively charged electrons around the atom. However, in atoms with several electrons, the core electrons positioned between the valence electrons and the positively charged nucleus provide some moderation of the full force attracting the valence electrons.

The core electrons introduce a shielding constant S that counteracts part of the attraction of the total **nuclear charge Z** (ie, the total number of protons) attracting the valence electrons. As a result, the valence electrons are partially shielded from the nucleus and experience an **effective nuclear charge** Z_{eff}, which is less than the full nuclear charge. A diagram of the effect of electron shielding in an atom is shown in Figure 1.32.

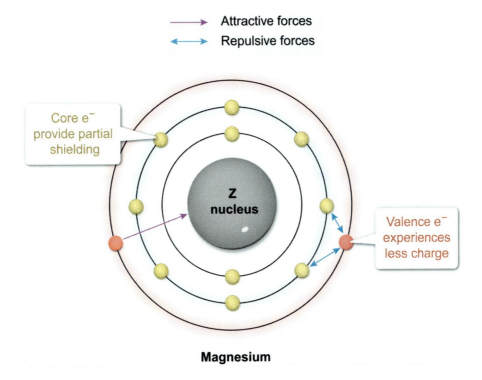

Figure 1.32 Electron shielding in a magnesium atom.

The effective nuclear charge for a specific atom or ion is calculated as:

$$Z_{\text{eff}} = Z - S$$

Precisely calculating S is difficult and depends on the electron density of a given electron orbital configuration. However, a rough, first-order approximation can be made by assuming that S is approximately equal to the number of core electrons. Based on this approximation, when comparing atoms within the same period of the periodic table, Z increases as the number of core electrons (shielding) stays constant. As such, Z_{eff} increases from left to right across a period.

Because this approximation does not fully account for electron density or orbital distance, it is less useful for precisely comparing Z_{eff} between atoms in the same group. Although the number of core electrons increases down a group, Z_{eff} generally increases slightly because the electron shells become larger and the core electrons are spread farther apart (ie, lower core electron density), which shields the nucleus less effectively as the nuclear charge increases. Periodic trends for Z_{eff} are shown in Figure 1.33.

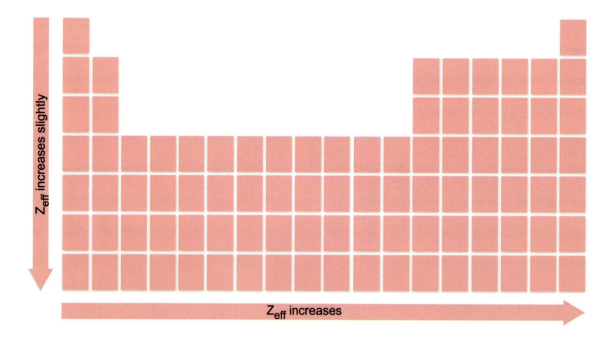

Figure 1.33 Periodic trends in effective nuclear charge Z_{eff}.

1.5.02 Size of Atoms (Atomic Radius)

An **atomic radius** is defined as the distance between the center of an atom's nucleus to its outermost shell containing the valence electrons, as shown in Figure 1.34.

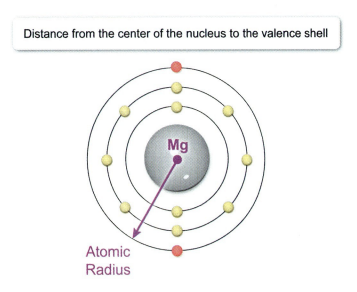

Figure 1.34 Definition of atomic radius.

Trends in atomic radius can be predicted based on the elements' positions on the periodic table, and this allows qualitative comparisons of the relative sizes of atoms. The relative sizes and general periodic trend of atomic radii are summarized in Figure 1.35.

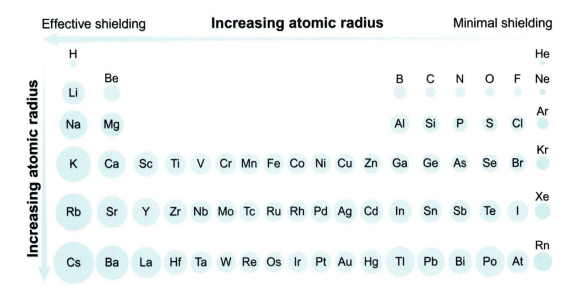

Figure 1.35 Periodic trends of atomic radii.

Within each group, the atomic radius generally increases from top to bottom because each square down the column adds another shell of electrons farther from the nucleus. The increased number of core electrons also partly shields the outer valence electrons from the attraction of the protons in the nucleus. As such, the increased number of shells causes the outer shell of electrons to be positioned farther from the nucleus.

Within each period, the atomic radius tends to decrease from left to right (with some exceptions). Each square to the right adds one proton to the nucleus and one electron to the outer shell, but the total number of shells remains constant across the row. The added proton increases the Z_{eff}, which causes the outer electrons to be attracted more strongly and pulls the electron shells closer to the nucleus.

> **Concept Check 1.7**
>
> Se, Br, Cl, S
>
> Of the four elements listed, which element has the smallest atomic radius?
>
> **Solution**
>
> *Note: The appendix contains the answer.*

1.5.03 Size of Atomic Ions (Ionic Radii)

As discussed in Concept 1.5.02, the atomic radius is the measurement for neutral atoms that retain all their electrons (ie, atoms in their elemental state with no net charge). In contrast, the **ionic radius** is the measurement for atoms that have gained or lost electrons (often an entire electron shell) and have acquired a net charge. As such, periodic trends in these two measurements are distinct and must not be treated equivalently.

Compared to the neutral atom of a given element, its cation is smaller and its anion is larger. Losing electrons to form a cation often empties the entire outer shell and leaves only the closer inner shells. In addition, the remaining electrons experience less repulsion due to fewer neighboring electrons, which pulls the electrons closer to the nucleus.

Conversely, gaining electrons to form an anion often fills a previously vacant shell farther from the nucleus. Adding electrons produces greater electronic repulsion, which also pushes electrons farther from the nucleus. **Isoelectronic** ions have the same number of electrons but because the number of protons is different in each ion, the electrons experience a different effective nuclear charge. In an isoelectronic series of ions, ionic radii decrease as the atomic number increases.

Therefore, ionic radii tend to decrease across a period (left to right) and increase down a group (top to bottom). This trend for metal cations resets and repeats for anions beginning near the division between metals and nonmetals, past which anions tend to preferentially form. These periodic trends in ionic radii are shown in Figure 1.36.

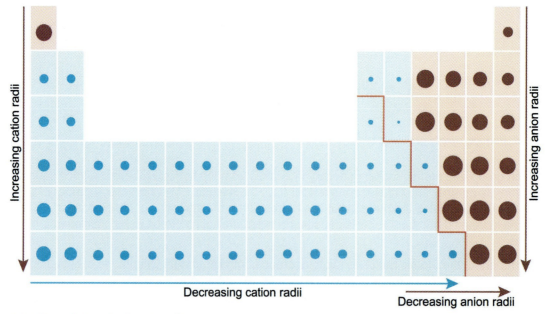

Figure 1.36 Periodic trends of ionic radii.

Chapter 1: Structure and Properties of Atoms

 Concept Check 1.8

A neutral (uncharged) atom of S has a smaller radius than a neutral atom of Mg. However, a S^{2-} ion has a larger radius than a Mg^{2+} ion. Why?

Solution

Note: The appendix contains the answer.

1.5.04 Electronegativity

Electronegativity is a measure of how strongly valence electrons are attracted to an atom within a chemical bond to another atom. The electronegativity of an atom depends on its effective nuclear charge Z_{eff} and the distance of its valence electrons from the nucleus. Atoms with both a larger Z_{eff} and fewer electron shells (ie, a smaller atomic radius) have greater electronegativities. Electronegativity values are typically reported using the **Pauling scale**, which consists of dimensionless relative numbers that range from 0 to 4.

Electronegativity values tend to increase from left to right in a period and from bottom to top in a group, with fluorine being the most electronegative element (4.0) and francium being the least (0.70). The periodic trends in electronegativity are shown in Figure 1.37.

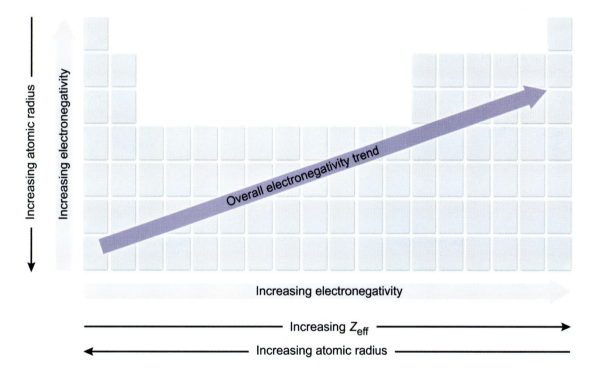

Figure 1.37 Periodic trends in electronegativity.

> **✓ Concept Check 1.9**
>
> As, In, Sr, N
>
> List the elements shown above in order of decreasing electronegativity values (from most to least electronegative).
>
> **Solution**
>
> *Note: The appendix contains the answer.*

1.5.05 Electron Affinity

Electron affinity assesses how readily a neutral atom of an element accepts an additional electron. The electron affinity of an element is measured quantitatively as the change in energy resulting from **adding an electron** (e⁻) to a neutral atom of an element X in the gas state to form an anion with a −1 charge, as illustrated in Figure 1.38.

$$X(g) + e^- \rightarrow X^-(g)$$

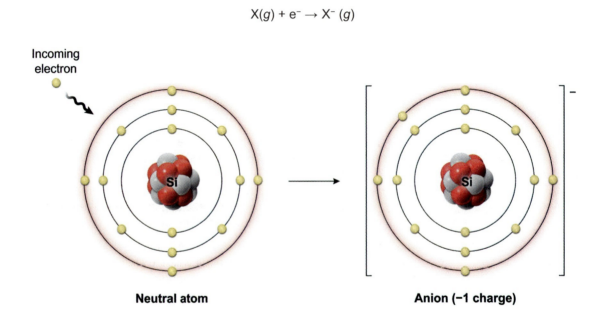

Figure 1.38 Electron affinity assesses the energetic favorability of an atom accepting an additional electron.

When a stable anion is formed, energy is released, which results in a negative value for the change in energy (ie, an exothermic e⁻ addition). The more strongly the valence electrons of a neutral atom are attracted to the nucleus (ie, the greater the effective nuclear charge), the greater the energy release when an incoming electron is gained. As such, elements with a more negative electron affinity more readily accept the addition of an electron.

If a less stable anion is formed, a significant amount of *energy input* is required for an atom to accept an electron, resulting in a positive value for the change in energy (ie, an endothermic e⁻ addition).

Periodic trends show that electron affinity values generally become more negative from left to right across a period (with some intermittent exceptions), as shown in Figure 1.39.

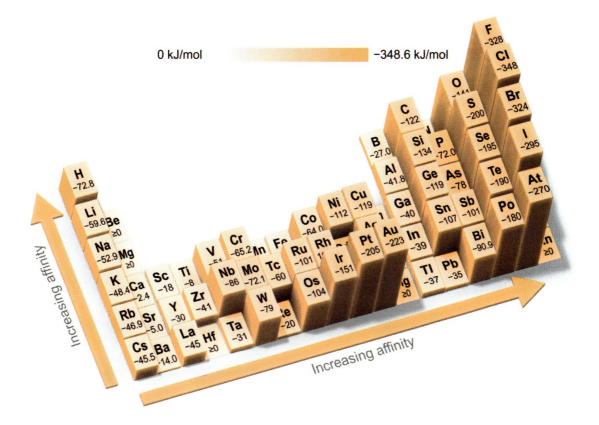

Figure 1.39 Trends in the magnitude of electron affinity.

Conversely, electron affinity values generally become less negative down a group on the periodic table because any additional electron is added to an orbital at an increased distance from the nucleus, which causes the nuclear attraction to be weaker due to increased distance and core electron shielding. The exception to this trend is when an atom is very small (eg, most of Period 2). Small atoms such as oxygen and fluorine have several electrons crowded around a small nucleus, resulting in greater electron-electron repulsion, and adding another electron increases repulsion forces even further.

Noble gases (Group 18) do not follow the periodic trends discussed here (ie, they possess an electron affinity value of nearly zero) because these elements already have a full valence shell, which makes it energetically unfavorable to accept an additional electron. The general periodic trends for the electron affinities of the elements are summarized in Figure 1.40.

 Concept Check 1.10

Which element—vanadium, germanium, or barium—has the least exothermic electron affinity?

Solution

Note: The appendix contains the answer.

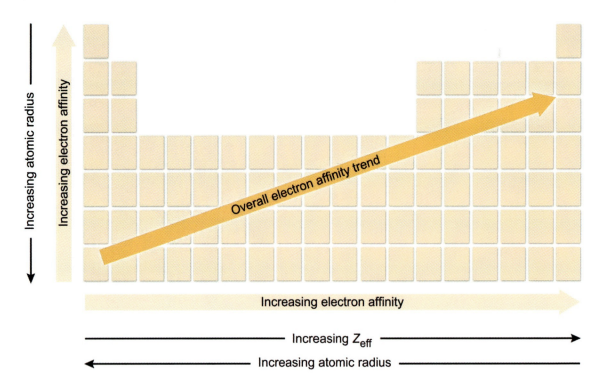

Figure 1.40 Periodic trends of electron affinity measured in terms of exothermicity.

1.5.06 Ionization Energy

Ionization energy is the minimum amount of energy a neutral atom or ion in the gas state must absorb for an electron (e⁻) to be removed. The **first ionization energy** refers to the removal of the first, most loosely bound valence electron from a neutral atom of an element to form a cation with a +1 charge, as illustrated in Figure 1.41.

$$X(g) + energy \rightarrow X^+(g) + e^-$$

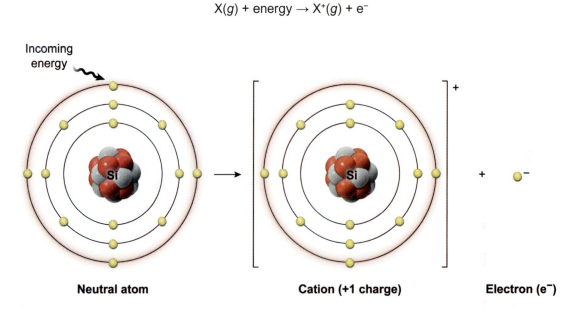

Figure 1.41 Ionization energy is the minimum amount of energy required to remove an electron from a neutral atom.

Group 1 metals have the lowest first ionization energies of all the element groups (ie, less energy is needed to remove an electron) because removing a single electron gives the atom a stable octet configuration (ie, ns^2np^6) like the noble gases (Group 18). In contrast, the noble gases have the highest first ionization energies (ie, more energy is needed to remove an electron) as they already possess a stable octet configuration.

The amount of energy needed to remove a valence electron from an atom is also influenced by the effective nuclear charge Z_{eff} of the atom and the electron's distance from the nucleus (ie, atomic radius). Greater values of Z_{eff} result in a stronger electrostatic force that pulls the electrons more tightly toward the positively charged nucleus, which takes more energy to overcome. In contrast, an electron that is farther from the nucleus requires *less* energy to remove because the electrostatic force between the nucleus and electron decreases with increasing atomic radius (ie, a greater distance).

Therefore, the first ionization energy generally increases across a period and decreases down a group, as shown in Figure 1.42.

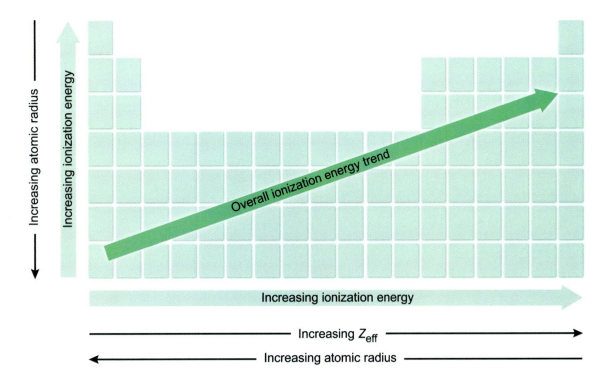

Figure 1.42 Periodic trends of the first ionization energies.

The energy needed to remove a second electron from an atom to form a cation with a +2 charge is referred to as the second ionization energy:

$$X^+(g) + \text{energy} \rightarrow X^{+2}(g) + e^-$$

Similarly, the third, fourth, and fifth ionization energies are the energies required to remove the third, fourth, and fifth electron, respectively. Each subsequent ionization energy is always greater than the last because Z_{eff} increases with the removal of each electron.

> **Concept Check 1.11**
>
> List the elements S, K, and Se in order of increasing first ionization energies (least to greatest).
>
> **Solution**
>
> *Note: The appendix contains the answer.*

Lesson 1.6

Measurements and Chemical Composition

Introduction

In the medical field, the use of measurements is ubiquitous, both in diagnostic testing and in treatment. The care used in taking measurements can mean the difference between the life and death of a patient, or between a correct diagnosis and a faulty one. This lesson reviews the system used for scientific measurements, as well as a number of its applications in the field of chemistry.

1.6.01 Metric Units

Scientific measurements must be expressed with appropriate units to be meaningful for calculations and comparisons. The system universally accepted by scientists is the **International System of Units** (SI), which is based on the **metric system**. Metric prefixes scale a unit value by powers of 10, allowing both large and small quantities to be expressed more concisely. Listed in Table 1.3 are some common metric prefixes used in chemistry.

Table 1.3 Common metric prefixes and their meaning.

Prefix	Symbol	Meaning
kilo–	k	10^3
deci–	d	10^{-1}
centi–	c	10^{-2}
milli–	m	10^{-3}
micro–	µ	10^{-6}
nano–	n	10^{-9}
pico–	p	10^{-12}

Seven fundamental physical quantities have standard SI base units, which are listed in Table 1.4.

Table 1.4 Standard SI base units.

Physical quantity	Unit	Symbol
Length	Meter	m
Mass	Kilogram	kg
Time	Second	s
Electric current	Ampere	A
Temperature	Kelvin	K
Number of particles	Mole	mol
Luminous intensity	Candela	cd

Units for all other quantities are derived from two or more of these seven SI base units. For example, **volume** is derived from length extended in three dimensions (eg, a cube with sides each 1 m in length). Accordingly, the SI base unit of volume is a cubic meter (m^3):

$$\text{Volume} = \text{length (m)} \times \text{width (m)} \times \text{height (m)} = m^3$$

In chemistry, volumes smaller than cubic meters are usually more convenient. As such, cubic meters can be converted to cubic decimeters (dm^3) given that 1 m = 10 dm:

$$1 \, m^3 = (1 \, m)^3 = (10 \, dm)^3 = 1{,}000 \, dm^3$$

Based on the cubic decimeter and cubic centimeter, two smaller units of volume commonly used in chemistry, the liter (L) and the milliliter (mL), can be defined:

$$1 \, L = (1 \, dm)^3 = (10 \, cm)^3 = 1{,}000 \, cm^3$$

$$1 \, L = 1{,}000 \, cm^3 = 1{,}000 \, mL$$

$$1 \, mL = 1 \, cm^3$$

Other derived units involve a ratio of two or more SI base units. For example, **density** (SI unit of kg/m^3) indicates the mass of a sample per unit volume, which is discussed further in Concept 1.6.02.

$$\text{Density} = \frac{\text{Mass}}{\text{Volume}}$$

A unit ratio can also be used to describe the composition of a mixture by quantifying the amount of one component in a mixture relative to the total amount of the mixture. For example, a mass ratio can be expressed as a **mass percent** using a one-hundred-unit basis, as discussed further in Concept 1.6.04.

$$\text{Mass percent} = \frac{\text{Mass of component}}{\text{Mass of mixture}} \times 100$$

Similarly, for very small amounts of a substance in a mixture, a larger one-million-unit basis (ie, 1,000,000 = 10^6) can be used to express the composition in **parts per million (ppm)**. The ppm calculation is numerically equivalent to the mass of the component in milligrams divided by the total mass of the mixture in kilograms:

$$\text{ppm} = \frac{\text{Mass of component (g)}}{\text{Mass of mixture (g)}} \times 10^6 = \frac{\text{Mass of component (mg)}}{\text{Mass of mixture (kg)}}$$

Alternatively, the composition of a mixture can be expressed in terms of its **molarity**, which is a ratio of the number of particles (ie, atoms or molecules) of one component per liter of a mixture (eg, a solution), as discussed further in Concept 1.6.08.

$$\text{Molarity (M)} = \frac{\text{Moles (mol)}}{\text{Volume (L)}}$$

1.6.02 Density

As previously mentioned, the **density** ρ of a substance is defined as the mass m it contains per unit of volume V that it occupies:

$$\rho = \frac{m}{V}$$

Because materials tend to expand when heated and contract when cooled, the volume that a substance occupies can undergo small changes with temperature. As a result, the exact density of a substance can vary slightly with temperature. However, when the density of a substance is known at a given temperature, the density equation can be rearranged and used to determine the mass or volume of a sample of that substance.

When ρ and V are known: $m = V\rho$

When ρ and m are known: $V = \dfrac{m}{\rho}$

Chemistry calculations often convert between the mass and volume of a substance by using the density directly as a **conversion factor** (ie, a ratio oriented to yield the correct cancellation of units when multiplied by another value). For example, if a sample has 2.0 g of mass in each cubic centimeter of volume, that substance has a density that can be written as a conversion factor in two ways:

$$\frac{2.0 \text{ g}}{1 \text{ cm}^3} \quad \text{or} \quad \frac{1 \text{ cm}^3}{2.0 \text{ g}}$$

The density ratio can be used as a conversion factor to convert between a given sample volume and the corresponding amount of mass present in that volume. As such, if a sample with a density of 2.0 g/cm³ has a volume of 4.0 cm³, multiplying by the density ratio above shows the mass contained in the sample:

$$4.0 \text{ cm}^3 \times \frac{2.0 \text{ g}}{1 \text{ cm}^3} = 8.0 \text{ g}$$

Similarly, if another sample with a density of 2.0 g/cm³ is known to have a mass of 10.0 g, multiplying by the inverted form of the density ratio mathematically cancels the units of grams and shows the volume of the sample:

$$10.0 \text{ g} \times \frac{1 \text{ cm}^3}{2.0 \text{ g}} = 5.0 \text{ cm}^3$$

The use of density as a conversion factor is applied extensively for calculations in Lesson 2.5.

> **Concept Check 1.12**
>
> The density of ethanol at 35 °C is 0.77 g/mL. What is the volume in mL of a 2.5 g sample of ethanol at 35 °C?
>
> **Solution**
>
> *Note: The appendix contains the answer.*

1.6.03 Molecular Weight

Because a molecule is a collection of atoms, the **molecular weight** (also called the molecular mass) of a molecule is the sum of the atomic masses of each atom in the molecule. However, the molecules in a sample are assembled from a variety of different isotopes.

As such, the molecular weight of a molecule or ionic formula unit is calculated using the average atomic masses reported in atomic mass units (amu) on the periodic table. The mass of each element is multiplied by the number of atoms of that element present in the chemical formula of the molecule, and the resulting masses are then summed together.

> **Concept Check 1.13**
>
> What is the molecular weight of $(NH_4)_2CO_3$?
>
> **Solution**
>
> *Note: The appendix contains the answer.*

1.6.04 Mass Percent

The elemental composition of a molecule or the concentration of a given component within a mixture can be expressed as percentages. However, a variety of different ratios can be chosen as the basis for calculating a percentage (eg, an atom ratio, a volume ratio, a mass ratio).

When a mass ratio is used to calculate a percentage, it is called a **mass percent** (denoted as "% *m/m*"), which is defined as the component mass divided by the total mass of the substance (ie, the combined mass of each component) expressed as a percentage:

$$\% \, (m/m) = \frac{\text{Component mass}}{\text{Total mass}} \times 100$$

> **Concept Check 1.14**
>
> What is the mass percent of fluorine in the compound PF_3?
>
> **Solution**
>
> *Note: The appendix contains the answer.*

As shown in the next concept, if the mass percent of each element in a compound is known, then the empirical formula of a compound can be determined by calculating the atom ratio of each element in the compound.

1.6.05 Empirical and Molecular Formulas

A **chemical formula** lists the elements that make up a compound and gives the ratios of the elements to each other. The **law of definite proportions** states that in any pure sample of a chemical compound, the mass ratio of one type of atom to another is constant no matter how the compound is obtained or prepared. For example, the mass ratio of hydrogen to oxygen atoms in pure water (H_2O) is always 1:8, as shown in Figure 1.43. Any pure compound that contains hydrogen and oxygen atoms in a mass ratio other than 1:8 (eg, H_2O_2 has a mass ratio of 1:16) is not water.

The **ratio of the masses** of the constituent elements in different samples of the same pure compound is always the same.

H_2O
2 H atoms = 2.02 amu
1 O atom = 16 amu
H:O **mass ratio is 2:16**, which reduces to **1:8**.

Pure samples of water from different sources have the same **1:8** hydrogen to oxygen mass ratio.

Figure 1.43 The composition of water adheres to the law of definite proportions.

Empirical and molecular formulas are two types of chemical formulas. An **empirical formula** gives the lowest whole-number ratio of the atoms of each unique element in a compound (eg, CH_2O). In contrast, a **molecular formula** lists the actual number of atoms of each type of element in one molecule of a compound and is an integer multiple (m) of the empirical formula:

$$\text{Molecular formula} = (\text{Empirical formula})m$$

For example, the molecular formula of CH_2O can be written as:

$$\text{Molecular formula} = (CH_2O)m = C_m H_{2m} O_m$$

If $m = 2$, then the molecular formula of the compound is $C_2H_4O_2$.

If the mass percent of each element in a compound is known, the empirical formula of the compound can be determined by evaluating the atom ratio of each element in the compound.

Concept Check 1.15

If an unknown compound has a molecular mass of 92.01 amu and contains 30.4% nitrogen and 69.5% oxygen by mass, what are the empirical and molecular formulas of the compound?

Solution

Note: The appendix contains the answer.

1.6.06 Avogadro's Number and the Mole

In a sample of a substance, individual particles of matter (eg, atoms, ions, molecules) are too small to be counted separately. To relate atomic-scale masses (amu) of individual particles to macroscale masses (grams) of samples containing trillions of particles, a counting unit called the mole (abbreviated "mol") is used.

A **mole** is defined as the amount of a substance that contains as many particles, atoms, or ions as the number of atoms in 12.00 g of ^{12}C. Experimental studies have demonstrated that 12.00 g of ^{12}C contains 6.022×10^{23} atoms, a constant known as Avogadro's number:

$$1 \text{ mole} = 6.022 \times 10^{23} \text{ particles}$$

$$6.022 \times 10^{23} \; ^{12}C \text{ atoms} = 1.00 \text{ mol } ^{12}C \text{ atoms} = 12.00 \text{ g } ^{12}C \text{ atoms}$$

Because each ^{12}C atom contains 12.00 amu and 1 mol of ^{12}C atoms contains 12.00 g, the ratio of amu to grams is 1:1 (ie, 1 mol of amu yields a mass of 1 g).

$$\frac{12.00 \text{ amu}}{1 \; ^{12}C \text{ atom}} = \frac{12.00 \text{ g}}{1 \text{ mol } ^{12}C \text{ atoms}}$$

Consequently, the mass of any single atom or molecule in amu numerically matches its molar mass, which is defined as the mass in grams of atoms present in 1 mole of the substance (g/mol). Although using Avogadro's number to define a mole results in the atomic mass and molar mass having the same numerical value, the quantities have different units to indicate atomic scale versus macroscale.

For example, the mass of ^{90}Zr is 90 amu and its molar mass is 90 g/mol, as shown in Figure 1.44.

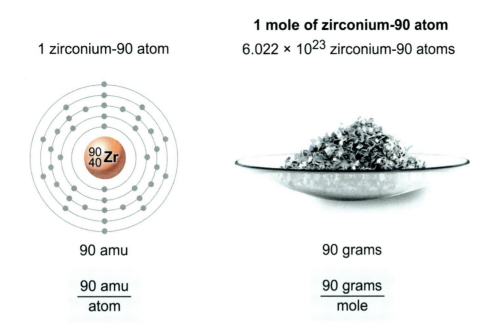

Figure 1.44 The atomic mass and molar mass of zirconium.

It is important to recognize the difference in atomic scale versus macroscale in both mass and quantity of atoms. One atom of zirconium has a mass of 90 amu, which is a very small mass (equivalent to 1.66×10^{-24} g), whereas 6.022×10^{-23} atoms (1 mol) of zirconium have a mass of 90 g.

Avogadro's number and the molar mass of a substance are important as conversion factors between the moles (number of particles) and the mass of a sample in chemical calculations. The number of moles of one chemical species (eg, atom, ion, molecule) relative to the moles of another chemical species can be found using the mole ratios from a balanced reaction equation or chemical formula. The use of these quantities as conversion factors is applied extensively in Lesson 2.5.

 Concept Check 1.16

If a sample of $CaCl_2$ has a mass of 0.75 g, how many $CaCl_2$ molecules are in the sample?

Solution

Note: The appendix contains the answer.

1.6.07 Compound Stoichiometry

The stoichiometry of a compound describes the quantitative relationship (ie, fixed ratios) among the various elements in the compound. As discussed in Concept 1.6.05, a chemical formula indicates how many atoms of an element (denoted as a subscript) are present in one molecule or formula unit of a substance (eg, 3 H atoms in H_3PO_4).

However, the subscripts in a chemical formula can also represent the number of moles of each element present in 1 *mole* of a molecule or formula unit of a substance because the proportions (ie, the elemental ratios) are constant. For example, 1 mol of H_3PO_4 includes 3 mol of H atoms, which can be written as a stoichiometric mole ratio in two ways:

$$\frac{3 \text{ mol H}}{1 \text{ mol H}_3\text{PO}_4} \quad \text{or} \quad \frac{1 \text{ mol H}_3\text{PO}_4}{3 \text{ mol H}}$$

The molar proportions given by stoichiometric mole ratios can be used as conversion factors between a given number of moles of a compound and the corresponding number of moles of an element present in that amount of the compound. As such, if a sample contains 2.0 mol of H_3PO_4, multiplying by the mole ratio above shows the number of moles of H in the sample:

$$2.0 \text{ mol H}_3\text{PO}_4 \times \frac{3 \text{ mol H}}{1 \text{ mol H}_3\text{PO}_4} = 6.0 \text{ mol H}$$

Similarly, if another sample of H_3PO_4, is known to contain 12.0 mol of H, multiplying by the inverted form of the mole ratio above mathematically cancels the moles of H and shows that the sample must consist of 4.0 mol of H_3PO_4.

> ☑ **Concept Check 1.17**
>
> How many grams of oxygen atoms are in a 0.50 mol sample of H_3PO_4?
>
> **Solution**
>
> *Note: The appendix contains the answer.*

1.6.08 Molarity

The concentration of a solution refers to the amount of solute (ie, dissolved compound) per unit volume of solution (ie, a type of homogeneous mixture). The molar concentration (molarity) of a solution is defined as the moles of dissolved solute per liter of solution:

$$\text{Molarity (M)} = \frac{\text{Moles of solute}}{\text{Solution volume } V \text{ (in liters)}}$$

When the molarity and either the volume or the moles of solute are known for a solution, the molarity equation can be rearranged to solve for the unknown quantity.

When M and V are known: $\quad Moles = MV$

When M and $moles$ are known: $\quad V = \dfrac{Moles}{M}$

However, chemistry calculations often convert between the moles and volume of a solution by using the molarity as a conversion factor. For example, if a solution has a molar concentration of 3.0 mol/L, there are 3.0 mol in 1 L of solution, which can be expressed as a conversion factor in two ways:

$$\frac{3.0 \text{ mol}}{1 \text{ L}} \quad \text{or} \quad \frac{1 \text{ L}}{3.0 \text{ mol}}$$

As such, if a solution with a molarity of 3.0 mol/L has a volume of 0.50 L, multiplying the molarity by the volume shows the moles of solute contained in the sample:

$$0.50 \text{ L} \times \frac{3.0 \text{ mol}}{1 \text{ L}} = 1.5 \text{ mol}$$

Similarly, if another solution with a molarity of 3.0 mol/L contains 12.0 mol of solute, multiplying the reciprocal of molarity by the solution volume mathematically cancels the units of moles, yielding the volume of the solution:

$$12.0 \text{ mol} \times \frac{1 \text{ L}}{3.0 \text{ mol}} = 4.0 \text{ L}$$

Because molarity expresses concentration in terms of moles, mole ratios from the formula of an ionic compound can be easily used to relate the molar concentration of one ionic species to the molar concentration of another ionic species in a solution.

For example, each formula unit of the ionic compound K_2S forms two K^+ ions and one S^{2-} ion when dissolved in water (ie, a 2:1 mole ratio of K^+ to K_2S). Accordingly, a 2.0 mol/L K_2S solution has a K^+ concentration that is two times higher. This is demonstrated mathematically by multiplying the mole ratio from the compound by the solution molarity:

$$\frac{2.0 \text{ mol } K_2S}{1 \text{ L}} \times \frac{2 \text{ mol } K^+}{1 \text{ mol } K_2S} = \frac{4.0 \text{ mol } K^+}{1 \text{ L}} = 4.0 \text{ M } K^+$$

The use of molarity as a conversion factor in chemical calculations is further applied in Lesson 2.5.

Concept Check 1.18

The ionic compound $AlCl_3$ dissolves in water to form Al^{3+} and Cl^- ions. What is the molar concentration of Cl^- ions in a 0.2 M solution of $AlCl_3$?

Solution

Note: The appendix contains the answer.

Chapter 1: Structure and Properties of Atoms

END-OF-UNIT MCAT PRACTICE

Congratulations on completing **Unit 1: Atomic Theory and Chemical Composition**.

Now you are ready to dive into MCAT-level practice tests. At UWorld, we believe students will be fully prepared to ace the MCAT when they practice with high-quality questions in a realistic testing environment.

The UWorld Qbank will test you on questions that are fully representative of the AAMC MCAT syllabus. In addition, our MCAT-like questions are accompanied by in-depth explanations with exceptional visual aids that will help you better retain difficult MCAT concepts.

TO START YOUR MCAT PRACTICE, PROCEED AS FOLLOWS:

1) Sign up to purchase the UWorld MCAT Qbank
   IMPORTANT: You already have access if you purchased a bundled subscription.
2) Log in to your UWorld MCAT account
3) Access the MCAT Qbank section
4) Select this unit in the Qbank
5) Create a custom practice test

Unit 2 Interactions of Chemical Substances

Chapter 2 Chemical Bonding, Reactions, and Stoichiometry

2.1 Chemical Bonding

2.1.01	Lewis Symbols and the Octet Rule	
2.1.02	Octet Rule Exceptions	
2.1.03	Ionic Bonds	
2.1.04	Covalent Bonds	
2.1.05	Multiple Bonding	
2.1.06	Bond Dissociation Energy	
2.1.07	Coordinate Covalent Bonds	
2.1.08	Lewis Structures	
2.1.09	Formal Charge	
2.1.10	Polyatomic Ions	
2.1.11	Electron Delocalization and Resonance Structures	

2.2 Bond Polarity and Molecular Polarity

2.2.01	Polar and Nonpolar Covalent Bonds	
2.2.02	Covalent Versus Ionic Character	
2.2.03	Molecular Geometry	
2.2.04	Orbital Hybridization	
2.2.05	Molecular Polarity	

2.3 Intermolecular Forces

2.3.01	London Forces	
2.3.02	Dipole Interactions	
2.3.03	Hydrogen Bonding	
2.3.04	Relative Strengths of Intermolecular Forces	
2.3.05	Properties Influenced by Intermolecular Forces	

2.4 Chemical Reactions

2.4.01	Types of Reactions	
2.4.02	Oxidation-Reduction (Redox) Reactions	
2.4.03	Oxidation States	
2.4.04	Reaction Prediction	
2.4.05	Balancing Chemical Reactions	
2.4.06	Balancing Redox Reactions	

2.5 Stoichiometry of Chemical Reactions

2.5.01	Relationships Between Species in Chemical Reactions	
2.5.02	Molarity and Density as Conversion Factors	
2.5.03	Limiting Reactant	
2.5.04	Theoretical, Actual, and Percent Yields	

Lesson 2.1
Chemical Bonding

Introduction

In the electron configuration of an atom, the outermost shell (ie, the shell with the highest value of the principal quantum number n) contains the valence electrons. Atoms are generally more stable (ie, less reactive) when their outermost (valence) shell of electrons is full, like the noble gases (Group 18).

Consequently, the valence electrons primarily determine the **chemical properties** of atoms (ie, how atoms interact with each other to form new substances) because atoms can achieve a more stable (ie, energetically favorable) state by gaining, losing, or sharing valence electrons via interactions with other atoms. In the process, these interactions form **chemical bonds** (ie, links between atoms) and produce new substances.

This lesson examines the role of valence electrons in the formation of chemical bonds and the types of bonds that can form.

2.1.01 Lewis Symbols and the Octet Rule

In an electron configuration, the outermost shell contains the valence electrons and the lower shells contain the core electrons. Because the valence electrons primarily determine the chemical properties of atoms, **Lewis symbols** use dots placed around an element's chemical symbol to represent the number of valence electrons held by an atom of that element.

As noted in Lesson 1.3, elements in the same column of the periodic table have the same number of valence electrons. These configurations are vividly illustrated by placing the Lewis symbols for the representative elements onto a periodic table layout, as shown in Figure 2.1.

Chapter 2: Chemical Bonding, Reactions, and Stoichiometry

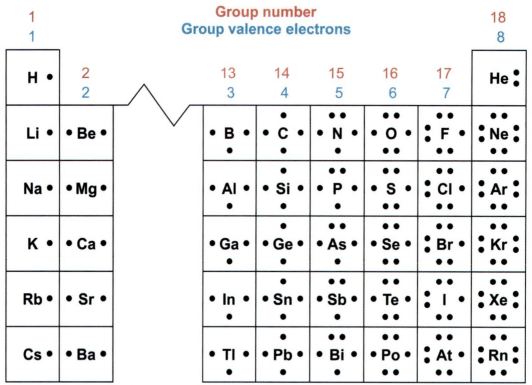

Figure 2.1 Lewis symbols showing the number of valence electrons in the representative elements.

Having the same number of valence electrons causes the elements within the same column to have similar chemical properties (ie, reactivity, bonding), whereas elements across a row have different chemical properties because the number of valence electrons differs. The reason for this pattern in reactivity is that atoms tend to be more stable when they possess a full outer shell of electrons, like the noble gas configurations (Group 18). Except for helium, the noble gases all possess a full valence shell of 8 electrons (an octet) with a valence electron configuration of ns^2np^6.

Accordingly, the **octet rule** states that atoms participate in chemical reactions with other atoms by gaining, losing, or sharing electrons to achieve a full outer shell of electrons (usually an octet).

- For atoms with 1 to 3 valence electrons (Groups 1, 2, and 13), it is energetically more favorable to lose electrons and keep the fully occupied lower shells (core electrons) than to gain several additional electrons to complete an octet in a partially filled higher shell. As such, an atom that loses electrons acquires an electron configuration like the noble gas preceding it on the periodic table.
- For atoms with 5 to 7 valence electrons (Groups 15, 16, and 17), gaining additional electrons for an octet in the outer shell is more favorable than losing several electrons. An atom that gains electrons acquires an electron configuration like the noble gas following it on the periodic table.

An example of two atoms each achieving an octet via gaining or losing electrons is shown in Figure 2.2.

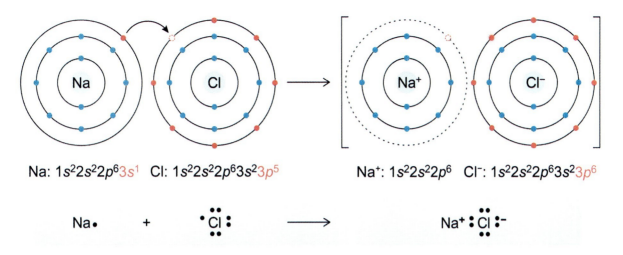

Figure 2.2 Achieving an octet by an electron transfer between two atoms.

When two atoms interact, the tendency of each atom to gain, lose, or share valence electrons relates to the relative difference in electronegativity between the atoms. A large difference in electronegativity causes an electron transfer from the atom with lower electronegativity to the atom with higher electronegativity, which produces oppositely charged atomic ions held together by an ionic bond.

In contrast, a small (or moderate) difference in electronegativity prevents either atom from taking electrons from the other atom, which causes the two atoms to obtain an octet by sharing electrons as part of a covalent bond. Ionic bonds and covalent bonds are discussed in greater detail later in this lesson.

2.1.02 Octet Rule Exceptions

Although the octet rule states that atoms tend to gain, lose, or share electrons to form bonds that achieve a full outer shell containing 8 electrons (an octet), four primary exceptions to this rule exist:

- **Elements in Period 1**. Both hydrogen and helium are small atoms that have only an s sublevel as a valence shell. Because an s sublevel has a maximum capacity of two electrons, H and He both achieve a full valence shell with only 2 electrons instead of 8.
- **Electron-deficient species**. Beryllium and boron are capable of forming stable compounds with fewer than 8 valence electrons. For example, beryllium can form compounds such as $BeCl_2$, in which the Be atom has only 4 electrons in its valence shell. Similarly, boron can form compounds such as BCl_3 in which B is surrounded by only 6 valence electrons.
- **Hypervalent species**. Atoms in Period 3 and beyond with d sublevels that can be used in bonds to other atoms may sometimes be hypervalent (ie, have more than 8 electrons in their valence shell). Although the representative elements tend to follow the octet rule, two elements that frequently have expanded octets are phosphorus (10 electrons) and sulfur (12 electrons). Hypervalency is also common in the transition metals, but elements in Periods 1 and 2 cannot be hypervalent because only s and p subshells are present in their valence shells.
- **Free radicals**. Species with an odd number of valence electrons cannot follow the octet rule because one unpaired electron (ie, a free radical) always remains in the valence shell. This configuration results in an incomplete octet around one or more atoms.

When determining the bonding arrangement for an atom in a chemical formula, it is usually best to start by following the octet rule and the exception for elements in Period 1. If a reasonable bonding pattern still cannot be produced for all atoms in the formula, the remaining exceptions should then be considered to determine if any of them apply to the atoms being evaluated.

The next sections of this lesson examine the types of bonds that can form between atoms and how these bonds result in the formation of chemical structures.

2.1.03 Ionic Bonds

An **ionic bond** forms when an atom with a relatively low electronegativity (usually a metal) transfers its valence electrons to another atom with a relatively high electronegativity (usually a nonmetal) to produce two oppositely charged ions held together by strong electrostatic attractions.

Atoms that lose electrons form **cations** (ie, positively charged ions) and atoms that gain electrons form **anions** (ie, negatively charged ions). The magnitude of the charge of a **monatomic ion** (ie, an ion made from a single atom) is equal to the number of electrons gained or lost.

A summary of the tendency of the representative elements to gain or lose electrons to achieve an octet is shown in Figure 2.3.

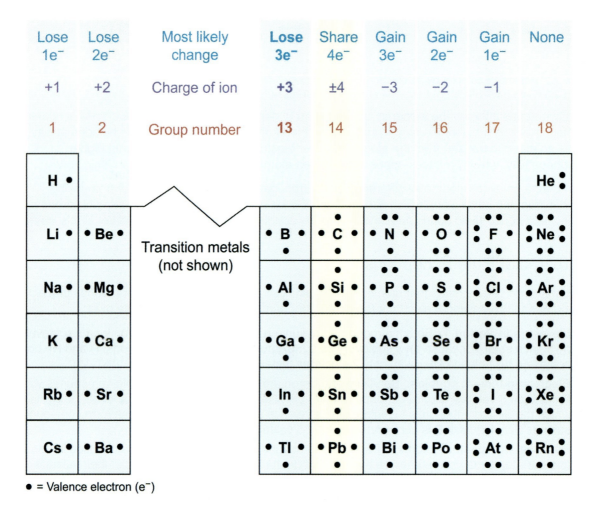

Figure 2.3 Trends in the tendency of atoms to gain or lose valence electrons.

When ionic bonds form, the oppositely charged ions assemble in ratios that give a formula unit in which the positive and negative charges are counterbalanced (ie, the resulting formula unit has a net charge of zero). For example, Na^+ and S^{2-} ions assemble in a 2:1 ratio to form a Na_2S formula unit because a S^{2-} ion requires two Na^+ ions to counterbalance the −2 charge (ie, −2 + 1 + 1 = 0), as illustrated in Figure 2.4.

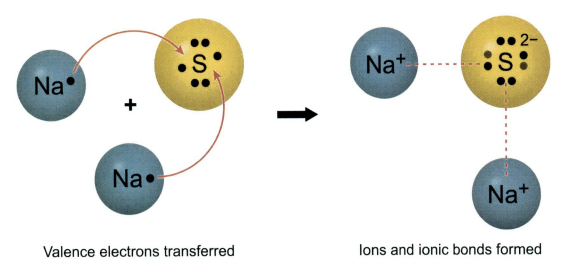

Valence electrons transferred   Ions and ionic bonds formed

Figure 2.4 Formation of Na^+ and S^{2-} ions and a Na_2S formula unit from atoms of the component elements.

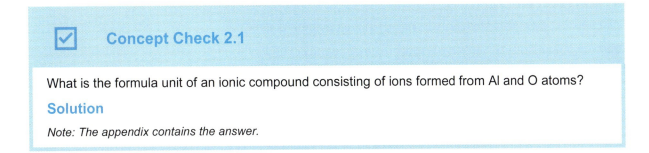

Concept Check 2.1

What is the formula unit of an ionic compound consisting of ions formed from Al and O atoms?

Solution

Note: The appendix contains the answer.

As a rough approximation, ionic bond lengths can be estimated as the sum of the ionic radii of the joined ions. By using this estimate while considering the general trends in ionic radii, relative comparisons of ionic bond lengths can be predicted without measurements. However, this method is imperfect and does introduce some error because the actual three-dimensional packing of ions within ionic solids can result in different ion spacing than predicted.

Ionic compounds are usually electrically conductive when dissolved in water because they dissociate into freely moving, charged ions that assist the movement of electrons (ie, electric current) through the mixture. As such, dissolved ionic compounds typically function as electrolytes.

2.1.04 Covalent Bonds

A **covalent bond** is formed when two atoms (usually nonmetals) share two or more valence electrons to achieve full valence shells (ie, an octet). Because nonmetal atoms have a nearly complete octet of valence electrons, it is energetically unfavorable to lose several electrons to another atom to achieve an octet.

In addition, most pairs of nonmetal atoms have only a small or moderate difference in electronegativity, which prevents either atom from exerting enough attraction to take the needed electrons from the other atom. As a result, the two atoms can achieve an octet only by sharing some of the electrons between them, as illustrated in Figure 2.5.

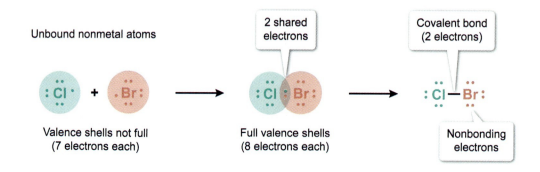

Figure 2.5 Nonmetal atoms achieve full valence shells by sharing electrons to form a covalent bond.

The **bonding electrons** (ie, the shared pair of electrons) between the atoms form a covalent bond (often represented by a line), which links the atoms together and allows both atoms to achieve full valence shells without either atom losing any valence electrons. The remaining **nonbonding electrons** (ie, the unshared electron pairs) are retained in orbitals around each atom. The increased stability achieved by obtaining an octet through the formation of a covalent bond explains why the halogens exist as diatomic molecules (ie, pairs of atoms: F_2, Cl_2, Br_2, I_2) in their elemental state.

Relative comparisons of the bond lengths between atoms joined by a single covalent bond can made by approximating the bond length as the sum of the atomic radii of the bonded atoms. For example, based on the trends in atomic radii, a Cl–Br bond is expected to be longer than a Cl–F bond because the atomic radius of a bromine atom is greater than the atomic radius of a fluorine atom (Figure 2.6).

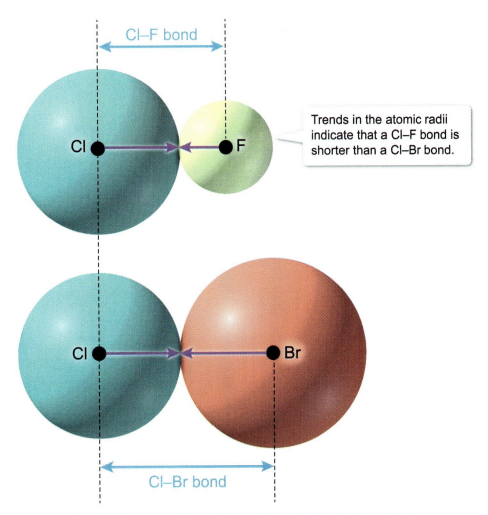

For relative comparisons, the bond length of a single covalent bond can be roughly estimated as the sum of the atomic radii.

Figure 2.6 Relative comparisons of single covalent bond lengths made using atomic radii.

Molecules formed from covalently bonded atoms are typically not electrically conductive when dissolved in water because the compounds are uncharged and the bonding electrons are localized (ie, confined within the covalent bonds). These characteristics do not assist the movement of electrons (ie, electric current) through the mixture. As such, molecular compounds are not electrolytes.

2.1.05 Multiple Bonding

When two atoms share valence electrons to form a covalent bond, sharing one pair of electrons (ie, forming one bond) is sometimes insufficient to provide both atoms with full valence shells (eg, an octet) of electrons. In such cases, the atoms may form **multiple bonds** (eg, double or triple bonds) to achieve a full valence shell. An example of a **double bond** is seen in the disulfur (S_2) molecule shown in Figure 2.7.

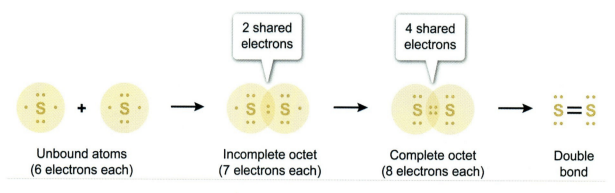

Figure 2.7 Double bond in a disulfur molecule formed by sharing two pairs of valence electrons.

In other pairs of atoms, forming a double bond still does not achieve a full valence shell, and forming a **triple bond** is necessary to achieve an octet. An example of a triple bond is seen in the dinitrogen (N_2) molecule shown in Figure 2.8.

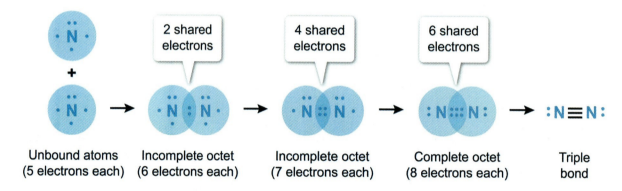

Figure 2.8 Triple bond in a dinitrogen molecule formed by sharing three pairs of valence electrons.

To understand how double and triple bonds are formed, it is necessary to consider the **atomic orbitals** involved because it is the overlap of atomic orbitals enables the sharing of electrons. Covalent bonds made by the end-to-end overlap of atomic orbitals are called **sigma (σ) bonds**, and covalent bonds made by the side-to-side overlap of *p* orbitals are called **pi (π) bonds**.

Although some atoms form only a single covalent bond (ie, σ bond), other atoms can participate in multiple bonds by the addition of one or more π bonds. As such, a double bond consists of one σ bond and one π bond, and a triple bond consists of one σ bond and *two* π bonds, as illustrated in Figure 2.9.

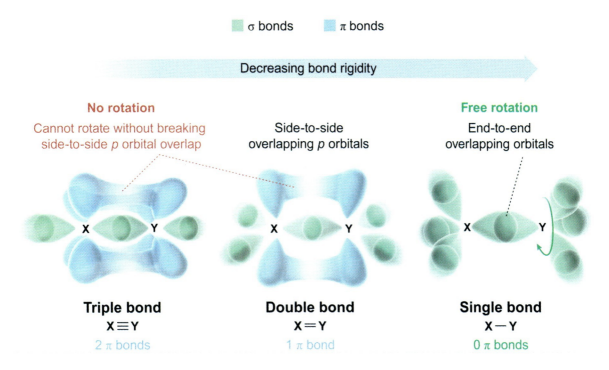

Figure 2.9 Overlap of atomic orbitals in the formation of σ and π bonds between atoms X and Y.

The orbital models in Figure 2.9 show how the end-to-end orbital overlap of σ bonds permits two σ-bonded atoms to freely rotate around the bond axis while maintaining the orbital overlap (ie, the bond can rotate without breaking). Conversely, free rotation is not possible around the axis between two π-bonded atoms because full rotation would disrupt the side-to-side *p* orbital overlap between the atoms and break the π bond. As a result, π bonds are more rigid than σ bonds due to the lack free rotation.

The covalent **bond length** (ie, the distance between the two bonded nuclei) depends on both the atomic radius of each atom and on the **bond order** (ie, single, double, or triple). Bonds between atoms with smaller radii are shorter because the nuclei of the atoms are closer together. Moreover, increasing the number of bonds shortens the bond length because the bonding electrons are confined more densely in orbitals between (instead of around) the atoms, which pulls the atoms closer together.

However, relative comparisons between the lengths of single, double, and triple covalent bonds can be made only if the bonds being compared involve the same two elements. An example of decreasing bond length with increasing bond order is shown in Figure 2.10.

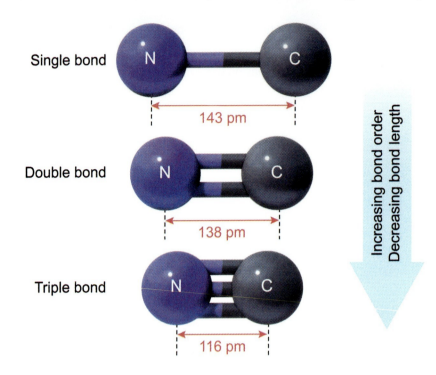

Figure 2.10 Comparison of bond length and bond order in covalently bonded nitrogen and carbon atoms.

2.1.06 Bond Dissociation Energy

The strength of a covalent bond can be measured by its bond dissociation energy, which is the amount of energy required to break the bond and separate the bonded atoms. This energy depends on the types of atoms (ie, the two elements) bonded and on the type of bond formed because atoms of various elements have orbitals that can overlap in different ways with a particular amount of energy. As such, different orbital combinations have different dissociation energies.

As discussed in Concept 2.1.05, atoms can form covalent bonds from their *s* and *p* orbitals either as **σ bonds**, produced from an end-to-end overlap of two orbitals, or as **π bonds**, produced from the side-to-side overlap of two *p* orbitals. When the σ and π bonds are compared individually, π bonds require less energy to break (ie, they have a smaller dissociation energy) than σ bonds because the end-overlap of the atomic orbitals in σ bonds is more efficient and more difficult to separate than the side-overlap of the orbitals in π bonds. This difference is seen in the graphs of the potential energy of σ and π bonds, as shown in Figure 2.11.

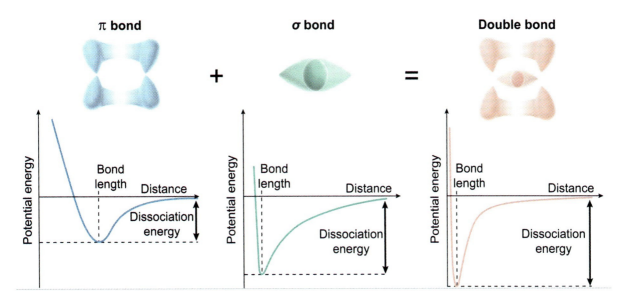

Figure 2.11 Comparison of the orbital configurations and bond dissociation energies of σ bond, π bond, and their combination (a double bond).

A double or triple bond involves a *combination* of σ and π bonds: A double bond consists of one σ bond and one π bond (Figure 2.11), and a triple bond consists of one σ bond and *two* π bonds (recall Figure 2.9). As such, the bond dissociation energy of a double or triple bond is the *total energy* required to break both the σ bond and any π bonds.

Therefore, when comparing covalent bonds involving the same two types of atoms, triple bonds have a greater bond dissociation energy than double bonds, and double bonds have a greater bond dissociation energy than single bonds. Each additional π bond increases the total amount of energy necessary to separate the atoms.

Increasing the bond order (ie, the number of bonds) not only tends to decrease the bond length (as discussed in Concept 2.1.05) but also increases the bond dissociation energy. A summary of the trends in bond length, strength, and dissociation energy in relation to the bond order is given in Figure 2.12.

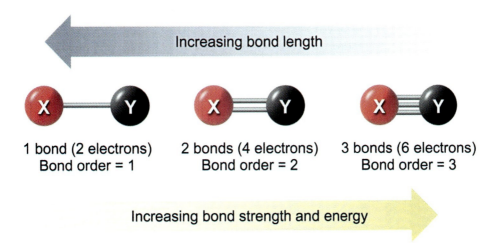

Figure 2.12 Trends in bond length, strength, and dissociation energy in relation to the bond order between atoms X and Y.

When comparing bonds of the same order, longer bonds tend to be weaker and have a lower bond dissociation energy, whereas shorter bonds tend to be stronger and have a higher bond dissociation energy. Calculations involving measured bond dissociation energies are covered in Lesson 3.5.

2.1.07 Coordinate Covalent Bonds

A coordinate covalent bond (also called a dative bond) is a special type of covalent bond formed when a pair of nonbonding electrons from an electron donor (ie, a ligand) is shared with a metal cation. Unlike other covalent bonds in which both bonded atoms contribute electrons, in a coordinate covalent bond, both shared electrons are donated from the same atom in a ligand. A metal cation may form several coordinate covalent bonds with various ligands, and the ensemble of the cation and its associated ligands is called a **coordination complex**.

In a coordinate covalent bond, a lone pair of electrons from an electron-rich atom in the ligand "plug into" a vacant orbital of an electron-deficient metal cation. Although these electrons are shared between the ligand and the metal cation, the ligand retains possession of the electrons. As such, the ligand and the electrons can "unplug" from the coordinate complex and be replaced by another ligand in a process called **ligand exchange**.

An electrical plug analogy for coordinate covalent bonding and an example of a coordination complex is shown in Figure 2.13.

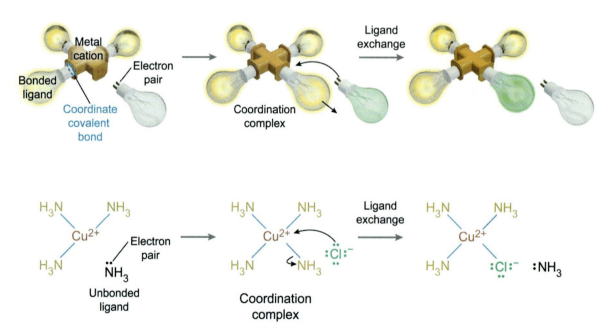

Figure 2.13 Electrical plug analogy for coordinate covalent bonding in a coordination complex.

It is worth noting that the formation of a coordinate covalent bond is a Lewis acid-base interaction in which a ligand, acting as a **Lewis base** (ie, an electron pair donor), coordinates with a metal cation, acting as a **Lewis acid** (ie, an electron pair acceptor). The theory of Lewis acids and bases is examined fully in Concept 8.1.03.

The **coordination number** of a coordination complex refers to the **number of coordinate covalent bonds** formed between the central metal ion and its nearest neighboring atoms. When all the atoms forming coordinate bonds are from *separate* molecules or ions, the number of ligands is equal to the coordination number. However, if two or more of the coordinating atoms are joined to the *same* ligand (ie, the ligand can form more than one coordinate bond), then the number of ligand units is not equal to the coordination number. In both cases, the number of *coordinating atoms* is the same but the number of *ligands* is different.

A ligand that can form two or more coordinate covalent bonds acts as a single unit called a **chelate** (a term derived from a Greek word meaning "crab's claw"). The **denticity** refers to how many atoms in the ligand can "bite" the metal cation (eg, a bidentate ligand forms two bonds, and a tetradentate ligand forms four bonds). Examples of coordination complexes with the same coordination number but different numbers of ligand units are given in Figure 2.14.

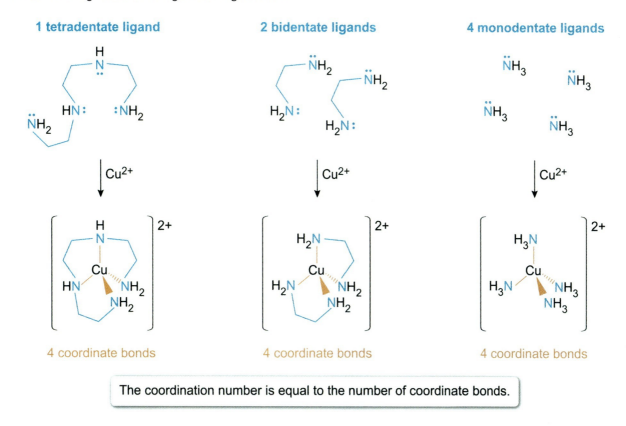

Figure 2.14 Copper complexes that have the same coordination number but different numbers of ligands.

Although a coordinate covalent bond involves a metal cation, the bond is not an ionic bond because it is not held together by electrostatic attraction between ionic charges. Unlike ionic compounds, which have a balance of positively and negatively charged ions and no net charge per formula unit, the ligands in a coordination complex are not necessarily charged ions and are often neutral (ie, uncharged) molecules. As such, the net charge of a coordination complex is equal to the sum of the charges of the cation and its ligands.

When the ligands in a coordination complex are charged, these negative charges partly or fully counterbalance the positive charge of the cation. When the ligands are neutral molecules (or do not fully counter the charge of the metal cation), the complex carries a net charge and is called a **complex ion**. Examples of three possible cases for the net charge of a coordination complex are given in Figure 2.15.

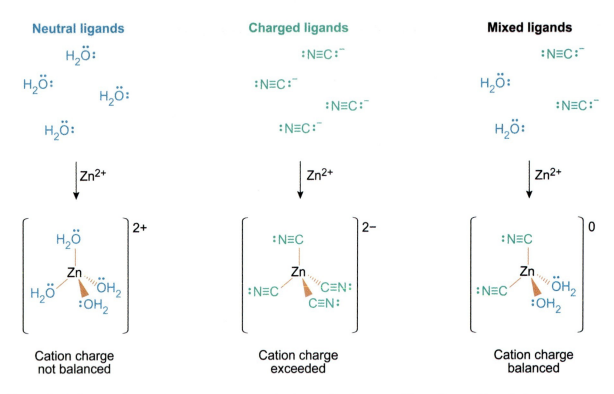

Figure 2.15 The net charge of a coordination complex is equal to the sum of the cation and ligand charges.

2.1.08 Lewis Structures

Lewis structures are used to represent the covalent bonds and bonding configuration of the atoms in a molecule. Dots placed between two element symbols represent shared electrons involved in covalent bonding, whereas dots placed around an element symbol represent unshared, nonbonding electrons. Each shared pair of electrons forms one bond (often shown as a line).

Lewis structures for a chemical formula are constructed using the following steps:

1. Determine the number of valence electrons for every atom in the molecule. (This can be done based on the location of the element on the periodic table.)

2. If the chemical formula has a net charge, add 1 extra electron for each unit of negative charge, or remove 1 electron for each unit of positive charge. (Any added electrons should be given to the most electronegative atom, and any removed electrons should be taken from the least electronegative atom.) The total number of valence electrons found in Steps 1 and 2 should not be different from the total found after the structure is completed.

3. Begin drawing the structure of the compound with the atom that has the lowest electronegativity (excluding hydrogen) positioned in the center, and then connect the atoms initially with single bonds. (Remember that each bond represents 2 shared electrons.)

4. Distribute the remaining nonbonding electrons (represented as dots) around the atoms so that each atom has a full valence shell (typically 8 electrons) according to the octet rule. If complete octets cannot be achieved for all atoms using only single bonds, move some of the nonbonding electrons to form double or triple bonds to the central atom.

5. If full valence shells still cannot be achieved by following the octet rule, consider an exception to the octet rule for the central atom (Concept 2.1.02). If a valid exception is possible (eg, atoms in Period 3 and higher), additional bonds or nonbonding electrons may be drawn around the central atom to incorporate more than 8 valence electrons.

6. Evaluate the formal charge of each atom in the structure, as given by the formula:

$$\text{Formal charge} = \begin{pmatrix} \text{Group} \\ \text{valence} \end{pmatrix} - \begin{pmatrix} \text{Nonbonding} \\ \text{electrons} \end{pmatrix} - \frac{1}{2}\begin{pmatrix} \text{Bonding} \\ \text{electrons} \end{pmatrix}$$

The sum of the formal charges in the structure should be equal to the net charge of the chemical formula. When more than one valid bonding configuration can be drawn, the best structure is the one that gives the fewest number of charged atoms possible. A full discussion of formal charge is found in Concept 2.1.09.

An example of drawing a Lewis structure using the steps above is given in Figure 2.16.

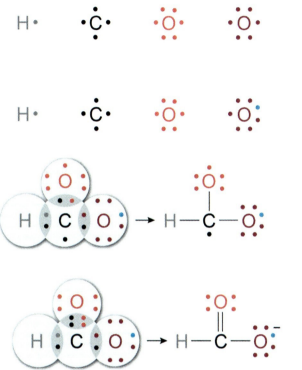

Step 1.
Lewis symbols evaluated from the periodic table for the HCO_2^- formula.

Step 2.
One extra electron added to the O atom due to the net negative charge in the HCO_2^- formula.

Step 3.
Lewis structure attempted with only single bonds, but the C atom has an incomplete octet.

Steps 4 and 6.
One double bond formed (no exceptions to the octet rule), and the net formal charge matches the HCO_2^- formula.

Note: Step 5 was unnecessary because the structure could be made without exceptions to the octet rule.

Figure 2.16 Drawing the Lewis structure of the formula HCO_2^-.

 Concept Check 2.2

What is the Lewis structure for the formula $HCONH_2$?

Solution

Note: The appendix contains the answer.

> **✓ Concept Check 2.3**
>
> Structures 1 and 2 shown below are incorrect Lewis structures for the formula H₃PO₄.
>
> Structure 1   Structure 2
>
> What are the errors in each of these structures?
>
> **Solution**
>
> *Note: The appendix contains the answer.*

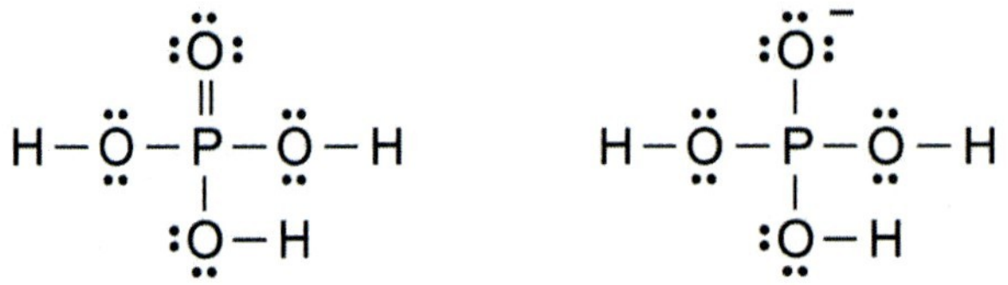

2.1.09 Formal Charge

Formal charges are values calculated for each atom in a Lewis structure to roughly estimate the distribution of electronic charge among the atoms within a molecule based on the bonding configuration of each atom. A formal charge is determined relative to an atom's group valence, which is the number of valence electrons held by an uncharged atom within a given group (column) on the periodic table.

The calculation of formal charges assumes that any nonbonding electrons around an atom are entirely unshared by the atom, whereas any electrons within a covalent bond with another atom are shared (ie, divided) equally between the atoms. These assumptions and the calculation of formal charge are expressed mathematically as:

$$\text{Formal charge} = \left(\text{Group valence}\right) - \left(\text{Nonbonding electrons}\right) - \frac{1}{2}\left(\text{Bonding electrons}\right)$$

The calculation of the formal charges of the atoms in the Lewis structure of nitromethane (CH₃NO₂) is shown in Figure 2.17.

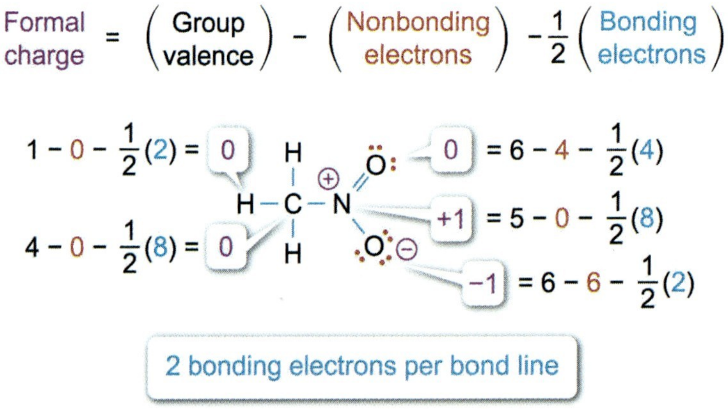

$$\text{Formal charge} = \left(\text{Group valence}\right) - \left(\text{Nonbonding electrons}\right) - \frac{1}{2}\left(\text{Bonding electrons}\right)$$

$1 - 0 - \frac{1}{2}(2) = 0$

$4 - 0 - \frac{1}{2}(8) = 0$

$0 = 6 - 4 - \frac{1}{2}(4)$

$+1 = 5 - 0 - \frac{1}{2}(8)$

$-1 = 6 - 6 - \frac{1}{2}(2)$

2 bonding electrons per bond line

Figure 2.17 Determination of the formal charges of the atoms in nitromethane.

In Lewis structures, atoms without a stated formal charge have an implied charge of zero. As such, any nonzero formal charges must be shown for all atoms. Charges of −1 or +1 may be stated simply using a minus sign or plus sign without expressly writing the number.

> ☑ **Concept Check 2.4**
>
>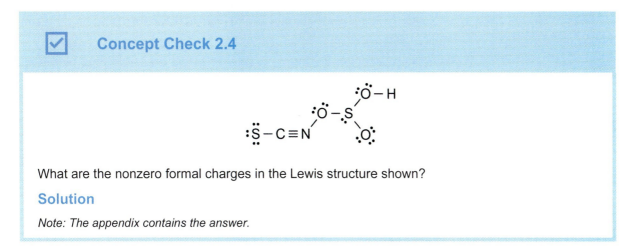
>
> What are the nonzero formal charges in the Lewis structure shown?
>
> **Solution**
>
> *Note: The appendix contains the answer.*

2.1.10 Polyatomic Ions

As previously discussed in Concept 2.1.03, **monatomic ions** are formed when an unbonded atom gains or loses one or more electrons and acquires a net charge; however, not all ions are formed from unbound atoms. This is demonstrated by **polyatomic ions**, which are groups of two or more covalently bonded atoms that have a net charge. If the total number of electrons exceeds the total number of protons in the group, the polyatomic ion is *negatively* charged, whereas a net excess of protons within the group results in a *positively* charged polyatomic ion.

Because the structure of a polyatomic ion has one or more charged atoms, polyatomic ions can form ionic bonds just like monatomic ions. However, polyatomic ion groups usually remain intact during ionic reactions due to the strong covalent bonds that hold the group of atoms together. As such, ionic compounds formed from polyatomic ions contain both ionic and covalent bonds in their structures, as seen in Figure 2.18.

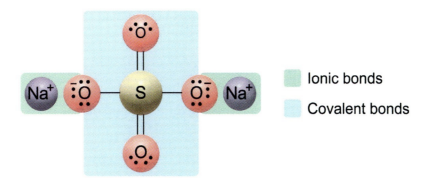

Figure 2.18 Lewis structure of Na_2SO_4, consisting of monatomic Na^+ cations and a polyatomic SO_4^{2-} anion.

Familiarity with the formulas and names of polyatomic ions is useful when working with written descriptions of chemical compounds and their reactions. A summary of some common polyatomic ions is given in Table 2.1.

Table 2.1 Common polyatomic ions sorted by net charge.

Charge	Ion formulas and names				
2+	Hg_2^{2+} dimercury				
1+	NH_4^+ ammonium				
1−	CH_3COO^- acetate	BrO_3^- bromate	OH^- hydroxide	CN^- cyanide	SCN^- thiocyanate
1−	MnO_4^- permanganate	ClO_4^- perchlorate	ClO_3^- chlorate	ClO_2^- chlorite	ClO^- hypochlorite
1−	$H_2PO_4^-$ dihydrogen phosphate	HCO_3^- hydrogen carbonate	HSO_4^- hydrogen sulfate	NO_3^- nitrate	NO_2^- nitrite
2−	HPO_4^{2-} hydrogen phosphate	CO_3^{2-} carbonate	SO_4^{2-} sulfate	SO_3^{2-} sulfite	$Cr_2O_7^{2-}$ dichromate
3−	PO_4^{3-} phosphate	PO_3^{3-} phosphite	AsO_3^{3-} arsenate		

The naming of polyatomic ions does not follow a fully unified system, but the names do follow some key patterns. Oxyanions (ie, polyatomic anions containing oxygen atoms) have names that end in either –ate or –ite. The –ate ending indicates the typical oxyanion of a given element, whereas the –ite ending indicates a less common oxyanion with the same net charge but one less oxygen atom. Examples of oxyanion names with –ite and –ate endings are compared in Table 2.2.

Table 2.2 Examples of oxyanion names with –ate or –ite suffixes.

Formula difference	Ion formulas and names				
1 more oxygen atom (–ate name ending)	NO_3^- nitrate	PO_4^{3-} phosphate	SO_4^{2-} sulfate	ClO_3^- chlorate
1 less oxygen atom (–ite name ending)	NO_2^- nitrite	PO_3^{3-} phosphite	SO_3^{2-} sulfite	ClO_2^- chlorite

Oxyanions of elements that can form bonds with up to four oxygen atoms may have names that include additional prefixes to further distinguish anions with the same net charge but a different number of oxygen atoms.

The prefix *per–* is used for an oxyanion that has one more oxygen atom than the oxyanion with a name ending in *–ate*. Similarly, the prefix *hypo–* is used for an oxyanion that has one less oxygen atom than the oxyanion with a name ending in *–ite*. A series of oxyanions demonstrating the use of these prefixes is given in Table 2.3.

Table 2.3 Example of a series of oxyanions that have names using the prefixes *per–* and *hypo–*.

	→ Decreasing number of oxygen atoms →			
Ion formulas and names	ClO_4^- perchlorate	ClO_3^- chlorate	ClO_2^- chlorite	ClO^- hypochlorite

2.1.11 Electron Delocalization and Resonance Structures

Some molecules cannot be fully represented using a single Lewis structure because some of the electrons can adopt more than one valid bonding configuration; this effect is known as resonance. In resonance, the arrangement of the atoms is unchanged, but some π bonds (eg, in double and triple bonds) and nonbonding electrons can "oscillate" from one atom to an adjacent atom in ways that do not violate the octet rule (or are a valid exception).

In effect, Lewis structures depict electrons as being localized at a particular site, but resonance actually causes certain electrons to be delocalized (ie, "spread out") among the participating atoms. An example of resonance is shown in Figure 2.19.

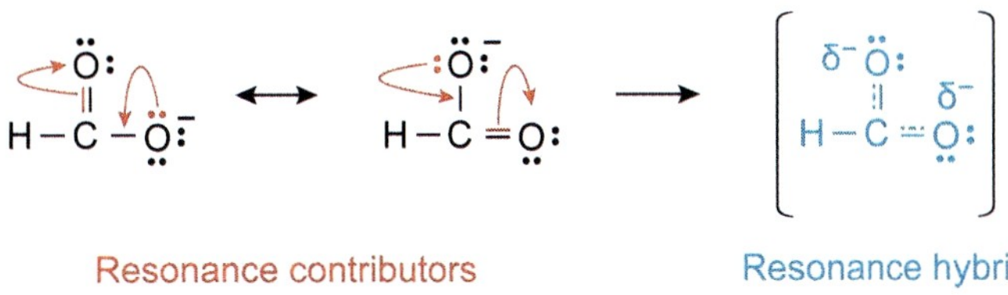

Figure 2.19 Resonance in the formate ion.

For the formate ion shown in Figure 2.19, the charge distribution and bond order cannot be fully represented using only one Lewis structure. Two valid Lewis structures can be drawn, and these structures predict one C=O double bond with an uncharged O atom and one C–O single bond with a charged O atom. However, experimental studies have found that the two carbon-oxygen bonds have an *equal* length that is shorter than a typical C–O single bond but longer than a typical C=O double bond, yielding a bond order of about 1.5 (ie, an average in between a single bond and a double bond).

In addition, the two oxygen atoms in the formate ion have been found to have equal *partial* negative charges (δ^-). These measurements demonstrate that the electrons are delocalized between the two oxygen atoms in the actual structure of formate. As such, resonance affects both bond length and the charge separation across a bond (ie, a bond dipole).

Consequently, for compounds that undergo resonance, a single Lewis structure represents only a resonance contributor (ie, the bond configuration at a moment in time). Because the electrons are in different places at different times, the true molecular structure is a resonance hybrid (ie, a weighted average) of all resonance contributors.

Resonance contributors that are more stable contribute more to the resonance hybrid than those contributors that are less stable (ie, the structure of the resonance hybrid more closely resembles the more stable contributors). In general, the most stable resonance contributor is the one that has the fewest number of atoms with a formal charge and places any negative charges on the more electronegative atoms in the structure.

Resonance is made possible by **conjugation**, which refers to a sequence of three or more adjacent *p* orbitals linked together that function as a single, continuous network along which electrons can be delocalized. These orbitals are usually involved in π bonding but may also involve nonbonding electrons or electron vacancies (ie, positive charges) at positions that can participate in the conjugated orbital sequence.

A conjugated system is often observed in a molecular structure as a chain of alternating single and multiple (ie, double or triple) bonds and can participate together in resonance. An example of a conjugated system is shown in Figure 2.20.

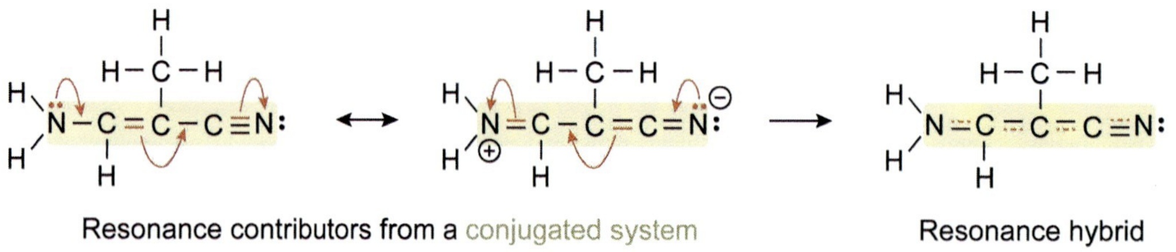

Resonance contributors from a conjugated system   Resonance hybrid

Figure 2.20 Resonance in a molecule with a conjugated system involving π bonds and nonbonding electrons.

When drawing resonance structures, the following rules should be observed:

- The octet rule must be obeyed for first- and second-row elements on the periodic table.
- All resonance structures must have the same total number of valence electrons.
- The number and position of the atoms must never change; only the electrons move.
- Only electrons in π bonds (eg, from double or triple bonds) or nonbonding lone pairs can move, not electrons in σ bonds (ie, single bonds).
- When going from one resonance structure to another, electron movement must be to adjacent atoms only.
- The overall (net) charge of the molecule must not change; however, the formal charges of the individual atoms can change.

Concept Check 2.5

Which of the four Lewis structures shown above are valid resonance structures of the structure given below? (Select all that are correct.)

Solution

Note: The appendix contains the answer.

Lesson 2.2
Bond Polarity and Molecular Polarity

Introduction

In chemistry, **polarity** (ie, the distribution of separated electrical charge) is typically examined in two categories: bond polarity and molecular polarity.

Bond polarity tends to be relatively simple and can be inferred from the difference in electronegativity between the two atoms sharing a bond. However, polarity is not just a binary phenomenon that can be reduced to simple categories of polar versus nonpolar; instead, polarity is a continuum spanning from the perfectly nonpolar H–H bond in hydrogen gas to polarity so extreme that atoms undergo a complete transfer of electrons to form an ionic bond, such as in NaF salt.

Molecular polarity describes an entire molecule and requires evaluating both the polarity of the bonds that make up a molecule as well as the orientations of those bonds (ie, the molecule's shape). For very large molecules, which are common in biological systems, both polar and nonpolar regions are often found on the same molecule. These concepts and related details are addressed in this lesson.

2.2.01 Polar and Nonpolar Covalent Bonds

As discussed in Concept 2.1.04, covalent bonds involve the sharing of electrons between two atoms; however, the nature of that sharing depends upon the **electronegativity** of each atom (Figure 2.21).

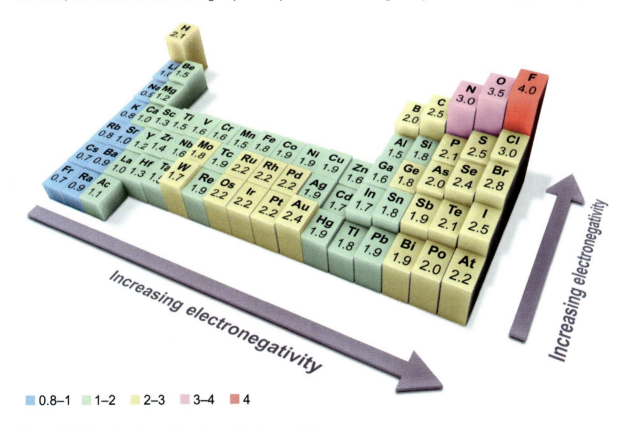

Figure 2.21 Trends and specific values of electronegativity.

Electronegativity is a measure of the degree to which an atom draws shared electrons to itself when bonded to other atoms. When bonded atoms have the same (or similar) electronegativity values (ie, an electronegativity difference of 0.4 or less), each atom's nucleus pulls on the shared electrons with roughly equal force, resulting in equal sharing of those electrons. This type of bond is a nonpolar covalent bond.

Conversely, if the electronegativity values are significantly different (ie, an electronegativity difference greater than 0.4 but less than 1.8), the nucleus of the more electronegative atom pulls more strongly on the shared electrons, resulting in a partial negative charge (δ⁻) on the more electronegative atom and a partial positive charge (δ⁺) on the less electronegative atom, as noted in Figure 2.22. This unequal sharing of electrons yields a polar covalent bond.

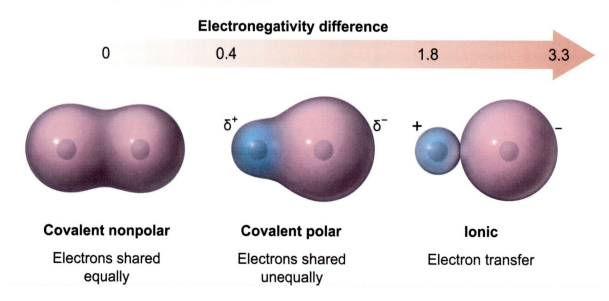

Figure 2.22 Electronegativity difference and bond polarity.

The degree of charge separation present in a polar bond is evaluated quantitatively by its **bond dipole moment (μ)**, which is a vector quantity (ie, has both magnitude and direction) determined by the following equation:

$$\mu = qr$$

where q is the magnitude of the partial charge on each atom and r is the distance between the atoms. Consequently, bond polarity (ie, the bond dipole moment) depends *not only* on electronegativity difference (which relates to the magnitude of the partial charge) *but also* on internuclear distance (ie, the bond length). Although resonance can affect the magnitude of a partial charge, a rough comparison of the polarities of different chemical bonds can be estimated using the electronegativity difference between the bonded atoms.

> ☑ **Concept Check 2.6**
>
> Which bond in 2-chloro-3-pentanol is the most polar?
>
>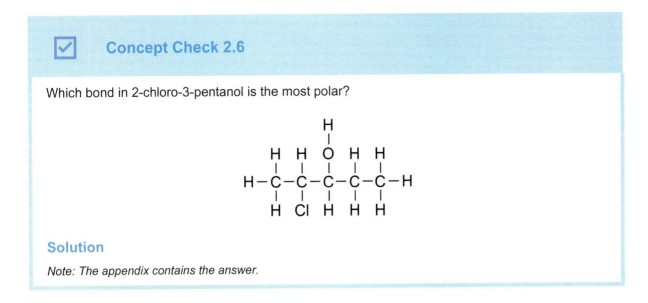
>
> **Solution**
>
> *Note: The appendix contains the answer.*

2.2.02 Covalent Versus Ionic Character

As explained in Concept 2.1.03, an ionic bond between two atoms results from the transfer of valence electrons from one atom to the other, which creates a cation and an anion with a mutual electrostatic attraction. When the electronegativity difference between two atoms is not sufficient to cause the complete transfer of electrons from one atom to the other, a polar covalent bond is formed; however, the bond still retains a certain degree of ionic character. The ionic character increases as the difference in electronegativity between two atoms increases (Figure 2.23).

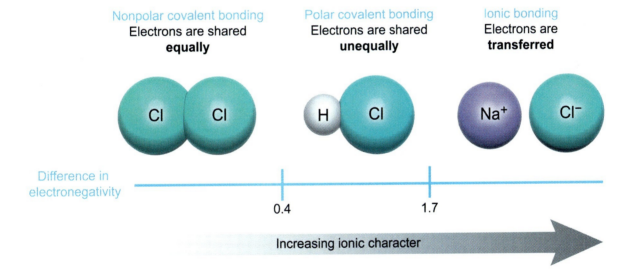

Figure 2.23 Polarity and ionic character increase as the difference in electronegativity difference between bonded atoms increases.

Just like polarity, the ionic or covalent character of a bond exists within a continuum rather than a hard binary distinction. Certain covalent compounds, such as HF, are made from two nonmetals but have a high degree of ionic character. Other compounds formed from metals and nonmetals that might be assumed to be ionic, such as $SnCl_4$, actually behave like covalent compounds.

The precise degree to which a substance exhibits ionic character can be determined by calculating its **percent ionic character**, according to the following equation:

$$\text{Percent ionic character of a bond} = \left(\frac{\mu_{\text{measured}}}{\mu_{\text{ionic (calculated)}}}\right) \times 100\%$$

where μ_{measured} is the experimentally measured dipole moment of the bond and $\mu_{\text{ionic (calculated)}}$ is the calculated value for what the dipole moment would be if the bond were fully ionic. In reality, no chemical bond is ever 100% ionic. Even bonded atoms with the most extreme electronegativity differences have at least some small degree of covalent character, as shown in Figure 2.24.

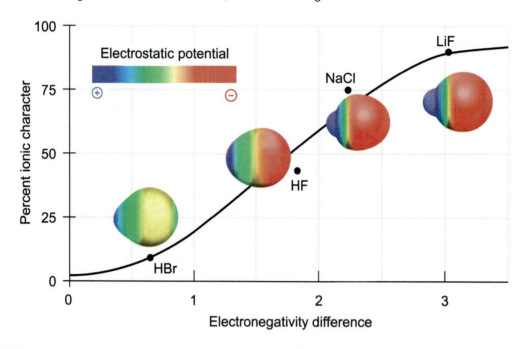

Figure 2.24 Percent ionic character versus electronegativity difference for various bonds.

Although a perfect correlation does not exist between ionic character and electronegativity difference, a relative evaluation of ionic character based on the electronegativity difference is often sufficient for an approximate analysis.

Concept Check 2.7

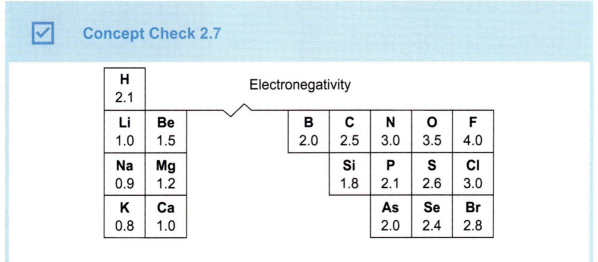

Consider the selected electronegativity values given here. Rank the following bonds in order of increasing ionic character: H–Cl, K–Cl, K–Br, S–Br.

Solution
Note: The appendix contains the answer.

2.2.03 Molecular Geometry

Before addressing the topic of molecular polarity, it is necessary to understand **molecular geometry** (ie, the shape formed by the spatial arrangement of atoms in a molecule). However, molecular geometry itself cannot be understood without knowing about electron geometry and the **valence-shell electron-pair repulsion (VSEPR) theory**.

According to VSEPR theory, **electron domains** (ie, regions of high electron density) around an atom arrange themselves in such a way as to maximize the space between the domains and minimize the repulsions between them. The spatial arrangement of electron domains around a central atom is referred to as its **electron geometry** (Table 2.4). The electron geometry around an atom determines the bond angles present in the molecule, which in turn helps determine the overall shape of the molecule.

Table 2.4 Electron geometries and their corresponding bond angles.

Number of electron domains	Arrangement of electron domains	Electron Geometry	Predicted bond angles
2		Linear	180°
3		Trigonal Planar	120°
4		Tetrahedral	109.5°
5		Trigonal Bipyramidal	120° 90°
6		Octahedral	90°

The number of electron domains around a molecule's central atom can be discerned from its Lewis structure. A single bond, a double bond, a triple bond, and a lone pair of nonbonding electrons are each considered to be one electron domain (Table 2.5).

Table 2.5 Types of electron domains and their symbols.

Type of electron domain	Lewis structure symbol
Single bond	—
Double bond	=
Triple bond	≡
Lone pair of electrons*	••

*Free radicals (ie, single unpaired electrons) are generally not considered electron domains for the purposes of VSEPR.

Concept Check 2.8

Potassium nitrate, an important component of chemical fertilizers, is represented by the Lewis structure shown:

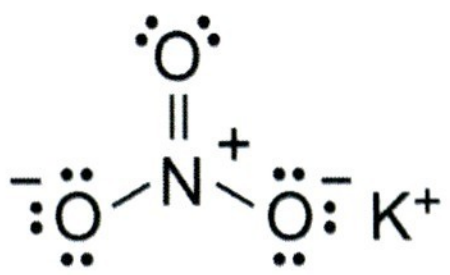

$$:\ddot{\underset{..}{O}}-\overset{\overset{\displaystyle :\ddot{O}:}{\|}}{\underset{}{N^+}}-\ddot{\underset{..}{O}}:^- \;\; K^+$$

What is the electron geometry around the nitrogen atom of this compound?

Solution

Note: The appendix contains the answer.

The electron geometry and the molecular geometry (with their associated angles) are both determined by the same arrangement of electron domains in a molecule. The key difference between the two is that molecular geometry considers only the shape resulting from the domains that form chemical bonds.

For example, if a molecule has no lone pairs of electrons around the central atom, the electron geometry and the molecular geometry are the same. In contrast, when a molecule has lone-pair electrons on the central atom, these lone pairs affect bond angles by contributing to the number of electron domains; however, the lone pair domains are excluded when determining the molecular geometry (ie, the shape made by the bonds), as illustrated in Table 2.6.

Table 2.6 Electron and molecular geometries.

Number of electron domains	Electron pair geometry	Molecular geometry		
		0 lone pairs	1 lone pair	2 lone pairs
 2	Linear	 Linear		
 3	Trigonal planar	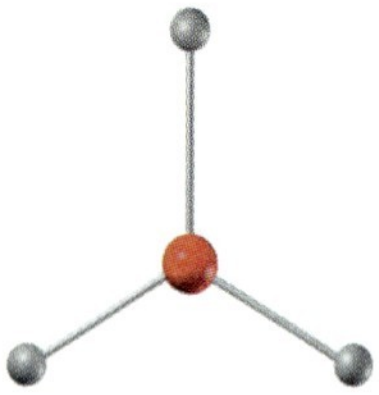 Trigonal planar	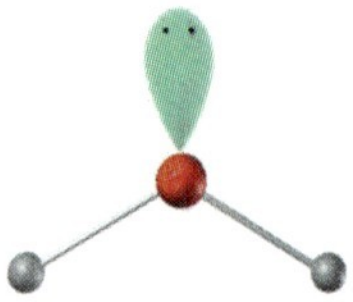 Bent	
 4	Tetrahedral	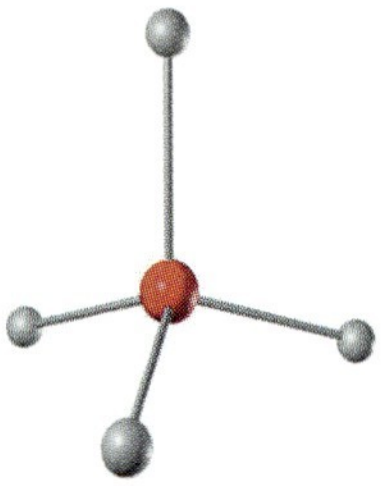 Tetrahedral	 Trigonal pyramidal	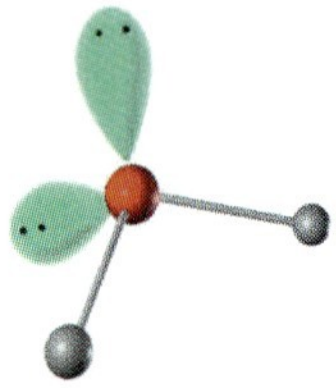 Bent
 5	Trigonal bipyramidal	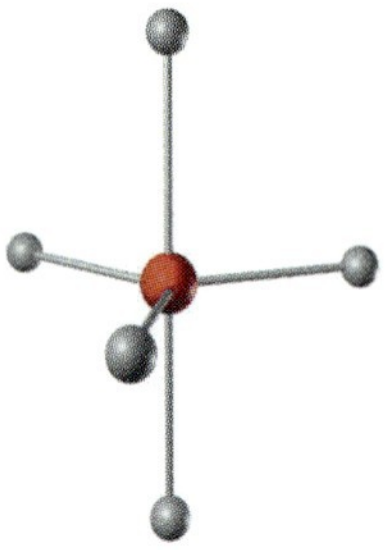 Trigonal bipyramidal	 See-saw	 T-shaped
 6	Octahedral	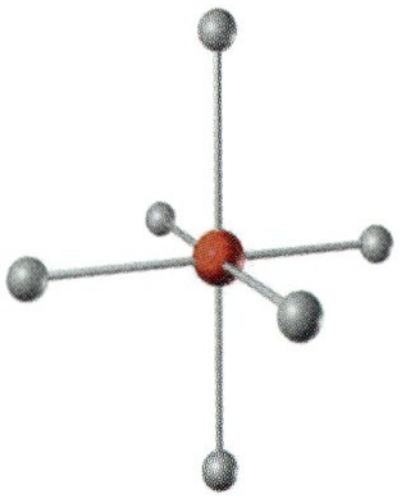 Octahedral	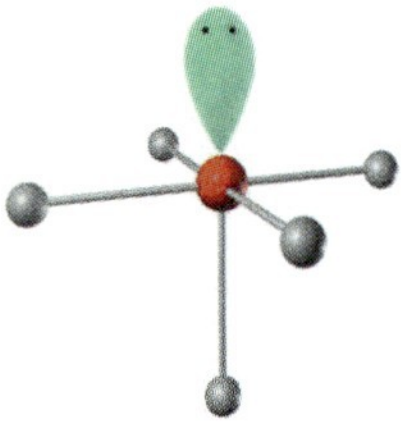 Square pyramidal	 Square planar

When determining the molecular geometry for a given compound (or the geometry around a particular atom in part of a larger compound), the following approach can be helpful:

1. Draw the Lewis structure for the compound (if not given).
2. Identify the number of electron domains around the central atom.
3. Determine the electron geometry based on the number of domains.
4. Exclude the lone-pair electrons from the electron geometry to determine the shape taken by the bonds (ie, the molecular geometry).

When considering bond angles for a molecule, it is important to note that not all electron domains have the same effect in terms of repulsion. Lone-pair electron domains have significantly greater repulsive forces compared to bonding electron domains. This phenomenon leads to slightly smaller bond angles in molecules that have lone pairs of electrons on the central atom when compared to bond angles in molecules with no lone pairs on the central atom, as shown in Figure 2.25.

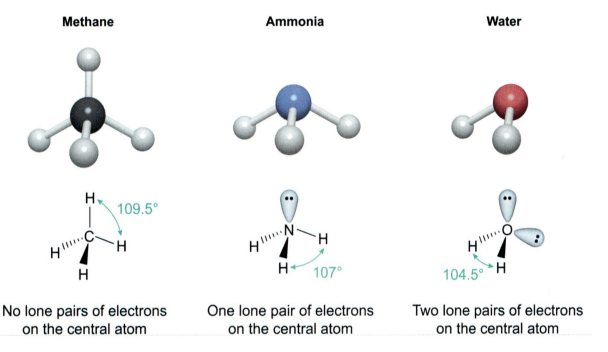

Figure 2.25 Effects of lone pairs of electrons on bond angles.

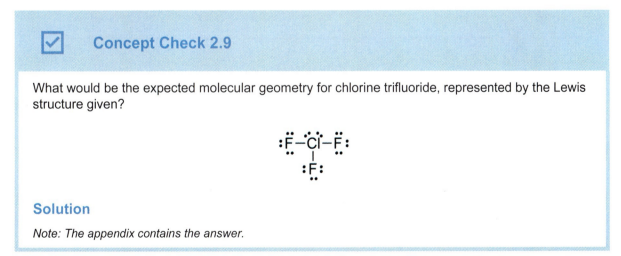

Concept Check 2.9

What would be the expected molecular geometry for chlorine trifluoride, represented by the Lewis structure given?

Solution

Note: The appendix contains the answer.

2.2.04 Orbital Hybridization

Covalent bonds are formed when the valence orbitals (eg, the outermost *s* and *p* orbitals) of two atoms overlap, allowing the electrons in those orbitals to be shared between the atoms. To create a more stable molecule, each atom in a molecule tends to hybridize its valence orbitals.

Hybridization is a process by which two or more atomic orbitals combine to form new **hybrid orbitals** of equivalent energy. Whenever atomic orbitals are hybridized, the total number of orbitals remains the same according to a principle known as **orbital conservation**. Although the total number of orbitals remains the same, the hybrid orbitals formed have different shapes than the original atomic orbitals.

Furthermore, to minimize electron repulsion, hybrid orbitals also adopt specific orientations determined by the number of hybrid orbitals formed. Figure 2.26 shows three common types of orbital hybridization.

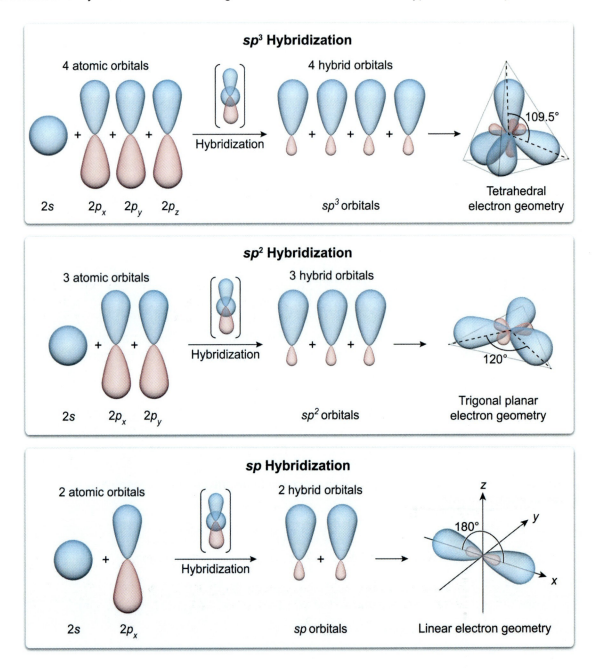

Figure 2.26 Hybridization of atomic orbitals.

Every atom capable of hybridization (hydrogen cannot hybridize) has three p orbitals in its valence shell. As such, sp and sp^2 hybridization result in an atom with one or two unhybridized p orbitals, respectively. These unhybridized p orbitals can participate in side-to-side overlap with unhybridized p orbitals from other atoms to form **pi (π) bonds** (Figure 2.27).

In contrast, the *end-to-end* overlap of hybrid orbitals results in a **sigma (σ) bond**. Ultimately, an atom's hybridization determines both the number and the type of bonds it can form.

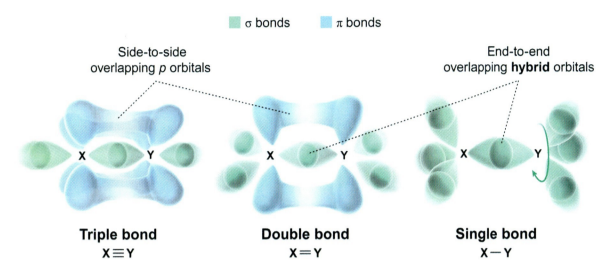

Figure 2.27 Orbital overlap in single, double, and triple bonds.

Hybridization has a stabilizing effect on a molecule because forming hybrid orbitals of equal energy lowers the overall energy of the atom, and thus the molecule. The number of hybrid orbitals that can form is always equal to the number of electron domains surrounding the atom, as summarized in Table 2.7.

Table 2.7 Hybridization of atoms based on the number of electron domains

Number of electron domains	2	3	4
Orbitals hybridized	s,p	s,p,p	s,p,p,p
Hybridization of atom	sp	sp^2	sp^3

Based on the pattern seen in Table 2.7, an easy approach to determine the hybridization of an atom is:

1. Draw the Lewis structure of the molecule (if not given).
2. Count the number of electron domains around the atom. The number of hybrid orbitals around the atom is equal to the number of electron domains surrounding that atom (ie, the sum of the superscripts of the hybridization type equals the number of electron domains).

For example, if an atom is surrounded by three electron domains, it must be sp^2 hybridized because the sum of the superscripts (ie, the number of hybridized orbitals) is:

$$sp^2 = s^1 p^2 = s + p + p \quad \Rightarrow \quad 1 + 2 = 3 \text{ hybridized orbitals}$$

$$3 \text{ hybridized orbitals} = 3 \text{ electron domains}$$

> **Concept Check 2.10**
>
> How many atoms are sp^3 hybridized in the following molecule?
>
> $$H-\underset{\underset{H}{|}}{\overset{\overset{H}{|}}{C}}-\underset{\underset{H}{|}}{\overset{\overset{H}{|}}{C}}-\overset{\overset{\ddot{\ddot{O}}}{||}}{C}-\underset{\underset{H}{|}}{\overset{\overset{H}{|}}{C}}-\ddot{\ddot{O}}-H$$
>
> **Solution**
>
> *Note: The appendix contains the answer.*

2.2.05 Molecular Polarity

A molecule is considered polar if it has a significant net dipole moment. Precise calculations of net dipole moments involve assessing vector quantities, which is a topic of the subject of physics and not discussed here. However, a qualitative assessment of whether a molecule possesses a significant net dipole moment can be made by considering the number of polar bonds present, the strength of the dipoles in each polar bond, the molecular geometry, and any resonance (when applicable).

As discussed in Concept 2.2.01, when two bonded atoms have significantly different electronegativities, the bond is polar and contains a dipole moment. Vector components of individual bond dipole moments that point in the same direction add together and increase the overall molecular dipole moment whereas those that point in opposite directions cancel each other, as seen in Figure 2.28. If all the bond dipole moments in three dimensions cancel each other, the molecule is nonpolar.

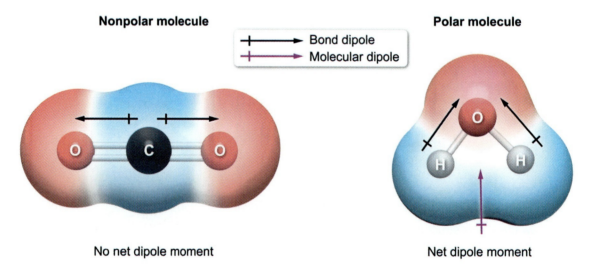

Figure 2.28 Effect of molecular geometry (bond orientations) on molecular dipole moments.

For molecules with a single central atom, these principles can be distilled into the following simple set of rules for determining molecular polarity:

1. If a molecule has one lone pair of electrons on the central atom, the molecule is polar. If a molecule has two or more lone pairs on the central atom, the molecule is polar *unless* those lone pairs are in geometrically opposite positions *and* all atoms bonded to the central atom are identical (Figure 2.29).

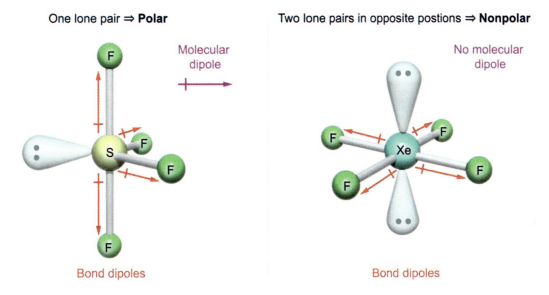

Figure 2.29 Effect of central atom lone pairs on molecular polarity.

2. If a molecule has no lone pairs and the atoms bonded to the central atom are all the same, the molecule is nonpolar. If a molecule has no lone pairs and the atoms bonded to the central atom are *not* the same, the molecule is generally polar (Figure 2.30).

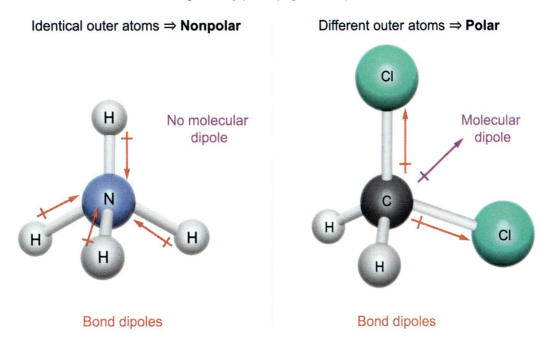

Figure 2.30 Effect of outer atoms on molecular polarity.

3. For any diatomic molecule, if the bond is polar, the molecule is polar (Figure 2.31).

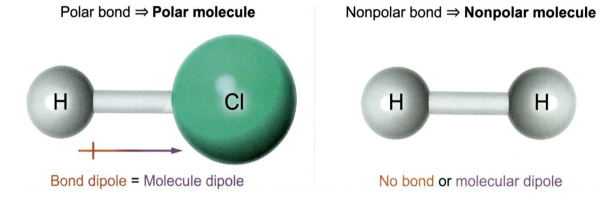

Figure 2.31 Polarity of diatomic molecules.

4. For molecules with resonance structures, the distribution of partial charges (δ⁺, δ⁻) can result in a bond dipole even in bonds between two atoms of the same element. In such cases, the polarity of the molecule can be evaluated from its resonance hybrid. If the partial charge dipoles in the resonance hybrid do not cancel, the molecule is polar; otherwise, the molecule is nonpolar (Figure 2.32).

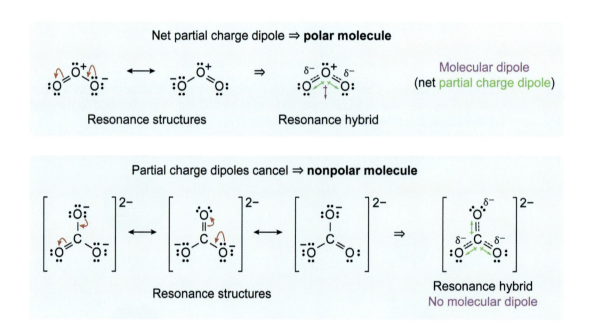

Figure 2.32 Effect of resonance on molecular polarity.

Bond polarity and molecular polarity are not the same thing. It is possible to have a nonpolar molecule with polar bonds (eg, CCl_4) or a polar molecule with seemingly nonpolar bonds that still form a net dipole due to resonance (eg, O_3). In addition, larger molecules can have both polar and nonpolar regions in their structures. Each of these situations can be evaluated by the same principles outlined here.

Concept Check 2.11

Determine whether the following molecule is polar or nonpolar:

$$:\!\ddot{\underset{..}{F}}\!-\!\overset{..}{\underset{..}{Xe}}\!-\!\ddot{\underset{..}{F}}\!:$$

Solution

Note: The appendix contains the answer.

Lesson 2.3

Intermolecular Forces

Introduction

As discussed in Concept 2.1.01, covalent molecules are held together internally by covalent bonds, which are strong forces that result from the sharing of electrons between atoms.

In contrast, **intermolecular forces** are weaker forces that result from *noncovalent* interactions between neighboring molecules. These types of noncovalent interactions can also be *intramolecular* (between different parts of the *same* molecule) in large molecules with extensive structures. Attractive intermolecular forces are responsible for causing molecules to associate with each other or to collectively aggregate, and this effect is stronger when molecules are close together. As such, the effects from intermolecular forces are particularly relevant for solids and liquids.

Noncovalent interactions between the dipoles of two or more neutral molecules are collectively called **van der Waals forces**. These forces include:

- Dipole-dipole interactions (ie, attractions between two permanent dipoles)
- Dipole–induced dipole interactions (ie, attractions between a permanent dipole and an induced dipole)
- London dispersion forces (ie, attractions between two induced dipoles)

This lesson discusses the different types of van der Waals forces, the nature and relative strengths of each type of interaction, and some selected properties influenced by these forces.

2.3.01 London Dispersion Forces

Due to the impossibility of determining both the position and the momentum of an electron (Concept 1.3.04), the most effective way to model the behavior of electrons is with probability distributions (ie, orbitals). It can be helpful to think of these orbitals as clouds of electrons.

These electron clouds can be distorted by the random motion of electrons within a molecule and collisions with (or proximity to) the electron clouds of other molecules. Such distortions cause weak, temporary dipoles to form in neighboring molecules, which attracts the molecules to one another. These weak attractive forces between induced, temporary dipoles are called London dispersion forces.

Figure 2.33 illustrates how a London force interaction might occur between two helium atoms.

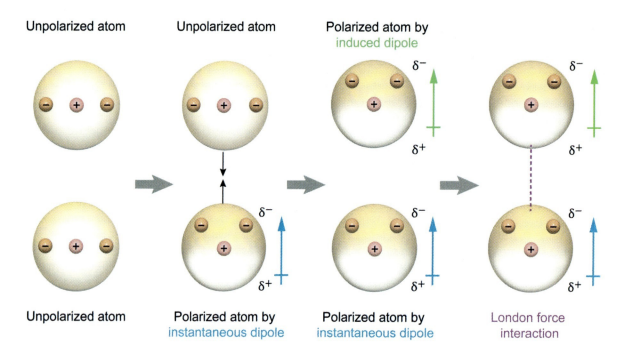

Figure 2.33 London dispersion force interaction between two adjacent helium atoms.

The strength of London dispersion forces depends on the polarizability of a molecule. Larger molecules have larger electron clouds that are more easily distorted. Likewise, longer molecules with fewer branches have more electron cloud "surface area" for interactions and are therefore more polarizable. Consequently, larger and longer molecules have stronger London dispersion forces than smaller and shorter molecules.

London dispersion forces are the only type of intermolecular force present in nonpolar molecules. However, London dispersion forces are also present in polar molecules and play a significant role in influencing the properties of all molecular compounds. While each individual London force interaction is relatively weak, the cumulative effect of these interactions can be incredibly strong, especially when considered in larger molecules.

Concept Check 2.12

Arrange the following molecules in order of increasing strength of London dispersion forces.

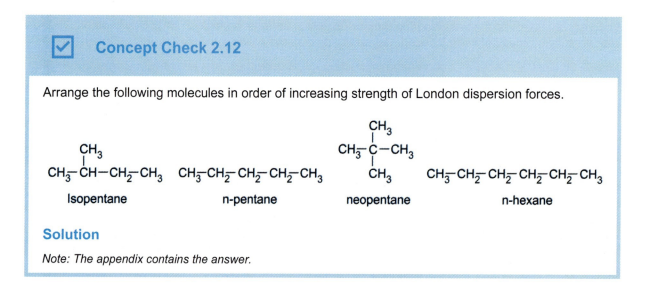

Solution

Note: The appendix contains the answer.

2.3.02 Dipole Interactions

A molecule with a significant net dipole moment is considered polar, having a partial positive charge (δ^+) on one region and a partial negative charge (δ^-) on another region of the same molecule (Concept 2.2.05). Several types of intermolecular forces involve noncovalent interactions with the permanent dipoles of polar bonds. This section examines a few ways that polar bond dipoles interact with other chemical species.

Dipole-Dipole Forces

Intermolecular forces resulting from the electrostatic attraction between the partially positive end of the permanent dipole of one polar molecule and the partially negative end of the permanent dipole of a neighboring polar molecule are called dipole-dipole forces (Figure 2.34).

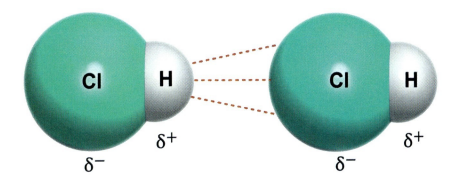

Figure 2.34 Dipole-dipole interaction of two HCl molecules.

Since the dipoles in polar molecules are permanent rather than instantaneous and temporary, the intermolecular forces between polar molecules tend to be significantly stronger than the London dispersion forces found in nonpolar molecules of comparable size. The strength of a dipole-dipole interaction is directly proportional to the magnitude of the net dipole moments of the molecules involved. Thus, larger net dipole moments result in stronger dipole-dipole forces.

> ☑ **Concept Check 2.13**
>
> Which of the following substances exhibits the strongest dipole-dipole forces?
>
>   H–Cl   H–Br   H–I
>
> **Solution**
>
> Note: The appendix contains the answer.

Dipole–Induced Dipole Forces

In addition to interactions between other polar molecules, the permanent dipoles of polar molecules can also interact with nonpolar molecules. These dipole–induced dipole interactions occur when the permanent dipole of a polar molecule induces a weak, temporary dipole in a neighboring nonpolar molecule, as shown in Figure 2.35. As such, a dipole–induced dipole interaction is weaker than a dipole-dipole interaction but stronger than an induced dipole–induced dipole interaction (ie, London dispersion forces).

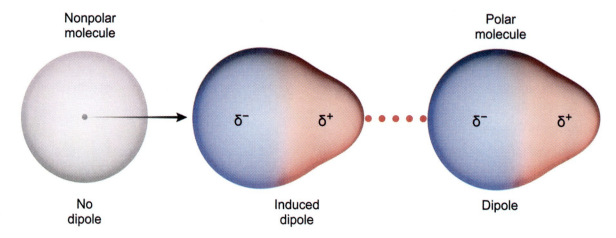

Figure 2.35 Dipole–induced dipole interaction.

Ion-Dipole Forces

Ion-dipole interactions involve the full charge of an ion being attracted to the opposite partial charge of a permanent dipole in a polar molecule, as shown in Figure 2.36. The attractive forces resulting from these interactions are the strongest of all the dipole interactions due to the relative magnitude of the charge on the ion.

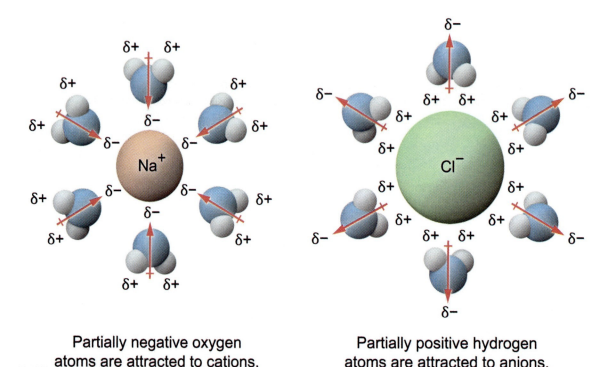

Figure 2.36 Ion-dipole forces observed when NaCl is dissolved in water.

Although ion-dipole forces are a type of dipole interaction, they are generally not considered one of the van der Waals forces because they involve fully charged ions. However, ion-dipole interactions are introduced here in the greater context of intermolecular forces due to their importance in dissolving ionic compounds to form solutions.

Concept Check 2.14

Indicate the type of intermolecular interactions that occur between the two different species in each of the following mixtures.

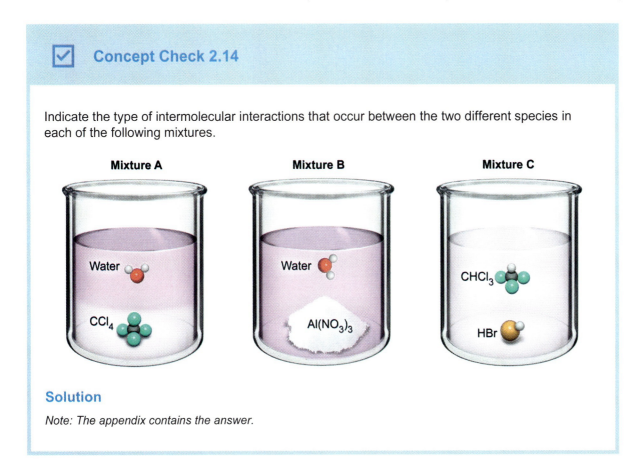

Solution
Note: The appendix contains the answer.

2.3.03 Hydrogen Bonding

Hydrogen bonding is a special type of dipole-dipole interaction that occurs between the particularly strong dipoles formed when a hydrogen atom is covalently bonded to the small, highly electronegative atoms of fluorine, oxygen, or nitrogen. The highly polar H–F, H–O, or H–N bonds result in hydrogen atoms with very large partial positive charges that act as **hydrogen bond donors** in these interactions. Conversely, the lone pairs of electrons on the fluorine, oxygen, and nitrogen atoms have a large partial negative charge and serve as **hydrogen bond acceptors** (Figure 2.37).

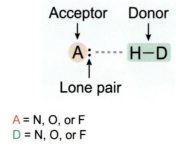

A = N, O, or F
D = N, O, or F

Figure 2.37 Hydrogen bonding donors and hydrogen bonding acceptors in a hydrogen bonding interaction.

Hydrogen bonding is the most powerful of the van der Waals forces. As such, the strength of these interactions has profound effects on the physical properties of compounds that are capable of hydrogen bonding interactions. These properties are discussed in more detail in Concept 2.3.05.

Concept Check 2.15

Consider the interactions labeled in the image shown.

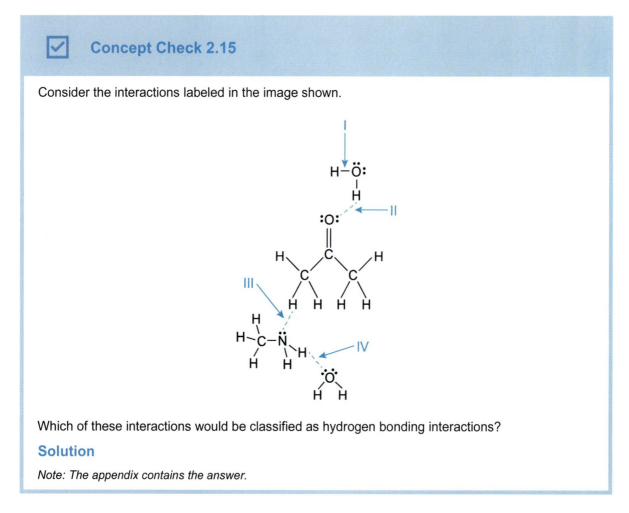

Which of these interactions would be classified as hydrogen bonding interactions?

Solution

Note: The appendix contains the answer.

2.3.04 Relative Strengths of Intermolecular Forces

All intermolecular forces involve an electrostatic attraction between opposite partial or full charges. The relative strength of each intermolecular force can be determined based on the relative magnitudes of those charges (Figure 2.38).

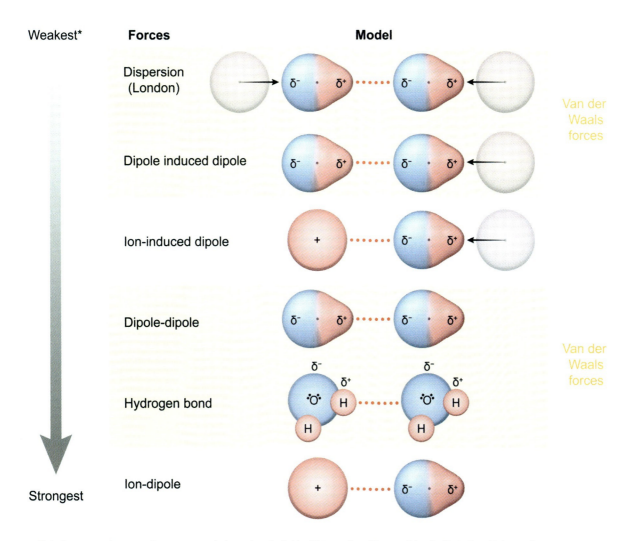

* Relative strength comparisons are made based on individual interaction. The combined effect of multiple weaker interactions can sometimes overpower stronger interactions

Figure 2.38 Relative strengths of intermolecular forces.

The weakest intermolecular forces are those that involve small, temporary dipoles (eg, dispersion forces). The stronger intermolecular forces are those that involve the stronger permanent dipoles of polar molecules or fully charged ions (eg, ion-dipole forces).

Concept Check 2.16

Arrange the interactions labeled in the image in order of increasing strength (ie, weakest to strongest).

Solution
Note: The appendix contains the answer.

2.3.05 Properties Influenced by Intermolecular Forces

The strength of the attractive forces between the molecules of a substance greatly influences its physical properties, including vapor pressure, boiling point, and hydrophobicity. When evaluating the effect that intermolecular forces have on physical properties, the types of intermolecular forces present and the number of interactions of each type must both be considered.

Vapor Pressure

For any liquid, the molecules in the sample with the highest energy can escape into the gas phase by a process called evaporation. These molecules can later condense and move back into the liquid phase. In a sealed system, these two processes eventually reach a **dynamic equilibrium**, the state in which the two opposing processes occur at equal rates such that the system has no net change. The pressure exerted by the small amount of vapor over a liquid in dynamic equilibrium is called the **vapor pressure** (Figure 2.39).

Chapter 2: Chemical Bonding, Reactions, and Stoichiometry

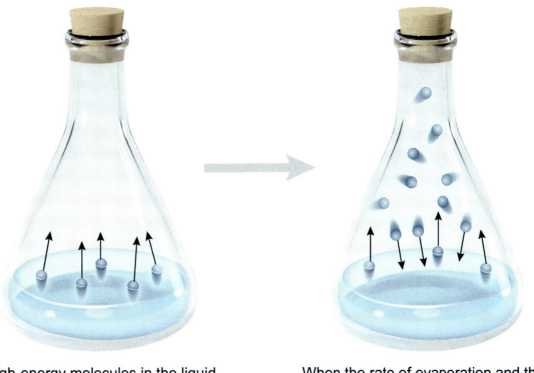

High-energy molecules in the liquid phase escape into the gas phase, creating pressure inside the container.

When the rate of evaporation and the rate of condensation are equal, dynamic equilibrium is established.

Figure 2.39 Vapor pressure in a closed system.

When intermolecular forces are stronger, it is more difficult (ie, requires more energy) for molecules to escape into the gas phase, and gas molecules are also more likely to be "captured" by intermolecular interactions and condense back into a liquid. Therefore, the stronger the intermolecular forces in a liquid, the lower the vapor pressure.

Boiling Point

The temperature at which the vapor pressure of a liquid is equal to the external pressure of the surroundings is the **boiling point**. Vapor pressure is dependent on temperature: Higher temperatures mean more molecules have enough energy to escape into the gas phase and are less likely to condense, resulting in a greater vapor pressure.

As temperature increases, the vapor pressure of a substance eventually equals the external pressure and boiling begins. The normal boiling point of a substance is the boiling point when the external pressure is 1.0 atm (ie, 760 mmHg). When intermolecular forces are greater, the vapor pressure is lower and much more heat must be added to the system to increase the vapor pressure, resulting in a higher boiling point. Thus, substances with stronger intermolecular forces have higher boiling points, as shown in Figure 2.40.

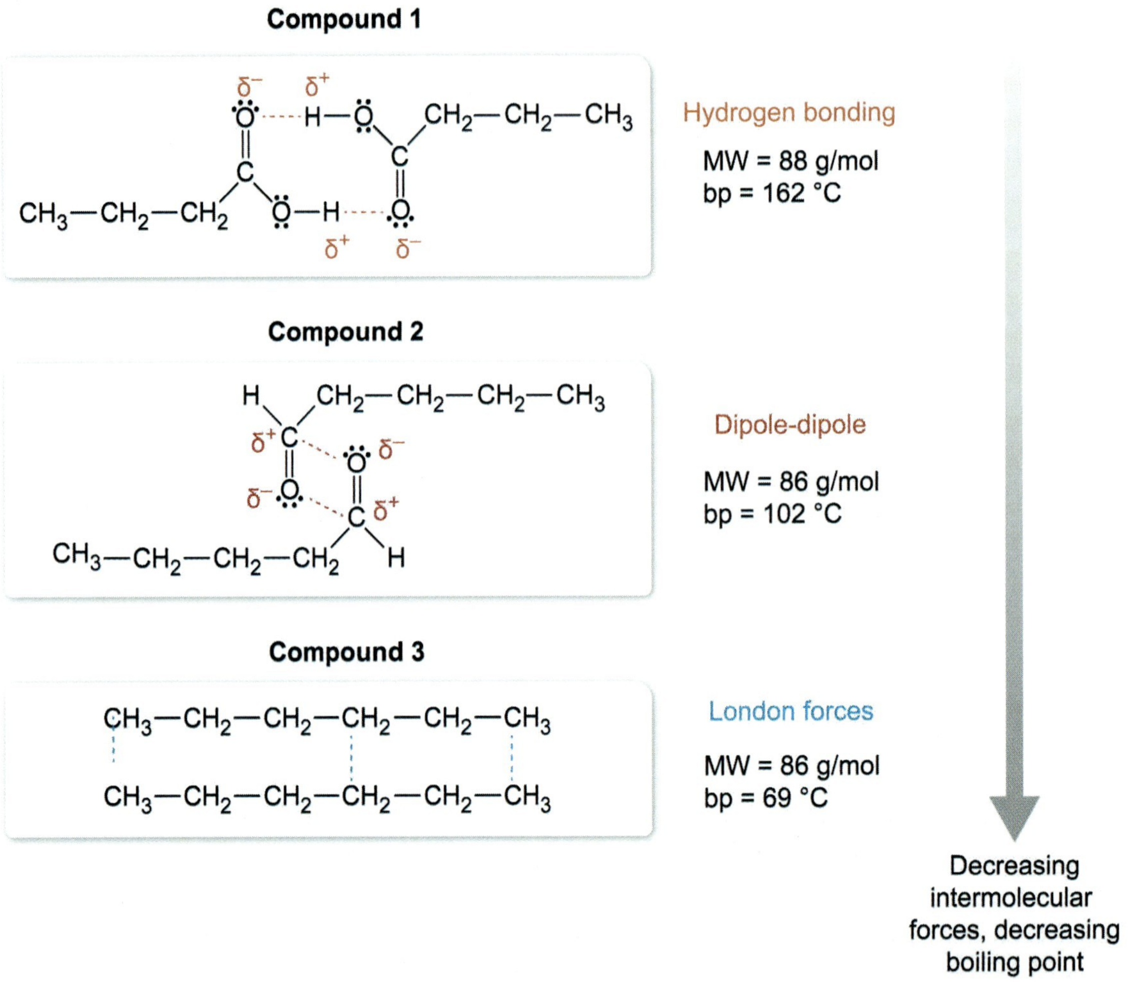

Compound 1

CH₃—CH₂—CH₂—C(=Ö⁻⋯H—Ö⁺)—Ö—H⋯:Ö:—C(=Ö)—CH₂—CH₂—CH₃

Hydrogen bonding

MW = 88 g/mol
bp = 162 °C

Compound 2

Structure showing two ester/ketone-like molecules with δ⁺C=Ö δ⁻ dipole-dipole interactions between CH₃—CH₂—CH₂—CH₂—C(H)=O and H—C(=O)—CH₂—CH₂—CH₂—CH₃

Dipole-dipole

MW = 86 g/mol
bp = 102 °C

Compound 3

CH₃—CH₂—CH₂—CH₂—CH₂—CH₃
CH₃—CH₂—CH₂—CH₂—CH₂—CH₃

London forces

MW = 86 g/mol
bp = 69 °C

Decreasing intermolecular forces, decreasing boiling point

Figure 2.40 Effect of intermolecular forces on the boiling point of compounds with similar molecular weights.

Since compounds with similar molecular weights have similar London dispersion forces, the difference in boiling points between compounds like H_2S and SiH_4 is almost entirely due to the dipole-dipole forces, which are present in H_2S but not in SiH_4 (Figure 2.41). However, in both polar and nonpolar compounds, the general upward trend in boiling points as molecular mass increases is due to the increase in London dispersion forces that occurs as a result of the greater polarizability of larger molecules.

The marked deviation of H_2O from this trend can be explained by the incredible strength of the hydrogen bonding interactions between water molecules.

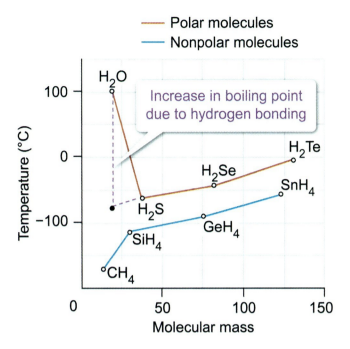

Figure 2.41 Boiling point of polar and nonpolar substances versus molecular mass.

Hydrophobicity

Molecules of H_2O, which are highly polar and engage in hydrogen bonding tend to be attracted to other polar substances due to the attraction between opposite partial charges. Polar substances that can easily interact with (and be dissolved in) water are called **hydrophilic** (ie, "water-loving").

In contrast, the weak temporary dipoles of nonpolar substances cannot effectively compete with the attraction water molecules have for each other, resulting in separation of the nonpolar molecules from the polar water molecules. Nonpolar substances such as these are referred to as **hydrophobic** (ie, "water-fearing"). Hydrophobic molecules tend to bunch together to limit their interactions with water molecules, as shown in Figure 2.42.

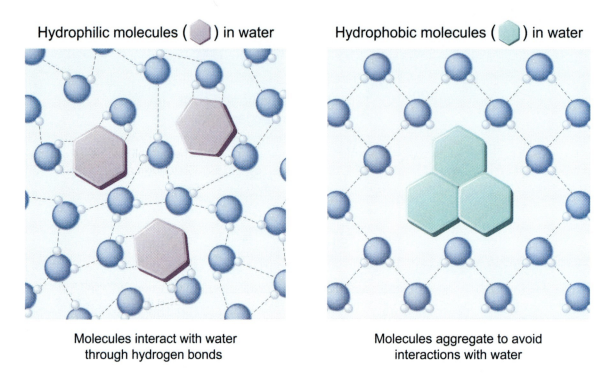

Figure 2.42 Behavior of hydrophilic and hydrophobic molecules in water.

Hydrophilic and hydrophobic behavior tends to be the result of the type of intermolecular forces present in a molecule rather than the magnitude of these forces. Compounds having only London dispersion forces are hydrophobic; compounds that have significant dipole-dipole forces or hydrogen bonding tend to be more hydrophilic. Large biological molecules often have both hydrophilic and hydrophobic regions within their structures, which affects the way they interact with other molecular species.

Concept Check 2.17

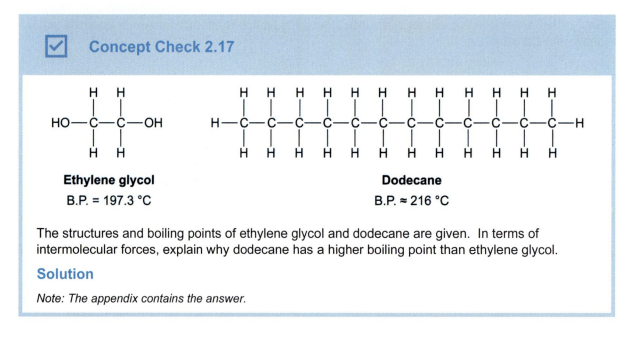

The structures and boiling points of ethylene glycol and dodecane are given. In terms of intermolecular forces, explain why dodecane has a higher boiling point than ethylene glycol.

Solution

Note: The appendix contains the answer.

Lesson 2.4
Chemical Reactions

Introduction

Chemical reactions are represented by chemical equations, which show the overall change of reactants into products. For example, the reaction between magnesium and oxygen to form magnesium oxide can be represented by the following chemical equation:

$$2 \text{ Mg}(s) + \text{O}_2(g) \rightarrow 2 \text{ MgO}(s)$$

A chemical equation has several key features:

- **Reaction arrow**: A single arrow ($\rightarrow$) separates the reactant species (left side of the arrow) from the product species (right side of the arrow). In some instances, a double arrow ($\rightleftharpoons$) is used to indicate that a reaction is reversible.
- **Reaction conditions**: When heat is added to a reaction, the word "heat" or the symbol "Δ" is often placed above the reaction arrow. If a catalyst is present during a reaction, its chemical formula is written above the reaction arrow. The addition of other compounds or the use of specific temperatures are also sometimes placed above or below the reaction arrow.
- **Stoichiometric coefficients**: The coefficients to the left of each chemical formula are whole number multipliers that indicate the number of each molecule or atom that reacts or is formed. For example, 2 Mg atoms react with 1 O_2 molecule to form 2 MgO molecules.
- **States**: The state of each reactant and product is often specified by an italicized letter inside parentheses placed to the right of the chemical formula. Four common states are abbreviated as solid (*s*), liquid (*l*), gas (*g*), and aqueous (*aq*) (ie, dissolved in water).

In this lesson, various aspects of chemical reactions are explored in detail, including the classification of reaction types and how to balance chemical equations.

2.4.01 Types of Reactions

Chemical reactions can be classified into different types based on certain patterns that they display. Five common types of chemical reactions are described as follows.

Combination Reactions

Combination reactions (also referred to as synthesis reactions) follow the pattern of *two* or more chemical species reacting to combine and form *one* product species. This type of reaction is shown in Figure 2.43.

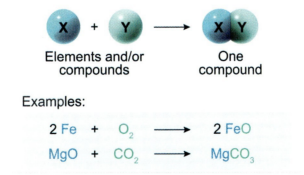

Figure 2.43 The general form of a combination reaction with selected examples.

Decomposition Reactions

A decomposition reaction involves the breaking down (decomposing) of a compound into simpler, more stable compounds or into its elemental constituents. This type of reaction usually requires the addition of energy (such as heat) to break the bonds of the original compound. As such, decomposition reactions follow the pattern of *one* reactant species decomposing into *two* or more product species, as shown in Figure 2.44. Note that a decomposition reaction can be viewed as the opposite of a combination reaction.

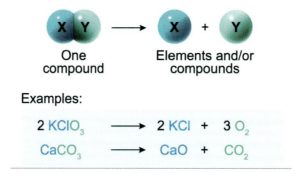

Figure 2.44 The general form of a decomposition reaction with selected examples.

Single Replacement Reactions

A single replacement reaction involves the replacement of one type of atom with another in a reaction between a *compound* and an *element* (ie, a substance consisting of one type of atom that is not ionized or bonded with other types of atoms). In these reactions, a sufficiently reactive atom in its elemental state loses or gains electrons and takes the place of one of the atoms in the compound. The replaced atom then loses or gains electrons and is ejected in its elemental state. As a result, a new compound and a new element are formed as the products, as demonstrated in Figure 2.45.

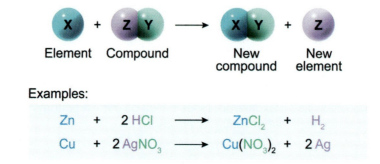

Figure 2.45 The general form of a single replacement reaction with selected examples.

Double Replacement Reactions

Double replacement reactions involve an exchange of constituent bonding partners between *two* reacting compounds, resulting in the formation of *two* new product compounds. For double replacement reactions between two ionic species, the cations and anions of the two compounds exchange counter ions. Frequently, in aqueous ionic solutions, one of the products that forms in a double replacement reaction is insoluble in water and manifests as a solid precipitate. This type of reaction is shown in Figure 2.46.

Chapter 2: Chemical Bonding, Reactions, and Stoichiometry

X Y + Z W ⟶ X W + Z Y

Two compounds   Two new compounds
                      (ion swap)

Examples:

AgNO₃ + NaCl ⟶ AgCl + NaNO₃

Pb(NO₃)₂ + 2 KI ⟶ PbI₂ + 2 KNO₃

Figure 2.46 The general form of a double replacement reaction with selected examples.

Combustion Reactions

Combustion reactions involve the burning of a fuel (often a hydrocarbon, an alcohol, or other flammable organic compound) in the presence of *oxygen* (often from the air). When the supply of oxygen is plentiful, complete combustion occurs to form *carbon dioxide* (CO_2) and *water*. This type of reaction is shown in Figure 2.47.

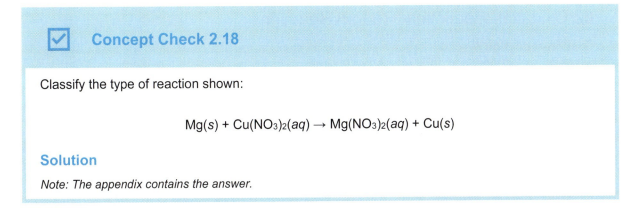

Hydrocarbon, alcohol or organic compound + O₂ ⟶ CO₂ + H₂O

Oxygen   Products of complete combustion

Examples:

C₃H₈ + 5 O₂ ⟶ 3 CO₂ + 4 H₂O

C₂H₆O + 3 O₂ ⟶ 2 CO₂ + 3 H₂O

Figure 2.47 The general form of a combustion reaction with selected examples.

☑ **Concept Check 2.18**

Classify the type of reaction shown:

$$Mg(s) + Cu(NO_3)_2(aq) \rightarrow Mg(NO_3)_2(aq) + Cu(s)$$

Solution

Note: The appendix contains the answer.

Although chemical reactions are often classified according to common reaction patterns, reactions can sometimes be more conveniently and descriptively classified by the type of chemical process involved or by the outcome of the reaction. As such, reactions can fall under more than one classification depending on the criteria used. Three common examples of this alternative approach to reaction classification are precipitation, chelation, and oxidation-reduction reactions.

Precipitation Reactions

Precipitation reactions include any reaction that produces an insoluble solid (precipitate) that drops out of the solution as the reaction proceeds. Although many double replacement reactions involving aqueous ionic solutions do form precipitates, precipitation reactions (as a broad classification) can also include other reactions that are not double replacement.

Chelation

The process of chelation involves a metal cation bound in a coordination complex by a single ligand that can form two or more coordinate covalent bonds in a pincer-like arrangement that results in the completion of a ring within the structure of the complex (Concept 2.1.07). Although chelation can be viewed broadly as a special case of a combination reaction, categorizing it as a chelation to indicate the chemical process and its outcome is more descriptive of the reaction.

Oxidation-Reduction (Redox) Reactions

Redox reactions are those in which the processes of oxidation and reduction take place. Because oxygen readily reacts with most elements, oxidation was historically defined as the formation of bonds with oxygen. Conversely, reduction was then taken to mean the loss of bonds to oxygen. Two redox reactions in which copper (Cu) atoms follow these definitions are:

Oxidation of Cu: $\quad 2\,Cu(s) + O_2(g) \rightarrow 2\,CuO(s)$

Reduction of Cu: $\quad CuO(s) + H_2(g) \rightarrow 2\,Cu(s) + H_2O(g)$

However, these historical definitions were ultimately found to be inadequate to identify all redox reactions because similar chemical processes are observed in reactions without oxygen atoms. In fact, all single replacement reactions (including those lacking oxygen) are also redox reactions. Redox reactions are examined more closely in Concept 2.4.02, where the modern definitions of oxidation and reduction are introduced.

2.4.02 Oxidation-Reduction (Redox) Reactions

An **oxidation-reduction reaction** (redox reaction) is an electron-transfer reaction in which one species is oxidized (loss of electrons) and another species is simultaneously reduced (gain of electrons). One way to remember the flow of electrons in a redox reaction is with the mnemonic "OIL RIG": oxidation is loss of electrons and reduction is gain of electrons. This mnemonic is illustrated in Figure 2.48.

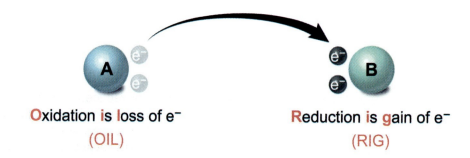

Oxidation is loss of e⁻   Reduction is gain of e⁻
(OIL)   (RIG)

Figure 2.48 Oxidation-reduction reactions involve the transfer of electrons.

During a redox reaction, the species containing the oxidized atom acts as the **reducing agent** because it *causes reduction* in another species as the reducing agent itself is oxidized in the process. Conversely, the species with the reduced atom functions as the **oxidizing agent** because it *causes oxidation* in another species as the oxidizing agent itself is reduced in the process. This interaction between the oxidizing and reducing agents is illustrated in Figure 2.49.

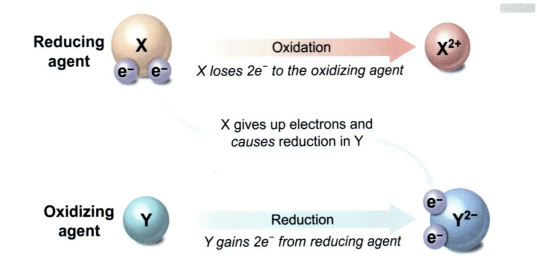

Figure 2.49 Reducing agents cause reduction, and oxidizing agents cause oxidation.

As discussed in Concept 2.4.01, the term *oxidation* was historically used to describe reactions that involve the formation of bonds to oxygen, while *reduction* was used to describe reactions that involve the loss of bonds to oxygen. Although these historical definitions for oxidation and reduction are limited, they can sometimes still be useful because oxidation and reduction in organic molecules can be identified by comparing the relative number of bonds to oxygen (O) before and after a reaction.

In general, carbon (C) atoms that gain bonds to O are oxidized whereas those that lose bonds to O are reduced. Because bonds to O that are lost are often replaced by bonds to H in organic compounds, redox reactions can also be identified by comparing bond to H atoms before and after a reaction (ie, losing C–H bonds often indicates oxidation, and gaining C–H bonds often indicates reduction).

☑ Concept Check 2.19

$$CH_4(g) + 2\,O_2(g) \rightarrow CO_2(g) + 2\,H_2O(g)$$

Which element undergoes oxidation and which element undergoes reduction during the reaction shown? What is the oxidizing agent, and what is the reducing agent?

Solution
Note: The appendix contains the answer.

In most redox reactions, one element is oxidized as a *different element* is reduced. In rare cases, a special type of redox reaction called a **disproportionation reaction** can happen in which both the oxidation and reduction occur to atoms of the *same element*. If a compound participates in disproportionation, atoms of a particular element undergo oxidation in some molecules of the compound while atoms of the same element undergo reduction in other molecules of the compound. As such, the reducing agent and oxidizing agent are both the same species in a disproportionation reaction.

Determining which substance is oxidized (ie, loses electrons) and which substance is reduced (ie, gains electrons) in a redox reaction can be achieved by comparing the **oxidation numbers** of each atom before and after the reaction.

2.4.03 Oxidation Numbers

Because redox reactions always involve electron transfer, a method of tracking the distribution of electron charge must be used to determine which atoms are gaining and losing electrons. The two primary methods of assessing charge distribution in chemical substances are formal charge and oxidation number.

Formal charge assesses the distribution of electron charges in Lewis structures by assuming that all bonding electrons are shared equally (ie, nonpolar covalent bonds). As such, the number of bonding electrons is divided in half when calculating formal charge:

$$\text{Formal charge} = \begin{pmatrix}\text{Group}\\\text{valence}\end{pmatrix} - \begin{pmatrix}\text{Nonbonding}\\\text{electrons}\end{pmatrix} - \frac{1}{2}\begin{pmatrix}\text{Bonding}\\\text{electrons}\end{pmatrix}$$

In contrast, the **oxidation number** (oxidation state) is the charge that an atom within a molecule would have if all bonding electrons are fully transferred to (and "owned" by) the most electronegative atom (ie, considering the bond as if it were an ionic bond). As such, when calculating the oxidation number, the bonding electrons are assigned to the most electronegative atom in a bond and *not divided* in half unless the bonded atoms have an equal electronegativity:

$$\text{Oxidation number} = \begin{pmatrix}\text{Group}\\\text{valence}\end{pmatrix} - \begin{pmatrix}\text{Nonbonding}\\\text{electrons}\end{pmatrix} - \begin{pmatrix}\text{Assigned}\\\text{bonding electrons}\end{pmatrix}$$

Accordingly, the oxidation number of an atom is conceptually different from its formal charge, and the two values are not necessarily equal. In reality, no chemical bond is fully covalent or fully ionic, and the actual electron density lies somewhere between the two extremes assumed by the different calculations of formal charge and oxidation number (Figure 2.50).

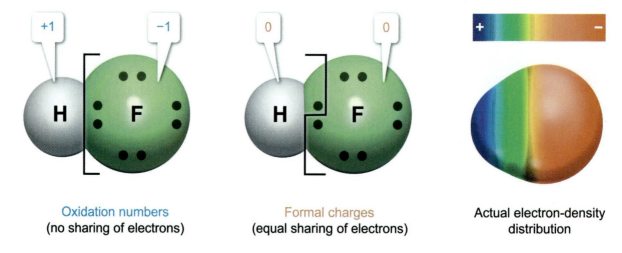

Figure 2.50 Oxidation number, formal charge, and electron-density models for an HF molecule.

Because redox reactions are fundamentally electron transfers, oxidation numbers can be used for tracking the movement of electrons. Comparing the oxidation number of an element in the reactants with the oxidation number of that same element in the products indicates whether the element has gained or lost electrons. An increased oxidation number indicates a loss of electrons (oxidation), whereas a decreased oxidation number signals a gain of electrons (reduction). A change of one unit occurs for each electron transferred.

Oxidation numbers may be assessed two different ways. The first method utilizes Lewis structures and the oxidation number equation to determine the hypothetical charge resulting from the number of electrons lost or gained relative to the number of valence electrons present in the atom's elemental state (ie, its group valence number). All bonding electrons are assigned to the most electronegative atom, but if the electronegativities are equal, the electrons are evenly divided between the atoms, as shown in Figure 2.51.

$$\begin{pmatrix} \text{Oxidation} \\ \text{number} \end{pmatrix} = \begin{pmatrix} \text{Group} \\ \text{valence} \end{pmatrix} - \begin{pmatrix} \text{Nonbonding} \\ \text{electrons} \end{pmatrix} - \begin{pmatrix} \text{Assigned bonding} \\ \text{electrons} \end{pmatrix}$$

Guideline 1: "Winner takes all"

Bonding electrons are assigned to the most electronegative atom in the bond.

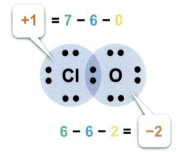

+1 = 7 − 6 − 0

6 − 6 − 2 = −2

Cl−O bond:
2 bonding electrons assigned to O (none assigned to Cl)

Guideline 2: "No winner, both share"

Bonding electrons are assigned to the most electronegative atom in the bond.

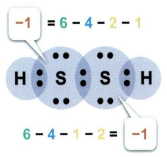

−1 = 6 − 4 − 2 − 1

6 − 4 − 1 − 2 = −1

S−S bond:
1 bonding electron assigned to each S (2 electrons, evenly shared)

S−H bond:
All electrons assigned to the S atom

Figure 2.51 Guidelines for assigning bonding electrons to determine oxidation numbers.

For compounds in which all atoms of a given element have the same oxidation number, oxidation numbers can be alternatively assessed using the compound formula and oxidation state rules. By this second method, atoms with consistent oxidation states (ie, those that rarely vary) are used to determine oxidation states for atoms that have variable states (eg, most metal atoms).

The following set of rules can be used to assign oxidation numbers to atoms in a compound:

- Free elements (ie, H_2, S_8, P_4, Mg) have an oxidation number of 0.
- Hydrogen usually has an oxidation number of +1. However, if hydrogen is bonded to a less electronegative element (eg, metals in Groups 1 and 2), its oxidation number is −1.
- Alkali metals (Group 1) always have an oxidation number of +1.
- Alkaline-earth metals (Group 2) always have an oxidation number of +2.
- Halogens (Group 17) have an oxidation number of −1 unless bonded to a more electronegative atom (eg, Cl = +1 in HOCl).
- Oxygen typically has an oxidation number of −2. However, oxygen has an oxidation number of −1 in peroxides (eg, H_2O_2) and +2 in OF_2.
- Monatomic ions (eg, Na^+, Mg^{2+}, F^-) have an oxidation number equal to their charge.
- The sum of the oxidation numbers of every atom in the structure of a compound or polyatomic ion must equal the net charge of the chemical formula.

For example, the oxidation number of sulfur in HSO₃⁻ can be determined using these rules. Since the oxidation states of oxygen (−2) and hydrogen (+1) are consistent, they can be used to find the value for sulfur algebraically:

$$HSO_3^- = -1 \text{ (net charge)}$$
$$H + S + 3\,O = -1$$
$$(+1) + S + 3(-2) = -1$$
$$1 + S - 6 = -1$$
$$S = +4$$

> ☑ **Concept Check 2.20**
>
> Consider the following redox reaction:
>
> $$Zn(s) + 2\,MnO_2(s) \rightarrow ZnO(s) + Mn_2O_3(s)$$
>
> Determine which element is oxidized and which element is reduced.
>
> **Solution**
>
> *Note: The appendix contains the answer.*

2.4.04 Reaction Prediction

As discussed in Concept 2.4.01, chemical reactions can often be broadly classified according to common reaction patterns or by the chemical processes involved (eg, precipitation, chelation, oxidation-reduction). Applying the associated patterns for these recognized reaction types along with other observations allows the products of a reaction to be predicted from known reactants.

Some examples of the prediction of reaction products using the five common reaction types are detailed here.

Combination Reactions

When predicting the product for a simple combination reaction involving two *elements*, consider the typical charges that each element adopts as ions to predict the ionic compound that can form. For example, consider the reaction between aluminum metal and oxygen gas:

$$Al(s) + O_2(g) \rightarrow \,?$$

Because aluminum typically acquires a +3 charge and oxygen typically acquires a −2 charge, the Al^{3+} and O^{2-} ions must combine in a 3:2 ratio to form a neutral ionic compound. Therefore, the product of the reaction is $Al_3O_2(s)$ and the balanced reaction is:

$$4\,Al(s) + 3\,O_2(g) \rightarrow 2\,Al_2O_3(s)$$

Furthermore, this combination reaction can *also* be classified as a redox reaction because Al is oxidized (ie, $Al \rightarrow Al^{3+} + 3\,e^-$) and O is reduced (ie, $O + 2\,e^- \rightarrow O^{2-}$). However, not all combination reactions can be classified as such.

Decomposition Reactions

The products resulting from a decomposition reaction can be difficult to predict. As such, the following general trends can be applied when predicting decomposition products:

- Metal carbonates can decompose to form a metal oxide and $CO_2(g)$:

$$CaCO_3(s) \xrightarrow{\Delta} CaO(s) + CO_2(g)$$

- Metal hydrogen carbonates can decompose to form a metal carbonate, $CO_2(g)$, and water:

$$2\ NaHCO_3(s) \xrightarrow{\Delta} Na_2CO_3(s) + H_2O(g) + CO_2(g)$$

- Some metal oxides can decompose to form $O_2(g)$:

$$2\ Ni_2O_3(s) \xrightarrow{\Delta} 4\ Ni(s) + 3\ O_2(g)$$

Notice that the decomposition of Ni_2O_3 is also a redox reaction because the oxidation state of Ni decreases from +3 to 0 (ie, a reduction) and the oxidation state of O increases from −2 to 0 (ie, an oxidation). However, not all decomposition reactions can be classified as such (eg, the decomposition reactions of $CaCO_3$ and $NaHCO_3$ are not redox reactions).

Single Replacement Reactions

If the reaction between a compound and an element is known to follow the pattern of a single replacement reaction, the products that form must be a new compound and a new element. For example, consider the reaction between $CaI_2(s)$ and $Cl_2(g)$:

$$CaI_2(s) + Cl_2(g) \rightarrow\ ?$$

Because chlorine atoms form negatively charged ions, Cl^- can replace I^- to form $CaCl_2$ (ie, one anion is switched with another anion):

$$CaI_2(s) + Cl_2(g) \rightarrow CaCl_2 + I_2(s)$$

Conversely, a cation can be switched with another cation in other cases, as demonstrated by the following reaction:

$$Zn(s) + 2\ HCl(g) \rightarrow ZnCl_2 + H_2(s)$$

All single replacement reactions are also classified as redox reactions because the neutral element that reacts either gains or loses electrons.

Double Replacement Reactions

Predicting the products resulting from double replacement reactions between two ionic compounds can be easily done by switching the partners of the cations and anions and determining the product formulas based on the ion charges (Figure 2.52). These counter ion exchanges frequently result in the formation of a new insoluble product (ie, a precipitate) and a soluble product.

Chapter 2: Chemical Bonding, Reactions, and Stoichiometry

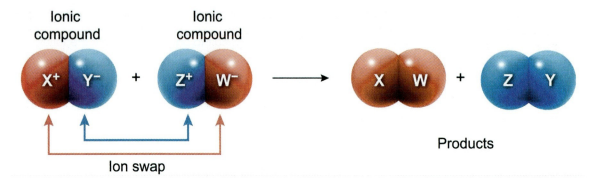

Figure 2.52 Predicting the products for a double replacement reaction.

> ☑ **Concept Check 2.21**
>
> FeS(s) undergoes a double replacement reaction in 1.0 M HCl(aq). What products are formed?
>
> **Solution**
>
> *Note: The appendix contains the answer.*

2.4.05 Balancing Chemical Reactions

For any chemical reaction, the **law of conservation of mass** states that atoms are neither created nor destroyed; they are only rearranged. Consequently, a chemical equation is balanced with respect to mass when the number of each type of atom in the chemical equation is the same on both sides of the reaction arrow (Figure 2.53).

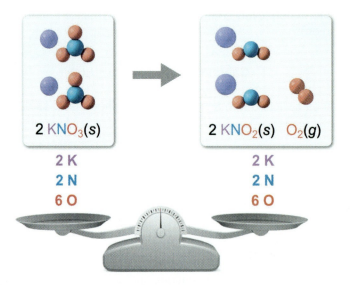

The same number of each type of atom is on each side of the reaction arrow.

Figure 2.53 A balanced chemical reaction follows the law of conservation of mass.

The stoichiometric coefficients and chemical formulas in a balanced reaction indicate the proportions of each type of atom in the reaction. However, the stoichiometric coefficients indicate how many of each type of molecule are present whereas the subscripts in the formulas indicate the internal composition (and identities) of the molecules.

As such, reactions are balanced by changing the stoichiometric coefficients, but the subscripts of the molecular formulas are never changed when balancing a reaction. Changing the subscripts would alter the *identity* of the molecule, not the *number* of molecules.

The following strategies can be helpful when balancing a chemical reaction:

- Balance elements present in only one compound on either side of the reaction arrow first.
- Balance compounds (eg, polyatomic ions) as a whole group if they appear on both sides of the reaction arrow.
- Balance the elements present in pure form (eg, $Br_2(l)$, $O_2(g)$, $C(s)$) last.

An example of the process of balancing a chemical reaction is shown in Figure 2.54.

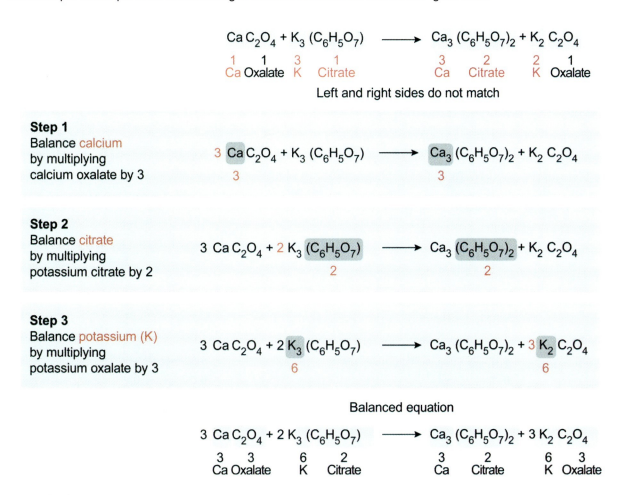

Figure 2.54 Balancing a chemical equation.

> ☑ **Concept Check 2.22**
>
> Balance the following chemical equation:
>
> $$C_2H_6(g) + O_2(g) \rightarrow CO_2(g) + H_2O(g)$$
>
> **Solution**
>
> *Note: The appendix contains the answer.*

2.4.06 Balancing Redox Reactions

For any reaction, the same number of atoms of each type must be present on both sides of the reaction arrow (ie, the law of conservation of mass must be obeyed). When balancing a redox reaction, an additional requirement is that the number of electrons lost must equal the number of electrons gained (ie, the **law of conservation of charge** must be obeyed). As such, the sum of the charges must be the same on both sides of the reaction, as demonstrated in Figure 2.55.

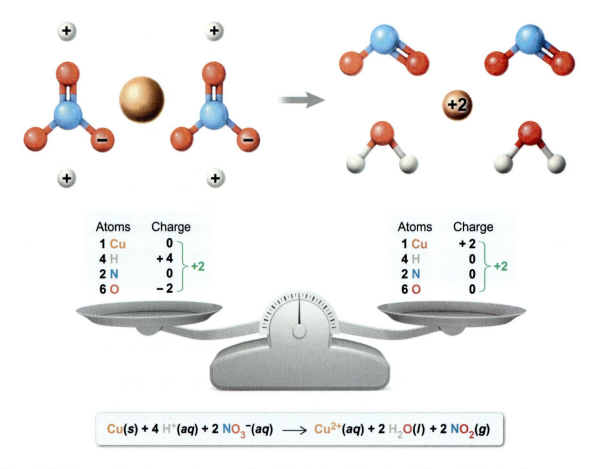

Figure 2.55 Conservation of mass and conservation of charge in a balanced oxidation-reduction reaction.

Simple redox reactions can generally be balanced by inspection. However, a more systematic approach is needed for balancing more complicated redox reactions (eg, those occurring in aqueous environments). One such approach is known as the **half-reaction method**.

Redox reactions can be broken into two **half-reactions** (one for oxidation and one for reduction) that add together to give the overall reaction. In a balanced redox reaction, the total number of electrons released from the oxidation half-reaction must *equal* the total number of electrons gained by the reduction half-reaction. Multiples of one or both half-reactions may be needed to achieve an electron balance in the overall summation.

For example, consider the unbalanced redox reaction:

$$Fe^{3+}(aq) + Sn^{2+}(aq) \rightarrow Fe^{2+}(aq) + Sn^{4+}(aq)$$

Expressing this reaction as two half-reactions, one for the oxidation of $Sn^{2+}(aq)$ and one for the reduction of $Fe^{3+}(aq)$, yields:

Oxidation half-reaction: $\quad Sn^{2+}(aq) \rightarrow Sn^{4+}(aq) + 2\,e^-$

Reduction half-reaction: $\quad Fe^{3+}(aq) + e^- \rightarrow Fe^{2+}(aq)$

Note that electrons are written as products in the oxidation half-reaction and as reactants in the reduction half-reaction. To cancel the electron terms in the overall reaction, the reduction half-reaction must be multiplied by 2 and then the two half-reactions added together:

Oxidation half-reaction: $\quad Sn^{2+}(aq) \rightarrow Sn^{4+}(aq) + \cancel{2\,e^-}$

2 × Reduction half-reaction: $\quad 2\,Fe^{3+}(aq) + \cancel{2\,e^-} \rightarrow 2\,Fe^{2+}(aq)$

Overall: $\quad 2\,Fe^{3+}(aq) + Sn^{2+}(aq) \rightarrow 2\,Fe^{2+}(aq) + Sn^{4+}(aq)$

In the overall reaction, the net charge of the reactants is 2(+3) + 1(+2) = +8 and the net charge of the products is 2(+2) + 1(+4) = +8.

However, when a redox reaction takes place in acidic or basic conditions, additional steps must be applied to balance the reaction. A modified version of the half-reaction method can be used to balance redox reactions occurring in acidic or basic conditions, as outlined by the following stepwise process:

1. Separate the unbalanced equation into oxidation and reduction half-reactions.
2. For each half-reaction:
   a) Balance all elements except hydrogen and oxygen first.
   b) Balance charge by adding H^+ if under *acidic* conditions or OH^- if under *basic* conditions.
   c) Balance oxygen last by adding H_2O.
3. If the number of electrons transferred in both the oxidation and reduction half-reactions are not equal, multiply one or both half-reactions by a multiple to equalize the number of electrons transferred.
4. Add both half-reactions together and cancel all identical species that appear in equal number on both sides of the reaction (eg, electrons, H_2O).
5. Check that all elements are balanced and that the net charge of the reactants is equal to the net charge of the products.

> **✓ Concept Check 2.23**
>
> The following unbalanced redox reaction takes place under basic conditions:
>
> $$Al(s) + NO_2^-(aq) \rightarrow AlO_2^-(aq) + NH_3(g)$$
>
> What is the balanced equation?
>
> **Solution**
>
> *Note: The appendix contains the answer.*

Lesson 2.5

Stoichiometry of Chemical Reactions

Introduction

According to the law of definite proportions, pure samples of the same substance consist of the same elements in the same proportions by mass. This observation holds true because atoms and molecules interact in particular ratios when reacting to form new compounds.

Stoichiometry is the application of the law of definite proportions to determine the quantitative relationships (ie, the amounts and ratios) of different substances in a chemical reaction. It is a powerful tool that can be used to predict various quantities such as the amount of reactant needed to produce a certain amount of product, or the amount of reactant consumed given the amount of product produced.

To correctly describe chemical systems, stoichiometry must simultaneously account for the proportions of chemical substances on three levels:

- **Particles** (eg, atoms, ions, molecules) are the fundamental units of matter in chemical reactions. Each reaction involves specific numbers of discrete particles in specific ratios. As such, rigorous systems of stoichiometry must be based on interactions at the particle level.
- **Moles** represent the very large numbers of particles present within chemical samples. The mole is a counting unit that provides a means of scaling up a calculation to account for the vast numbers of particles participating in a reaction while still maintaining the reaction ratios found in an individual reaction.
- **Mass** is a convenient measurement to relate the relative amounts of chemical substances in a reaction. Accordingly, the number of moles of a chemical substance are often correlated to an equivalent amount of mass of that substance.

The relationships between the three levels of stoichiometry outlined above are illustrated by the flowchart in Figure 2.56, in which X and Y represent any two substances within a chemical reaction.

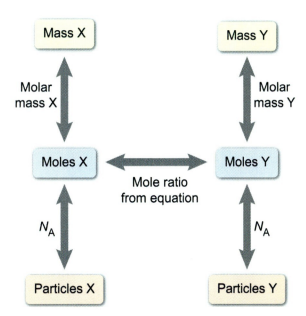

Figure 2.56 Flowchart summarizing the three levels of stoichiometry and the unit conversion steps that relate each level. Substances X and Y represent any two chemical species in a reaction.

Each stochiometric level is related to its neighboring level by a specific quantity or ratio (seen placed next to the arrow between the two levels). These values are the required conversion factors necessary to go from one level to the next. Particles must be scaled to moles (and vice versa) using **Avogadro's number** (N_A). The moles of a substance must be related to mass (and vice versa) using the **molar mass** (g/mol) of that substance. The only way to relate one substance to another is by a **mole ratio** from a balanced chemical reaction that specifies the interaction of the two substances.

Although the use of this flowchart is not required for solving stoichiometry calculations, it can provide a visual representation of the necessary calculation pathway. By identifying the starting point (ie, what quantity is known) and the ending point (ie, the quantity to be determined), the number of conversion steps required (and their order) can be easily seen. In this lesson, the application of stoichiometric calculations is examined in detail using this method.

2.5.01 Relationships Between Species in Chemical Reactions

As mentioned in Concept 1.6.04, elements combine in definite proportions when forming chemical compounds. For example, consider the balanced chemical equation for the reaction between unspecified species A and B to form C:

$$3A + B \rightarrow 2C$$

Suppose that 3.6×10^{24} particles of A are consumed during the reaction. What is the mass of product C that would be produced? Adapting the chart from Figure 2.56 to this system provides an outline for the calculation to determine the mass, as shown in Figure 2.57.

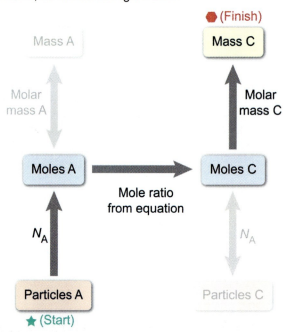

Figure 2.57 Flowchart summarizing the stoichiometric calculation steps necessary to determine the mass of product C formed from a given number of particles of reactant A.

Using Avogadro's number as a conversion factor, it is possible to convert a known number of particles of a substance to the number of moles of that substance (and vice versa). As such, if 3.6×10^{24} particles of A are consumed during the reaction, dividing the number of particles of A by Avogadro's number yields the number of moles of A:

$$3.6 \times 10^{24} \text{ particles A} \times \frac{1 \text{ mol A}}{6.022 \times 10^{23} \text{ particles A}} = 6.0 \text{ mol A}$$

Because elements combine in definite proportions when forming chemical compounds, the stoichiometric coefficients from the balanced equation can represent either individual particles or moles of particles. As such, the number of moles of one chemical species is related to the number of moles of another chemical species in a reaction by stoichiometric mole ratios made from any two coefficients.

The balanced reaction given above shows that 3 moles of A react with 1 mole of B to form 2 moles of C. Therefore, the mole ratio relating reactant A to product C can be written in two ways:

$$\frac{3 \text{ mol A}}{2 \text{ mol C}} \quad \text{or} \quad \frac{2 \text{ mol C}}{3 \text{ mol A}}$$

This molar proportion can be applied to determine the number of moles of C that can be produced from the 6.0 mol of A (determined as previously shown):

$$6.0 \text{ mol A} \times \frac{2 \text{ mol C}}{3 \text{ mol A}} = 4.0 \text{ mol C}$$

Note that any two chemical species in a reaction can *only* be directly related by moles (ie, not grams, not particles). However, it is common to need to convert from grams of a substance to moles (or vice versa) using its molar mass before a mole ratio can be applied (or after a ratio is applied). Therefore, once it has been determined that 4.0 mol of product C is formed during the reaction, the grams of C produced can be calculated by multiplying the moles of C by its molar mass (assume 8.0 g/mol) as shown:

$$4.0 \text{ mol C} \times \frac{8.0 \text{ g C}}{1 \text{ mol C}} = 32 \text{ g C}$$

In the preceding discussion, a series of separate calculations are given that demonstrate how to use Avogadro's number, mole ratios, and a compound's molar mass with the given number of particles of reactant A to determine the mass of product C formed in a reaction. Although these calculations can be done in separate steps (as above), all these calculations involve conversion factors and can be combined sequentially in a single operation to yield the same numerical result:

$$3.6 \times 10^{24} \text{ particles A} \times \frac{1 \text{ mol A}}{6.022 \times 10^{23} \text{ particles A}} \times \frac{2 \text{ mol C}}{3 \text{ mol A}} \times \frac{8.0 \text{ g C}}{1 \text{ mol C}} = 32 \text{ g C}$$

Concept Check 2.24

Consider the following reaction.

$$2 \text{ Cu}_3\text{FeS}_3(s) + 7 \text{ O}_2(g) \rightarrow 6 \text{ Cu}(s) + 2 \text{ FeO}(s) + 6 \text{ SO}_2(s)$$

If 170 g of $Cu_3FeS_3(s)$ reacts with excess $O_2(g)$, how many grams of Cu(s) are produced?

Solution

Note: The appendix contains the answer.

> ☑ **Concept Check 2.25**
>
> When heated, Ni$_2$O$_3$(s) decomposes to produce O$_2$(g):
>
> $$2\ Ni_2O_3(s) \rightarrow 4\ Ni(s) + 3\ O_2(g)$$
>
> If 1.8×10^{24} molecules of O$_2$(g) were produced during this decomposition reaction, how many grams of Ni$_2$O$_3$(s) decomposed?
>
> **Solution**
>
> *Note: The appendix contains the answer.*

2.5.02 Molarity and Density as Conversion Factors

Thus far, particle-mole, mole-mole, and mass-mole unit conversions have been applied to stoichiometry calculations. However, in some cases the volume of a reactant or a product is known and must be converted to moles before a stoichiometric mole ratio can be applied to relate one species to another in the chemical reaction. If the density or molarity of the substance is known, the volume of the substance can easily be converted into moles (or vice versa).

Recall from Concept 1.6.02 that the **density** ρ of a substance is defined as the mass m it contains per unit of volume V that it occupies:

$$\rho = \frac{m}{V}$$

The density of a substance at a given temperature can be used as a **conversion factor** between the volume and mass of that substance. For example, if a 12 mL sample of substance A has a density of 0.50 g/mL, multiplying the volume of the sample by its density shows that the sample has a mass of 6.0 g (note that the units of volume must be the same to cancel):

$$12\ \text{mL A} \times \frac{0.50\ \text{g A}}{1\ \text{mL A}} = 6.0\ \text{g A}$$

Dividing the mass of A by its molar mass (assume a molar mass of 4.0 g/mol) yields the number of moles of A in the 12 mL sample:

$$6.0\ \text{g A} \times \frac{1\ \text{mol A}}{4.0\ \text{g A}} = 1.5\ \text{mol A}$$

Combining both steps in one calculation yields the same numerical value:

$$6.0\ \text{mL A} \times \frac{0.50\ \text{g A}}{1\ \text{mL A}} \times \frac{1\ \text{mol A}}{2.0\ \text{g A}} = 1.5\ \text{mol A}$$

Similarly, if the number of moles is initially known, inverting the conversion factors allows a determination of the corresponding volume occupied by that number of moles. For example, a sample containing 4.0 mol of A would occupy a volume of:

$$4.0\ \text{mol A} \times \frac{2.0\ \text{g A}}{1\ \text{mol A}} \times \frac{1\ \text{mL A}}{0.50\ \text{g A}} = 4.0\ \text{mL A}$$

Although density is very useful to relate the mass and volume of a sample, **molarity** is used more frequently in chemical calculations involving moles because the molarity of a solution provides a direct relationship between the moles of dissolved solute and the liters of solution (Concept 1.6.08):

$$\text{Molarity (M)} = \frac{\text{Moles of solute}}{\text{Solution volume } V \text{ (in liters)}}$$

Accordingly, the molarity of a solution can be used as a conversion factor in stoichiometric calculations when either the moles of solute or the volume of a solution is known. For example, if a solution of substance A has a molarity of 0.10 M (ie, 0.10 mol/L), a 0.50 L sample of the solution contains 0.050 mol of A:

$$0.50 \text{ L} \times \frac{0.10 \text{ mol A}}{1 \text{ L}} = 0.050 \text{ mol A}$$

Similarly, if another sample of the solution contains 0.20 mol A, multiplying by the inverted molarity as a conversion factor shows that the sample must have a volume of 2.0 L:

$$0.20 \text{ mol A} \times \frac{1 \text{ L}}{0.10 \text{ mol A}} = 2.0 \text{ L}$$

Expanding the flowchart in the introduction to include the use of density and molarity as conversion factors in stoichiometric calculations gives the operational summary shown in Figure 2.58.

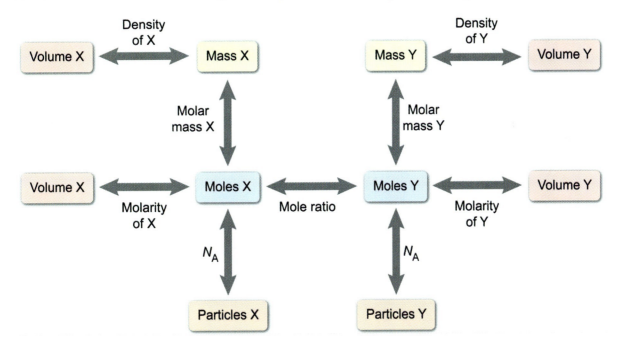

Figure 2.58 Flowchart summarizing the unit conversion steps involving density, molarity, molar mass, and Avogadro's number in stoichiometric calculations. Substances X and Y represent any two species in a chemical reaction.

> **Concept Check 2.26**
>
> $$2\ NaOH(aq) + 2\ Al(s) + 6\ H_2O(l) \rightarrow 2\ NaAl(OH)_4(aq) + 3\ H_2(g)$$
>
> Consider the balanced equation shown. For this reaction, how many liters of a 0.050 M NaOH(aq) solution are consumed to produce 1.26 L of $H_2(g)$ at 25 °C? (Note: Assume that the density of $H_2(g)$ at 25 °C is 0.090 g/L.)
>
> **Solution**
>
> *Note: The appendix contains the answer.*

2.5.03 Limiting Reactant

During a chemical reaction, the reaction continues until one of the reactants is fully consumed. The limiting reactant is the reactant completely consumed first, thus stopping the reaction and limiting the amount of product that can be produced. If a reactant is not fully consumed during the reaction, it is in excess and will not cause the reaction to stop.

The limiting reactant can be found by determining the number of moles of product that the stated amounts of each reactant could potentially form if completely consumed. This quantity is based on the number of moles of each reactant and the mole ratios of the stoichiometric coefficients in the balanced reaction equation. The limiting reactant is the one that results in the smallest theoretical amount of product.

For example, Li(s) reacts with $N_2(g)$ to form $Li_3N(s)$ according to the following balanced equation:

$$6\ Li(s) + N_2(g) \rightarrow 2\ Li_3N(s)$$

If 6 moles of Li(s) and 5 moles of $N_2(g)$ react, the amount of $Li_3N(s)$ produced if each reactant were completely consumed can be calculated using mole ratios:

$$6\ \cancel{mol\ Li(s)} \times \frac{2\ mol\ Li_3N(s)}{6\ \cancel{mol\ Li(s)}} = 2\ mol\ Li_3N(s)$$

$$5\ \cancel{mol\ N_2(g)} \times \frac{2\ mol\ Li_3N(s)}{1\ \cancel{mol\ N_2(g)}} = 10\ mol\ Li_3N(s)$$

Because Li(s) yields a smaller theoretical amount of product than $N_2(g)$—that is, Li(s) limits the amount of product that can form—Li(s) is the limiting reactant and $N_2(g)$ is in excess, as illustrated in Figure 2.59.

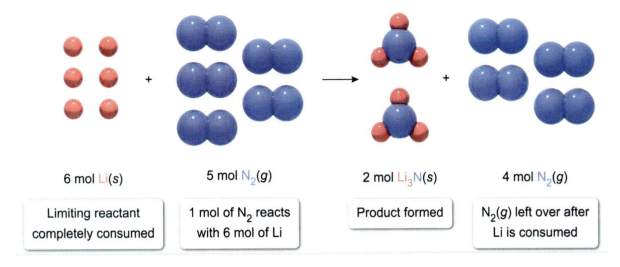

Figure 2.59 Example of a limiting reactant in a chemical reaction.

> ☑ **Concept Check 2.27**
>
> Consider the reaction shown:
>
> $$4\ Al(s) + 3\ O_2(g) \rightarrow 2\ Al_2O_3(s)$$
>
> If 162 g of Al(s) react with 96.0 g of $O_2(g)$, which reactant is completely consumed during the reaction? How many moles of the excess reactant are left unreacted?
>
> **Solution**
> *Note: The appendix contains the answer.*

2.5.04 Theoretical, Actual, and Percent Yield

The **theoretical yield** of a chemical reaction is the *maximum* amount of product that can be produced during the reaction from the given amounts of reactants initially present, as illustrated in Figure 2.60. The amount of theoretical yield is determined by the number of moles of the limiting reactant and the stoichiometric mole ratios from the balanced chemical reaction. Once the limiting reactant is gone, the reaction stops, and no more product can be formed.

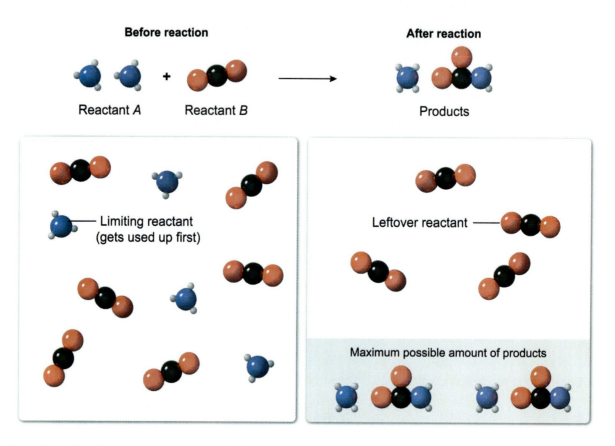

Figure 2.60 The theoretical yield of a reaction is the maximum amount of product that can be formed from the initial amount of reactants.

However, the amount of product produced during a reaction is seldom equal to the amount predicted by the theoretical yield. The **actual yield** of product obtained is usually less than the theoretical yield due to several factors including incomplete conversion, unwanted side reactions, experimental errors, and transfer losses (eg, losing product during a filtration process). The actual yield can sometimes *appear* greater than the theoretical yield in instances when an isolated product contains large amounts of impurities (eg, insufficiently dried product that has extra mass from water).

The extent to which reactants are converted into products and isolated during a reaction is often expressed as a percentage of the theoretical yield known as the **percent yield**. The percent yield for a given product is defined as the ratio of the product mass obtained from an experiment (the actual yield) to the calculated product mass (theoretical yield) reported as a percentage:

$$\% \text{ Yield} = \frac{\text{Actual yield}}{\text{Theoretical yield}} \times 100$$

For example, if a reaction produces 3.45 g of product instead of a predicted 6.20 g, the percent yield of the product would be:

$$\% \text{ Yield} = \frac{3.45 \text{ g}}{6.20 \text{ g}} \times 100 = 55.6\%$$

Concept Check 2.28

Consider the following reaction.

$$2\ Al(s) + 3\ CuSO_4(aq) \rightarrow 3\ Cu(s) + Al_2(SO_4)_3(aq)$$

If 80.0 g of Al(s) reacts with excess $CuSO_4(aq)$ and produces 150 g of Cu(s), what is the percent yield of Cu(s)?

Solution

Note: The appendix contains the answer.

END-OF-UNIT MCAT PRACTICE

Congratulations on completing **Unit 2: Interactions of Chemical Substances**.

Now you are ready to dive into MCAT-level practice tests. At UWorld, we believe students will be fully prepared to ace the MCAT when they practice with high-quality questions in a realistic testing environment.

The UWorld Qbank will test you on questions that are fully representative of the AAMC MCAT syllabus. In addition, our MCAT-like questions are accompanied by in-depth explanations with exceptional visual aids that will help you better retain difficult MCAT concepts.

TO START YOUR MCAT PRACTICE, PROCEED AS FOLLOWS:

1) Sign up to purchase the UWorld MCAT Qbank
   IMPORTANT: You already have access if you purchased a bundled subscription.
2) Log in to your UWorld MCAT account
3) Access the MCAT Qbank section
4) Select this unit in the Qbank
5) Create a custom practice test

Unit 3 Thermodynamics, Kinetics, and Gas Laws

Chapter 3 Thermodynamics

3.1 Zeroth Law and Temperature

- 3.1.01 Thermodynamic Systems and State Functions
- 3.1.02 Heat and Temperature
- 3.1.03 Modes and Direction of Heat Transfer
- 3.1.04 Thermal Equilibrium and the Zeroth Law of Thermodynamics

3.2 First Law of Thermodynamics

- 3.2.01 Conservation of Energy
- 3.2.02 Internal Energy, Heat, and Enthalpy
- 3.2.03 Heat Versus Enthalpy in Chemical Systems

3.3 Second Law of Thermodynamics

- 3.3.01 Entropy
- 3.3.02 Evaluating Entropy in Chemical Systems

3.4 Heat Transfer in Physical Changes

- 3.4.01 Phase Changes
- 3.4.02 Melting Point and Heat of Fusion
- 3.4.03 Boiling Point and Heat of Vaporization
- 3.4.04 Phase Diagrams
- 3.4.05 Heating Curves

3.5 Heat Transfer in Chemical Changes

- 3.5.01 Enthalpy of a Reaction
- 3.5.02 Hess' Law
- 3.5.03 Heats of Formation
- 3.5.04 Bond Dissociation Energy

3.6 Calorimetry

- 3.6.01 Experimental Measurement of Heat Transfer

3.7 Gibbs Free Energy

- 3.7.01 Gibbs Free Energy as a Thermodynamic Quantity
- 3.7.02 Enthalpic and Entropic Contributions to Gibbs Free Energy

Chapter 4 Kinetics

4.1 Reaction Mechanisms

- 4.1.01 Elementary Reactions and Net Reactions
- 4.1.02 Energy Considerations for Reactions
- 4.1.03 Rate-Determining Step
- 4.1.04 Thermodynamic Versus Kinetic Control

4.2 Reaction Rates

- 4.2.01 Reaction Rates and Rate Measurements
- 4.2.02 Rate Laws
- 4.2.03 Temperature Dependence of the Reaction Rate

4.3 Chemical Catalysis

- 4.3.01 Types and Mechanisms of Catalysis

Chapter 5 Chemical Equilibrium

5.1 Equilibrium in Reversible Reactions

5.1.01 Dynamic Equilibrium

5.2 Law of Mass Action

5.2.01 Introduction to the Law of Mass Action

5.3 Reaction Quotient

5.3.01 Equilibrium Constant
5.3.02 Effect of Catalysts
5.3.03 Gibbs Free Energy and the Equilibrium Constant

5.4 Le Châtelier's Principle

5.4.01 Effect of Concentration Changes
5.4.02 Effect of Pressure Changes
5.4.03 Effect of Temperature Changes

Chapter 6 Gas Laws

6.1 Pressure, Volume, and Temperature Relationships

6.1.01 Boyle's Law
6.1.02 Guy-Lussac's Law
6.1.03 Charles' Law
6.1.04 Combined Gas Law

6.2 Pressure and Volume Relationships to Molar Amount

6.2.01 Avogadro's Law
6.2.02 Dalton's Law of Partial Pressures

6.3 Ideal Gas Model

6.3.01 Ideal Gas Law
6.3.02 Molar Volume at STP
6.3.03 Deviations from Ideal Gas Behavior

Lesson 3.1
Zeroth Law and Temperature

Introduction

Energy is commonly defined as the ability to do work. Work always results in a change of some sort. Therefore, energy can be thought of as the "currency" of change. As such, work necessarily involves the transfer of energy, and often the transformation of energy from one form to another. The relationships between work and energy, and the changes that result from work, are all part of the field of **thermodynamics**. Applied specifically to chemical systems, this field of study is known as **thermochemistry**.

This lesson introduces the concept of thermodynamic systems and examines how they are defined and analyzed using state functions. This analysis includes an exploration of the modes and direction of energy transfer between systems, the relationship between heat and temperature in a system, and the relationship between systems at thermal equilibrium.

3.1.01 Thermodynamic Systems and State Functions

Investigations in thermodynamics typically involve separating the universe into two parts: the system and the surroundings. A **thermodynamic system** can be defined as any space being studied or observed, along with the matter and energy contained within that space. The **surroundings** are defined as everything outside the system (ie, the rest of the universe). Mathematically, this is expressed as:

Universe = System + Surroundings

In general, a system has a boundary of some sort with the surroundings. The nature of that boundary (ie, what can be exchanged through it) determines the type of system it is. As such, there are three common types of thermodynamic systems (Figure 3.1):

- **Open systems** are systems in which matter and energy can be freely exchanged with the surroundings.
- **Closed systems** are systems in which only energy can be exchanged with the surroundings.
- **Isolated systems** are systems in which neither matter nor energy can be exchanged with the surroundings.

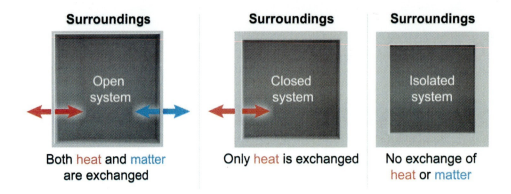

Figure 3.1 Types of thermodynamic systems.

One useful way of defining thermodynamic systems is with state functions. A state function describes a property of a thermodynamic system whose value depends only on the state of that system, not the process or path by which the state was achieved. Temperature, volume, and pressure are all examples of state functions.

In contrast, **path functions** describe properties that *are* dependent on the path taken. For example, the distance traveled on a trip from Los Angeles to New York depends on the path taken to get there. A direct flight results in a much shorter distance traveled than if two flights are taken by first going to Dallas and then getting a connecting flight to New York. Therefore, distance traveled is a path function. Table 3.1 lists examples of state and path functions that are applicable to the study of thermodynamics.

Table 3.1 Common state functions and path functions used in thermodynamics.

State Functions	Path Functions
Pressure (P)	Heat (q)
Volume (V)	Work (w)
Temperature (T)	
Enthalpy (H)	
Internal energy (U)	
Gibbs free energy (G)	
Entropy (S)	

To better understand state functions and path functions in thermodynamics, consider the following scenario. A system consists of a container partially filled with a gas and stoppered by a movable piston (Figure 3.2). Different ways (ie, paths) can be used to double the volume of this system. One way is to add heat to the gas in the container, causing the gas to expand and double its volume. Another way is to pull the piston (ie, use work to move the piston upward) and double the volume. In both cases, the outcome is the same: The volume is doubled.

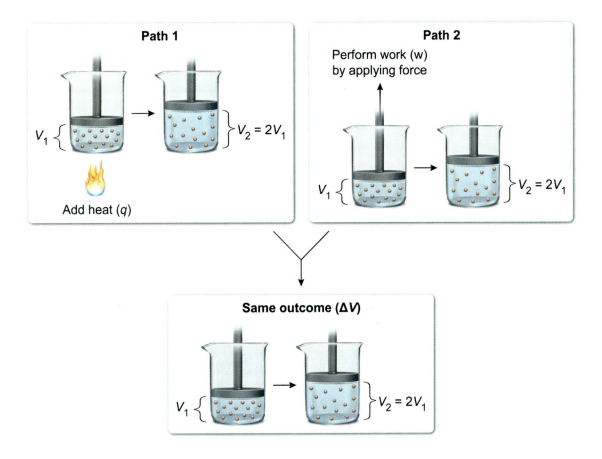

Figure 3.2 Heat and work as path functions.

The change in volume does not depend on the path, and therefore is a state function. However, depending on the path taken, the values for heat and work are different. This difference is why heat and work are path functions.

> ☑ **Concept Check 3.1**
>
> Indicate whether each of the following systems is best described as an open system, a closed system, or an isolated system.
>
> 1. The human circulatory system
> 2. An underground nuclear bunker encased in 6 feet of concrete
> 3. A glass jar sealed with an airtight lid
> 4. A teapot filled with water
>
> **Solution**
>
> *Note: The appendix contains the answer.*

3.1.02 Heat and Temperature

In thermodynamics, heat and temperature refer to different but related quantities. The link between the two is thermal energy. **Thermal energy** is the total kinetic energy of all particles that make up a system, whereas **temperature** is a measure of the average kinetic energy of the particles in the system. When two objects of different temperatures are brought into contact with one another, thermal energy flows from the object with a higher temperature to the object with the lower temperature. The transfer of thermal energy is called **heat** (Figure 3.3).

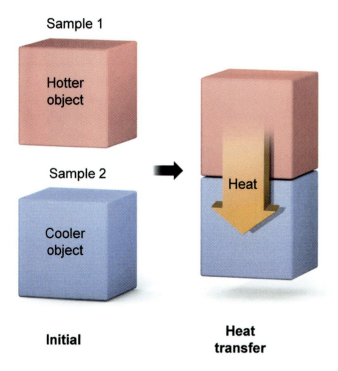

Figure 3.3 Heat is the flow of thermal energy from a hotter (ie, higher temperature) system to a colder (ie, lower temperature) system.

Although it is not possible to directly measure the thermal energy within a system, thermal energy can be measured as it moves between two systems as heat. Heat is typically measured based on the temperature change that occurs during heat transfer. For example, the calorie (cal) is a unit of heat that represents the amount of energy needed to increase the temperature of 1 gram of water by 1 °C. In chemistry, the more commonly used unit of energy is the joule (J). One calorie is approximately equal to 4.184 joules.

> ☑ **Concept Check 3.2**
>
> A cup of hot tea and a large tub of lukewarm water are placed in a walk-in freezer. A few hours later, the tea is frozen solid but the water in the tub is still in liquid form. Explain these observations.
>
> **Solution**
>
> *Note: The appendix contains the answer.*

Because temperature is a measure of the average kinetic energy of particles, logically a minimum temperature must exist at which particles would (at least theoretically) have no kinetic energy at all. This theoretical lowest possible temperature is referred to as **absolute zero**. Accordingly, the Kelvin temperature scale sets this temperature at a value at 0 K. Figure 3.4 presents the relationship between the Kelvin scale and two other commonly used temperature scales.

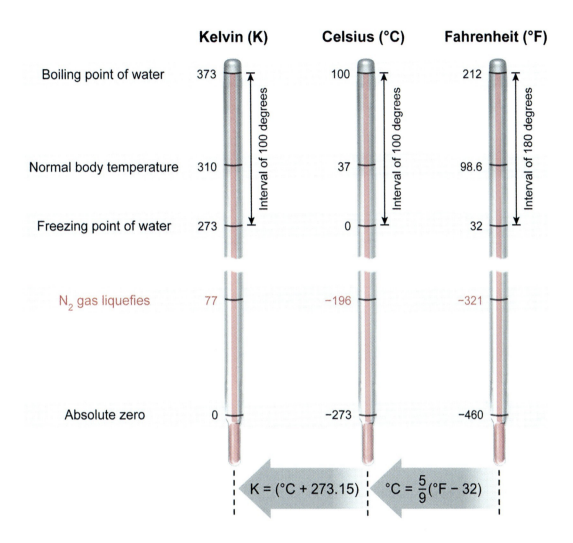

Figure 3.4 Comparison of Kelvin, Celsius, and Fahrenheit temperature scales.

Only a temperature scale based on absolute zero provides a directly proportional relationship between temperature and the average kinetic energy of particles (eg, a system cannot have negative kinetic energy, but some temperature scales allow for negative temperatures). As such, the Kelvin scale is standard for all calculations and relationships in thermochemistry.

> **Concept Check 3.3**
>
> Room temperature is generally considered to be about 22 °C. What would this temperature be on the Kelvin scale?
>
> **Solution**
>
> *Note: The appendix contains the answer.*

3.1.03 Modes and Direction of Heat Transfer

As explained in Concept 3.1.02, heat moves from systems at higher temperatures to systems at lower temperatures. The three ways or "modes" by which heat can be transferred between systems are conduction, convection, and radiation (Figure 3.5).

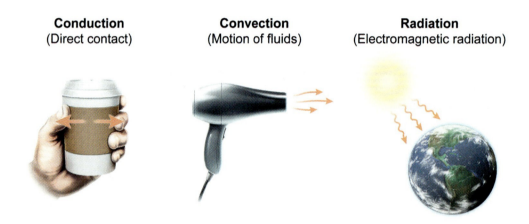

Figure 3.5 Modes of heat transfer.

Conduction

A system at a higher temperature has a greater proportion of molecules with higher kinetic energy, as shown in Figure 3.6. When such a system is brought into direct contact with a system at a lower temperature (ie, a system with a lesser proportion of molecules that have higher kinetic energy), the particles of the two systems undergo vibrational or translational collisions with one another.

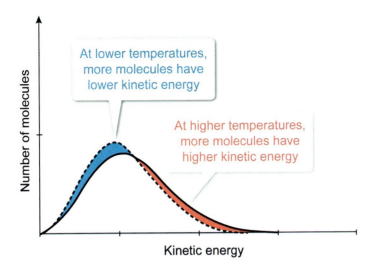

Figure 3.6 Energy distribution curves for the same system at two different temperatures.

As the particles of the two systems collide, kinetic energy spontaneously transfers from high-energy particles to low-energy particles. The transfer of heat from one system to another through these types of collisions is called **conduction** (Figure 3.7).

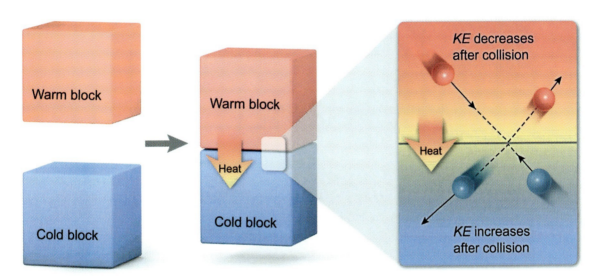

Figure 3.7 Heat transfer via conduction occurs through the exchange of kinetic energy during the collisions that take place between particles when objects are brought into direct contact with one another.

Convection

A second mode of heat transfer occurs when fluids (especially gases) expand while being heated. This expansion is caused by the increased average kinetic energy of particles that make up the fluid, which results in the fluid being less dense than it is at lower temperatures. The difference in density causes less dense, warmer fluids to rise and more dense, cooler fluids to sink. Heat transfer resulting from the movement of fluids in this way is called **convection**. Figure 3.8 shows a common example of convection that occurs when water is heated on a stove.

Figure 3.8 Heat is distributed throughout the water in a kettle by convection.

Radiation

A third mode of heat transfer that occurs through the absorption and emission of photons is known as **radiation**. Just as a photon of ultraviolet (UV) or visible light is emitted when an electron drops from a higher energy level to a lower energy level (Concept 1.3.02), a photon of infrared radiation is emitted when an atom or molecule drops from a higher kinetic energy state to lower kinetic energy state, as depicted in Figure 3.9.

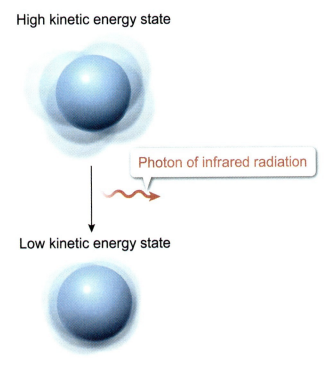

Figure 3.9 Photons of infrared radiation are released when particles go from higher kinetic energy states to lower kinetic energy states.

Conversely, photons of infrared radiation can be absorbed by atoms or molecules, causing them to jump from lower kinetic energy states to higher kinetic energy states. The analytical technique of spectroscopy uses various instruments to detect this mode of energy transfer between molecules and analyze the associated energy transitions in chemical samples. These analyses enable the identification of various structural components of complex organic molecules.

3.1.04 Thermal Equilibrium and the Zeroth Law of Thermodynamics

At temperatures above absolute zero, the atoms and molecules (ie, particles) of a sample of matter are in constant motion and have various amounts of kinetic energy. As Concept 3.1.02 explains, the average kinetic energy of particles in a sample is directly proportional to its Kelvin temperature.

When two systems are in **thermal contact** with one another, collisions between the particles of the two systems result in a net transfer of energy (ie, heat) from the higher-temperature system to the lower-temperature system until the two systems reach the same temperature, as shown in Figure 3.10.

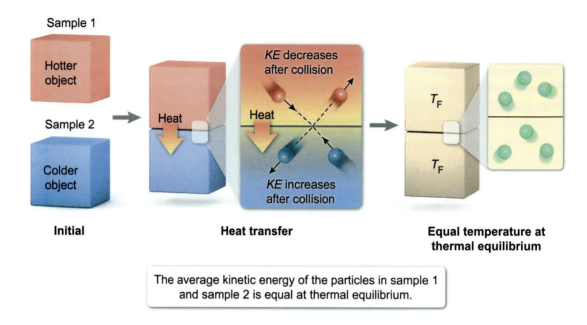

Figure 3.10 Net heat transfer between objects in thermal contact occurs until they reach the same temperature.

Even after the two systems reach the same temperature, collisions between particles continue to transfer energy from high-energy particles to low-energy particles. However, no *net* transfer of heat takes place. At this point, the two systems are said to be at **thermal equilibrium**.

If two systems are in thermal equilibrium and one of those systems is also in thermal equilibrium with a third system, then all three systems must be in thermal equilibrium with each other (ie, if A = B and B = C, then A = C). This principle is known as the **zeroth law of thermodynamics**, and it asserts that thermal equilibrium can exist between systems even if they are not in direct contact with one another provided they are at the same temperature (Figure 3.11).

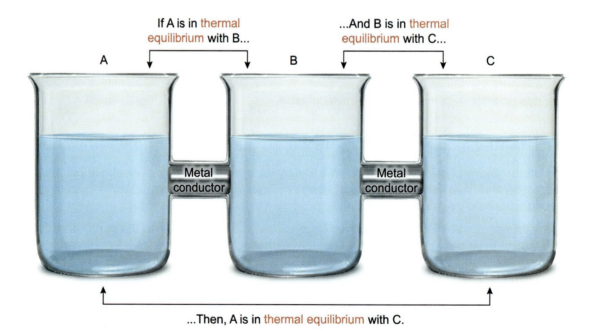

Figure 3.11 Zeroth law of thermodynamics.

Thus, the zeroth law of thermodynamics not only justifies the concept of temperature but also permits the conclusion that if two systems have the same temperature, no net heat transfer would occur if those systems were brought together.

Lesson 3.2
First Law of Thermodynamics

Introduction

In the study of thermodynamics, energy is tracked as it moves from one system to another and transforms from one form into another. Although tracking different forms of energy varies in difficulty, careful measurements confirm that the amount of energy at the beginning of a process is always equal to the amount of energy at the end of a process. This observation of the conservation of energy is the core principle behind the first law of thermodynamics, which is discussed in this lesson in terms of its relationship to chemical systems.

3.2.01 Conservation of Energy

The **first law of thermodynamics** states that the energy of the universe is constant, and therefore energy is neither created nor destroyed, although it can change from one form to another (Figure 3.12). Based on this law, every thermodynamic process can ultimately be defined by the following relationship:

$$\Delta U_{universe} = \Delta U_{system} + \Delta U_{surroundings} = 0$$

where $\Delta U_{universe}$, ΔU_{system}, and $\Delta U_{surroundings}$ are the change in internal energy of the universe, system, and surroundings, respectively.

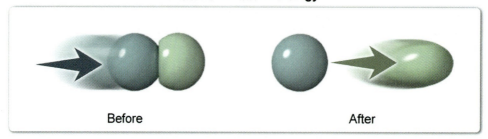

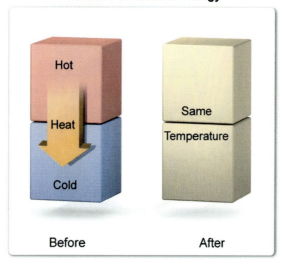

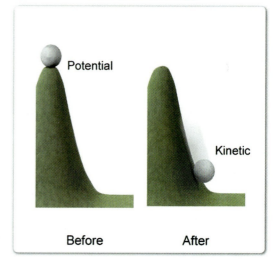

Figure 3.12 Energy can be transferred or transformed into different forms but is never created or destroyed.

From a thermodynamic perspective, this relationship means that every transfer of energy into a system must be equal to the energy taken out of its surroundings, an idea expressed mathematically as:

$$\Delta U_{\text{system}} = -\Delta U_{\text{surroundings}}$$

This important principle governs every thermodynamic analysis and allows for the evaluation of certain kinds of potential energy that cannot be measured directly in chemical systems.

3.2.02 Internal Energy, Heat, and Work

A system's **internal energy** U is the sum of the kinetic and potential energy of all particles within that system. As explained in Concept 3.1.02, heat is the transfer of thermal energy between systems. In contrast, work is the means by which all other forms of energy may be transferred. Since the first law of thermodynamics makes clear that energy is neither created nor destroyed, the only two ways to change the internal energy ΔU of a system are through heat (q) or **work (w)**, which is expressed mathematically as:

$$\Delta U = q + w$$

The work done in chemical systems is often limited to pressure-volume work (ie, a force per unit area applied through a three-dimensional displacement). When a system is compressed into a smaller volume ($-\Delta V$), the system gains energy by $+w$; when a system expands ($+\Delta V$) and pushes out against its surroundings, the system loses energy by $-w$. This relationship is expressed as:

$$w = -P\Delta V$$

The confusion that often arises in determining the signs for heat and work can be clarified by recognizing that the equation for ΔU takes the perspective of the system. If heat transfer causes the system to gain energy (ie, the system gets warmer), q is positive. If the system gets cooler, energy is lost, and therefore q is negative.

In the case of work, a system that does work loses energy whereas a system that has work done on it gains energy. For example, if work is done *by* the system *on* the surroundings, the system expands and loses energy to the surroundings (ie, w is negative). Table 3.2 summarizes these relationships.

Table 3.2 Sign Conventions for q, w, and ΔU.

Thermodynamic variable	Sign	Meaning
ΔU	+	Energy is gained by the system from the surroundings.
ΔU	−	Energy is lost by the system to the surroundings.
q	+	Heat is absorbed by the system from the surroundings.
q	−	Heat is released by the system to the surroundings.
w	+	Work is done on the system by the surroundings.
w	−	Work is done by the system on the surroundings.

> **Concept Check 3.4**
>
> As a system cools, it releases 178 J of heat. At the same time, the surroundings perform 59 J of work on the system, compressing the system to a smaller volume. What is the change in the internal energy of the system?
>
> **Solution**
>
> Note: The appendix contains the answer.

3.2.03 Heat Versus Enthalpy in Chemical Systems

The internal energy of a system (U) and the product of its pressure (P) and volume (V) combine to form a new state function known as **enthalpy H**, which is expressed by the equation:

$$H = U + PV$$

Although directly measuring either the internal energy or the enthalpy of a system is not feasible, the *change* in internal energy ΔU and *change* in enthalpy ΔH of a system can both be measured as energy moves into or out of the system. At *constant pressure*, the change in the enthalpy of a system is given by:

$$\Delta H = \Delta U + P\Delta V$$

Given that ΔU is equal to the sum of the heat (q) and work (w) associated with the system, the equation for ΔH can be rewritten as:

$$\Delta H = q + w + P\Delta V$$

Since $w = -P\Delta V$, this relationship simplifies to:

$$\Delta H = q + (-P\Delta V) + P\Delta V$$

$$\Delta H = q \text{ (at constant pressure)}$$

Therefore, the change in enthalpy is equal to heat at constant pressure. This relationship makes ΔH an extremely useful quantity when analyzing chemical systems because measuring heat transfer between systems is a relatively simple task. And although q is a path function, ΔH is a state function, which makes ΔH more useful for defining changes to a system.

In contrast, when volume is constant, $-P\Delta V = 0$ and the heat gained or lost by a system is equal to the change in its internal energy, as shown by the following:

$$\Delta U = q + w \quad \Rightarrow \quad \Delta U = q + (-P\Delta V)$$

$$\Delta U = q \text{ (at constant volume)}$$

In summary, q can be equal to either ΔH or ΔU depending on the conditions under which any heat transfer takes place. $\Delta H = q$ at constant pressure, and $\Delta U = q$ at constant volume.

Lesson 3.3

Second Law of Thermodynamics

Introduction

The more energy a sample of matter has, the more unstable it becomes. As such, isolated systems tend to progress toward configurations that have a more uniform distribution of energy. This phenomenon can be explained in terms of statistical probability. When energy can be exchanged freely among the particles of a system, the most probable distribution of energy is the most diffuse one. In other words, energy tends to spread out when given the chance. This tendency is a fundamental principle of the second law of thermodynamics.

The **second law of thermodynamics** states that for any spontaneous process, the **entropy** of the universe increases. An increase in entropy is often associated with increased disorder (ie, more random arrangements of particles), which results in a more diffuse (ie, uniform) distribution of energy. Understanding entropy, how and why it changes, and analyzing entropy changes in chemical systems is the focus of this lesson.

3.3.01 Entropy

A common definition for **entropy S** is a measure of the disorder or randomness of a system. For example, a crystal has a high degree of order (and therefore low entropy) because the atoms that compose it are aligned, spaced evenly, and adopt a single arrangement. In contrast, a gas is disordered and has high entropy because its individual molecules are randomly spaced, varously oriented, and can adopt many possible arrangements, as shown in Figure 3.13.

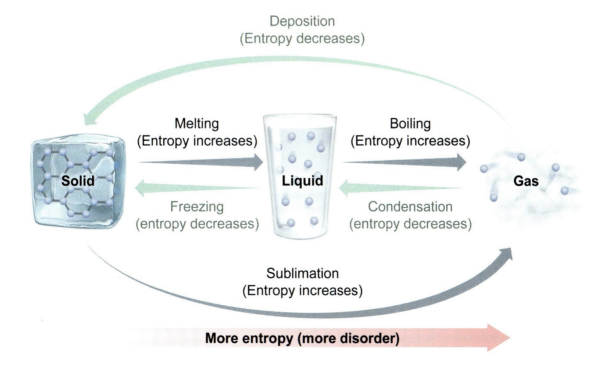

Figure 3.13 Entropy comparisons of different phases of matter.

Entropy and Microstates

The disorder associated with the entropy of a system is based on the number of possible microstates the system has. A **microstate** is one possible configuration of the particles that make up a system, including the positions, energies, and orientations of those particles (Figure 3.14).

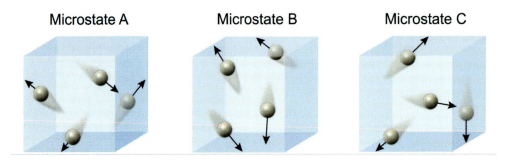

Figure 3.14 Three possible microstates of a four-particle system.

A larger number of microstates corresponds to more disorder (ie, greater entropy) in the system. The number of particles in a system, the volume of the system, the phase(s) of matter present, and the temperature of the system are all factors that affect the number of microstates a system can have. In general, more particles, larger volume, more energetic phases of matter (ie, $S_{solid} < S_{liquid} < S_{gas}$), and higher temperatures correspond to a greater number of possible microstates, and thus greater entropy (Figure 3.15).

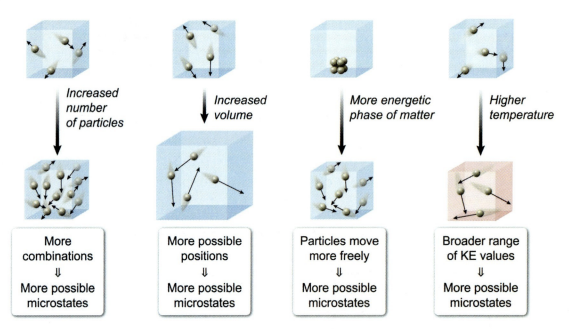

Figure 3.15 Factors that increase entropy.

The relationship between the entropy (S) of a system and the number of possible microstates (N) in a system is expressed mathematically as:

$$S = k \cdot \ln N$$

where k is Boltzmann's constant ($k = 1.38 \times 10^{-23}$ J/K).

Entropy and Dispersed Heat

Although entropy indicates the disorder of a system, entropy is more fundamentally a measure of how much energy in the system is unavailable to do work. In any thermal energy transfer, some of the energy of the system becomes dispersed in the random motion of the atoms and molecules. Disorder increases and part of the energy is made unavailable as dispersed heat that is no longer able to be converted to work. This reality is reflected in the mathematical representation of the second law of thermodynamics:

$$\Delta S_{universe} = \Delta S_{system} + \Delta S_{surroundings} > 0 \quad \text{(for any spontaneous process)}$$

Thus, spontaneous heat transfers out of a system (eg, the cooling of a hot cup of coffee) result in a lower entropy (ie, less unusable energy) for that system. However, the heat released into the surroundings causes the universe as a whole to have greater entropy (ie, more unusable energy).

Entropy of Reversible Processes

Reversible processes are those that, in theory, do not result in any energy being made unavailable to do work (ie, the entropy change of the universe is zero). Although no truly reversible processes exist, processes such as a phase change at a substance's corresponding melting or boiling point are close to being perfectly reversible. The change in the entropy of a system ΔS undergoing such a process can be estimated using the equation:

$$\Delta S = \frac{q_{rev}}{T} \quad \text{(at constant } T\text{)}$$

where q_{rev} is the heat transfer of the system for the reversible process, and T is the Kelvin temperature at which the process takes place. Since entropy is a state function, if an irreversible (ie, spontaneous) process and a reversible process both have the same initial and final states, both also have the same ΔS. Therefore, this equation can be used to calculate the ΔS of an irreversible process provided that q_{rev} is known.

Concept Check 3.5

Two ice cubes are placed in a glass of water on a warm, sunny day. In terms of their respective entropies, describe what happens next to:

1. the ice cubes.
2. the glass of water.
3. the universe.

Solution

Note: The appendix contains the answer.

3.3.02 Evaluating Entropy in Chemical Systems

In qualitative evaluations of entropy in chemical systems, the principles outlined in the previous concept can be applied. Table 3.3 provides a summary of general guidelines for qualitative comparisons of entropy in chemical systems.

Table 3.3 Factors affecting the entropy of chemical systems.

Factor	Trend	Reason
Number of particles	More particles ⇒ greater entropy	More particles allow more possible arrangements (ie, microstates).
Volume (V)	Increased V ⇒ greater entropy	More positions are possible, which increases the number of microstates.
Phase(s) of matter present	$S_{solid} < S_{liquid} \approx S_{aqueous} < S_{gas}$	Higher-energy phases of matter allow more possible positions and energies.
Temperature (T)	Increased T ⇒ greater entropy	Broader range of kinetic energy values are available at high T.
Atomic/molecular mass	Larger mass ⇒ greater entropy	Larger mass causes greater variation in energy level distribution.
Number of atoms in a molecule	More atoms ⇒ greater entropy	More atoms in a molecule allow for more internal variation of position.

A quick assessment of the entropy change for a chemical reaction can be performed by focusing specifically on the number of gas particles and the phases of matter present, as Figure 3.16 illustrates.

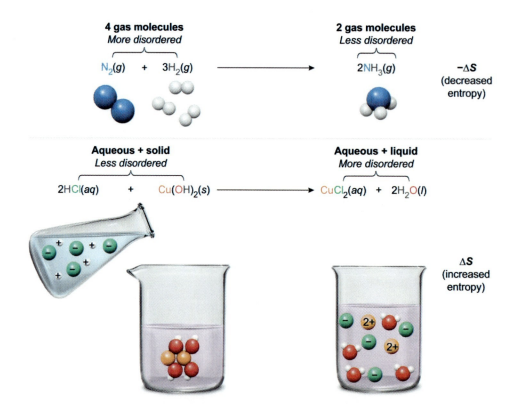

Figure 3.16 Qualitative evaluations of the entropy change of chemical reactions.

In general, chemical systems with more gas particles and more energetic phases of matter tend to have greater entropy.

> **Concept Check 3.6**
>
> Indicate whether the entropy is likely to increase or decrease for each of the following reactions:
>
> 1. $CaCO_3(s) \rightarrow CaO(s) + CO_2(g)$
> 2. $CaCl_2(s) \rightarrow CaCl_2(aq)$
> 3. $Pb(NO_3)_2(aq) + 2\ KI(aq) \rightarrow PbI_2(s) + 2\ KNO_3(aq)$
>
> **Solution**
>
> *Note: The appendix contains the answer.*

Standard Molar Entropy

Unlike enthalpy H and internal energy U, a definitive zero point exists for entropy S. At a temperature of absolute zero (ie, 0 K), a substance should theoretically be a crystalline solid with no motion. Under such conditions, that substance would have only one microstate and zero entropy (assuming a perfect crystal structure). From that starting point, the absolute entropies of chemical substances can be determined using experimental methods.

The **standard molar entropy $S°$** of a substance is the entropy of that substance under standard state conditions, in which all gases have a partial pressure of 1 atm and the concentrations of all species in solution are 1 M. Although temperature is not specified for the standard state in thermodynamics, most data tables use 298 K (25 °C) as a convenient reference.

Because entropy is a state function, the **standard entropy change $\Delta S°_{rxn}$** for any reaction can be calculated by taking the sum of the entropies of the products ($\Sigma S°_{products}$) and subtracting the sum of the entropies of the reactants ($\Sigma S°_{reactants}$):

$$\Delta S°_{rxn} = \Sigma S°_{products} - \Sigma S°_{reactants}$$

A table of experimentally determined standard entropy values is typically made available for calculations of this sort.

Chapter 3: Thermodynamics

 Concept Check 3.7

Substance	S° (J/mol·K)
$CO_2(g)$	213.8
$Fe(s)$	27.3
$Fe_2O_3(s)$	87.4
$HCl(aq)$	56.5
$H_2O(l)$	69.9
$NaCl(aq)$	115.5
$Na_2CO_3(s)$	135.0
$O_2(g)$	205.0

Given the table of standard entropies, determine the entropy change for the following chemical reactions:

1. $2\ Fe(s) + 3\ O_2(g) \rightarrow Fe_2O_3(s)$
2. $Na_2CO_3(s) + 2\ HCl(aq) \rightarrow 2\ NaCl(aq) + H_2O(l) + CO_2(g)$

Solution

Note: The appendix contains the answer.

Chapter 3: Thermodynamics

Lesson 3.4

Heat Transfer in Physical Changes

Introduction

The three phases (ie, states) of matter that exist under normal conditions are **solid**, **liquid**, and **gas**. Solids have a definite shape and volume, liquids have an indefinite shape (ie, take the shape of whatever container they are in) and a definite volume, and gases have an indefinite shape and volume (ie, expand to fill whatever container they are in), as Figure 3.17 illustrates.

Solid
- Particles closely packed
- Strong particle attractions
- Vibrational motion only
- Definite, fixed shape
- Definite, fixed volume

Liquid
- Particles in close contact
- Significant particle attractions
- Translational motion
- No definite fixed shape
- Definite, fixed volume

Gas
- Particles far apart
- Minimal particle attractions
- Highly translational motion
- No definite fixed shape
- Indefinite volume

Figure 3.17 Comparison of the particle models of solid, liquid, and gas phases.

Understanding the behavior of substances in each phase at the molecular level gives insight into how that behavior can be influenced by external factors, particularly with regard to phase changes.

Concept 2.3.05 discusses phase changes from the perspective of intermolecular forces, and Lesson 3.3 addresses phases of matter and phase changes with respect to entropy. This lesson focuses specifically on energy in the different phases of matter and examines the thermodynamics of phase changes along with other factors that may influence these changes.

3.4.01 Phase Changes

When enough energy (transferred as heat) is either absorbed or released by a substance, a physical process occurs in which the substance transitions from one phase to another (ie, a **phase change**). As summarized visually in Figure 3.18, six types of phase changes can occur among the solid, liquid, and gas states:

- **Melting** is a transition from a solid to a liquid.
- **Freezing** is the transition from a liquid to a solid (ie, the opposite of melting).
- **Vaporization** is the transition from a liquid to a gas.

- **Condensation** is the transition from a gas to a liquid (ie, the opposite of vaporization).
- **Sublimation** is the transition from a solid to a gas.
- **Deposition** is the transition from a gas to a solid (ie, the opposite of sublimation).

Figure 3.18 Phase changes among solids, liquids, and gases.

Molecular View of the Solid to Liquid to Gas Transition

To best illustrate the energetics of phase changes from a molecular point of view, consider the example of heat added to a perfect crystalline solid at 0 K until it reaches the gas phase. As discussed in Concept 3.1.02, a crystalline solid at 0 K should (theoretically) have no molecular kinetic energy. At this temperature, the particles of the solid are held together in a fixed arrangement by intermolecular forces.

As heat is added and the temperature increases, the individual particles gain kinetic energy and begin to vibrate but do not yet have enough energy to break free from the intermolecular forces holding them together. As the solid continues to be heated, the particles vibrate with more intensity until they can no longer be held in a fixed position by the intermolecular forces, and the particles begin to move around (Figure 3.19).

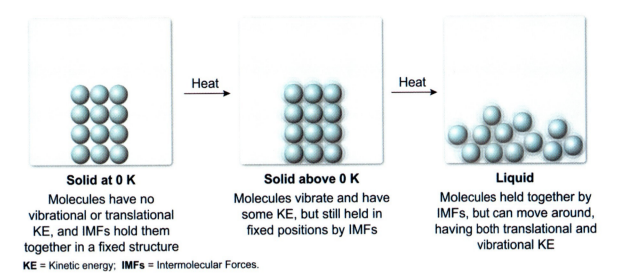

Figure 3.19 Particles at various stages of the solid-to-liquid phase transition.

After all the particles are moving freely and melting is complete, the molecules still experience significant attractions to each other; however, the attractions are weaker since the molecules of the liquid are continually moving around and are not in constant contact. As heat continues to be added, the molecules move faster. Eventually, some of the particles have enough kinetic energy to break completely free from the intermolecular forces holding them together and escape into the gas phase (Figure 3.20).

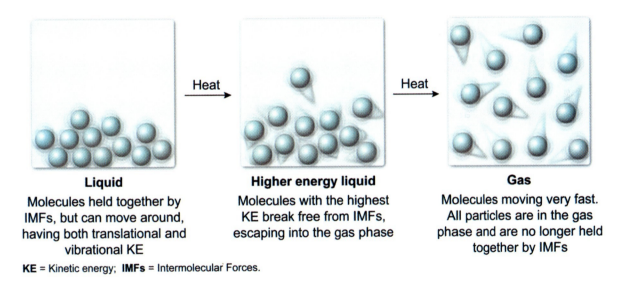

Figure 3.20 Particles at various stages of the liquid-to-gas phase transition.

Only the highest-energy molecules escape into the gas phase, leaving behind lower-energy molecules. If heat is continually added to the liquid, boiling continues until all the molecules gain enough energy to complete the transition from the liquid phase to the gas phase.

The process of removing heat from a gas until it transitions back to a liquid and then to a solid can be imagined as the reverse of the process just described. In the reverse process, condensation of the gas occurs followed by freezing (ie, solidification) of the condensed liquid.

Sublimination and Deposition

In sublimination and deposition, particles in the solid or gas phase bypass the liquid phase altogether. If the intermolecular forces in a solid are not strong enough to hold the particles in relatively close contact as the kinetic energy of the particles increases, the solid particles can escape directly to the gas phase (ie, sublimation). Conversely, when the forces and energy of particles in the gas phase are such that the particles are immediately held in a fixed position when enough heat is removed, the particles undergo deposition rather than passing through a more weakly interactive liquid phase.

Concept Check 3.8

For any given substance, deposition is the phase change that results in the release of the most energy. Explain why.

Solution

Note: The appendix contains the answer.

3.4.02 Melting Point and Heat of Fusion

As discussed previously, heat added to a solid causes the particles of that substance to gain kinetic energy, raising its temperature. When the solid reaches a temperature at which its particles have enough energy to break free from the intermolecular forces holding them in a fixed position, the substance is at its **melting point** (ie, the temperature at which both solid and liquid states coexist in equilibrium), as shown in Figure 3.21.

Chapter 3: Thermodynamics

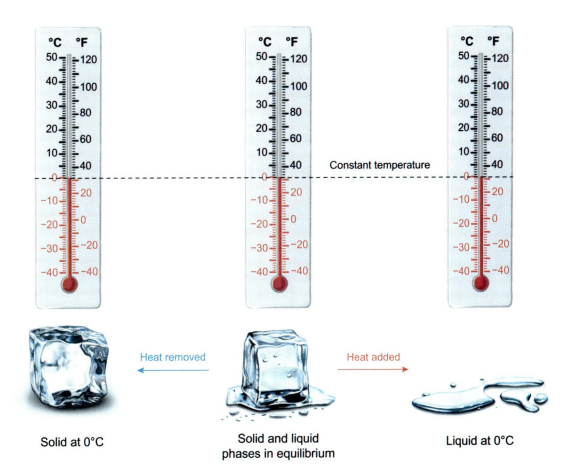

Figure 3.21 Solid phase and liquid phase coexist in equilibrium at the melting point.

Heat of Fusion

Heat energy absorbed by a solid at its melting point is used to break particles free from the attractive intermolecular forces holding them together but does not increase the temperature (ie, the average kinetic energy of particles). As such, the temperature during the melting process remains relatively constant for solids with a regular crystal structure.

The total amount of heat that must be added to a solid substance at its melting point to fully convert it to a liquid is its **latent heat of fusion ΔH_{fusion}**, or just heat of fusion (Figure 3.22).

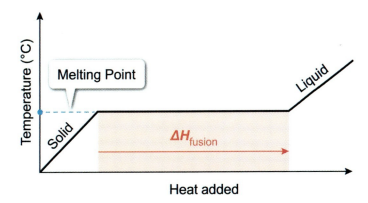

Figure 3.22 Heat of fusion is the heat necessary to fully transition a solid substance at its melting point to the liquid phase.

Both the melting point and **freezing point** of a substance correspond to the same temperature, with the difference being whether heat is added or removed. Consequently, when a substance in the liquid phase is cooled to its freezing point, an amount of heat equal to ΔH_{fusion} must be *released* as the liquid is converted to the solid phase. Therefore, the quantities of energy transferred for complimentary phase changes (eg, freezing and melting) are equal in magnitude but opposite in sign:

$$-\Delta H_{fusion} = \Delta H_{solidification}$$

Factors Affecting Melting Point

A major factor influencing the melting point of a substance is the strength of the intermolecular forces between its particles. Because melting involves increasing the kinetic energy of particles to overcome the attractive intermolecular forces between those particles, substances with *stronger* intermolecular forces have correspondingly *higher* melting points, as illustrated in Figure 3.23.

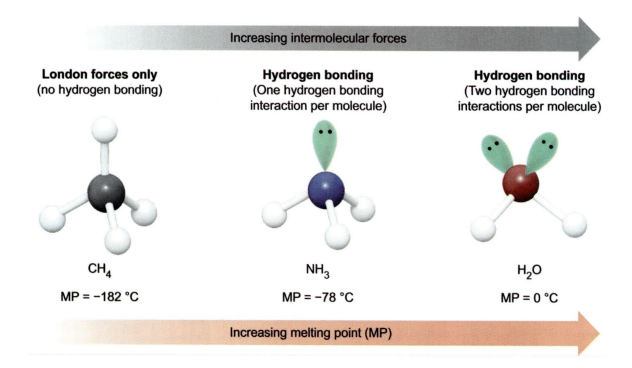

Figure 3.23 Effect of intermolecular forces on the melting point.

Another factor that influences melting point is the ability of a substance to form an organized crystal structure. Small, regularly shaped molecules fit together in an organized crystal structure far more easily than large, irregularly shaped molecules. The more efficient spacing of highly ordered solids maximizes the attractive forces between particles, leading to higher melting points. Conversely, **amorphous solids** (ie, solids with irregular crystal structures) tend to have lower melting points and greater melting point ranges (Figure 3.24).

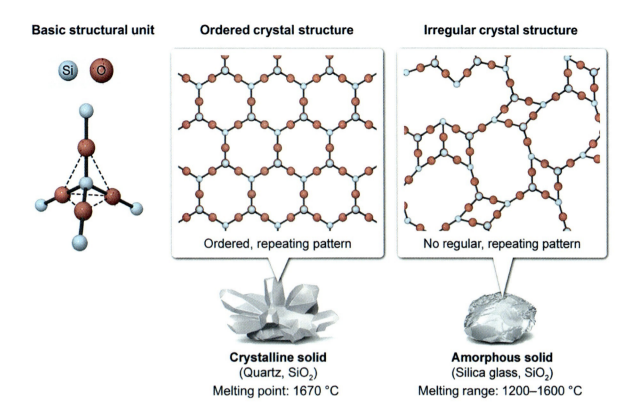

Figure 3.24 Melting points of highly ordered versus irregular crystal structures.

Impurities within a solid also have the effect of lowering the melting point and increasing the melting point range. As with irregularly shaped molecules, impurities limit the ability of a substance to form a highly ordered crystal structure, resulting in less-efficient spacing between particles and fewer intermolecular attractions. The effect depends on the concentration of impurities in the solid.

For similar reasons, a liquid with a significant number of dissolved particles has a lower freezing point than the liquid in its pure form. The dissolved particles interfere with the liquid molecules' ability to aggregate into a cohesive crystal structure by reducing the intermolecular interactions between the molecules.

In such cases, the kinetic energy of the liquid molecules must be reduced (ie, the temperature must be lowered) to enable the weaker crystal structure to form. Therefore, the greater the concentration of dissolved particles, the lower the freezing point, as Figure 3.25 illustrates.

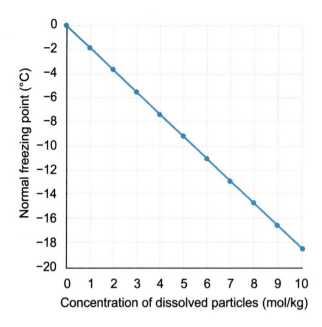

Figure 3.25 Normal freezing point of water versus molality of dissolved particles.

External pressure is a factor that affects the melting point differently depending on the relative densities of the solid and liquid phases. A higher external pressure always favors the denser phase. For water, the liquid phase is denser than the solid phase (ie, ice); therefore, putting pressure on ice causes it to melt. For most other substances, the solid phase is denser than the liquid phase. In such cases, added pressure causes the liquid to solidify to the denser solid phase. For example, high external pressure can solidify liquid nitrogen.

☑ Concept Check 3.9

From a molecular perspective, explain why salting roads in winter keeps them safer for travelers.

Solution
Note: The appendix contains the answer.

3.4.03 Boiling Point and Heat of Vaporization

At any given temperature, some of the molecules in a liquid have enough kinetic energy to overcome the attractive intermolecular forces holding them in the liquid phase. These higher-energy molecules can escape into the gas phase as vapors that are in equilibrium with the liquid phase. The resulting vapor pressure is significantly influenced by temperature. When the temperature is high enough for the vapor pressure to *equal* the external pressure exerted by the surroundings (ie, ambient pressure), the liquid boils. As such, this temperature is the **boiling point** of the liquid. The boiling point when the external pressure is exactly 1 atm is called the normal boiling point.

Heat of Vaporization

When a substance in the liquid phase is at its boiling point, the latent heat energy absorbed by the system is used to break the attractive intermolecular forces between the liquid molecules rather than cause an increase in temperature. Consequently, the energized liquid molecules can escape the surface of the liquid as vapor and enter the gas phase. The amount of heat that must be added to fully convert 1 mole of a substance at its boiling point from the **liquid phase to the gas phase** is that substance's **latent heat of vaporization** $\Delta H_{vaporization}$, which is illustrated in Figure 3.26.

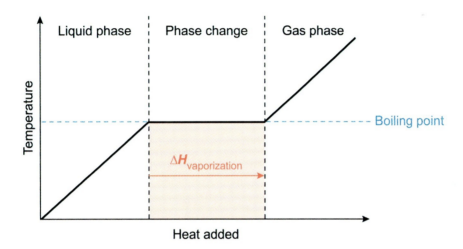

Figure 3.26 Heat of vaporization is the heat needed to fully convert a liquid substance at its boiling point to the gas phase.

The reverse process of vaporization is condensation (ie, going from gas to liquid) which *releases* energy. Consequently, the latent heat of condensation $\Delta H_{condensation}$ is equal in magnitude but has the opposite sign of the latent heat of vaporization. This relationship is expressed mathematically as:

$$\Delta H_{condensation} = -\Delta H_{vaporization}$$

Factors Affecting Boiling Point

Because the boiling point is the temperature at which the vapor pressure equals the external pressure, a greater external pressure results in a higher boiling point. Conversely, when a liquid is kept under a lower external pressure, a lower temperature is needed for the liquid's vapor pressure to equal the external pressure.

For example, at sea level where the air pressure is 1.0 atm, water boils at 100 °C, whereas at the top of Mt. Everest where the air pressure is much lower (0.336 atm), water boils at about 71 °C (Figure 3.27).

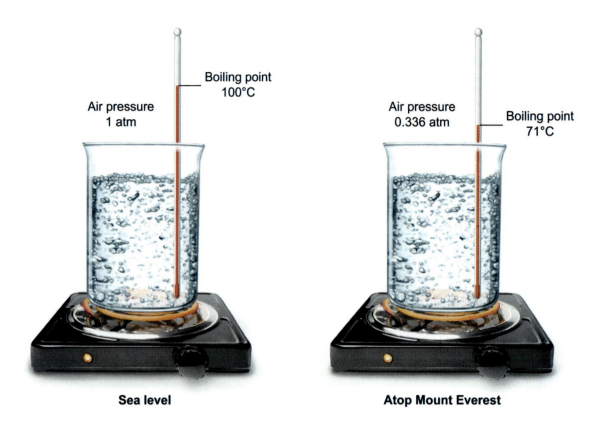

Figure 3.27 Effect of external pressure on boiling point.

The strength of the intermolecular forces holding liquid molecules together is another factor that influences the boiling point. As Concept 2.3.05 explains, substances with stronger intermolecular forces have higher boiling points because more energy is required to overcome these attractions between molecules during the phase change from a liquid state to a gas state (Figure 3.28).

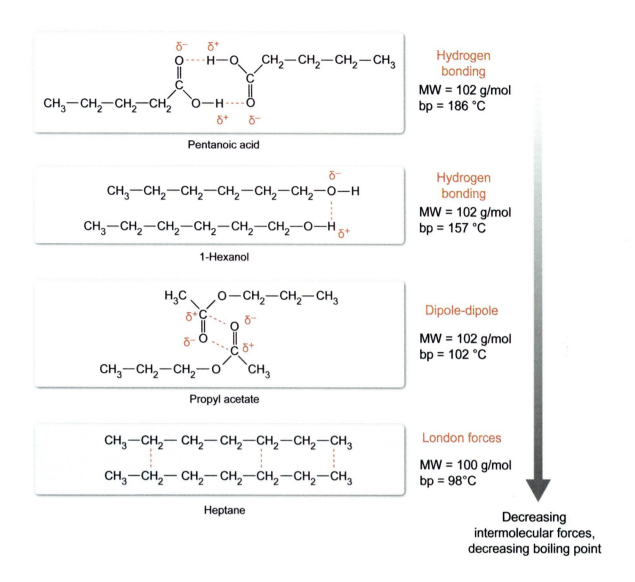

Figure 3.28 Effect of intermolecular forces on the boiling point of compounds with similar molecular weights.

Raoult's law states that when a nonvolatile solute (ie, a solute with no vapor pressure) is dissolved in a liquid, the vapor pressure of the liquid decreases. This phenomenon occurs because the solute molecules reduce the amount of surface area available for the liquid molecules to occupy and escape into the gas phase.

Consequently, the liquid containing the solute must be heated to a *higher temperature* for its vapor pressure to reach the same ambient pressure and begin to boil. As Figure 3.29 shows, a directly proportional relationship exists between the concentration of dissolved particles and the increase in the boiling point.

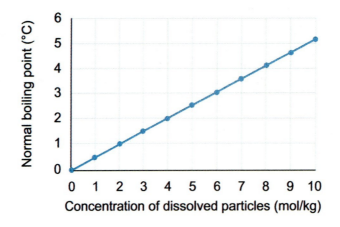

Figure 3.29 Normal boiling point of water versus the concentration of dissolved particles.

> **Concept Check 3.10**
>
> A pressure cooker is a specially designed cooking vessel that uses an air-tight seal to cook food under high pressure.
>
> 1. What effect does this device have on cooking time (compared to boiling the food in an open pot)?
> 2. Explain why this effect occurs in terms of the boiling point and vapor pressure.
>
> **Solution**
>
> *Note: The appendix contains the answer.*

3.4.04 Phase Diagrams

A **phase diagram** is a graph that illustrates a substance's stability in a given phase (solid, liquid, or gas) as a function of pressure (*y*-axis) and temperature (*x*-axis). The **boundary lines** between phases on a phase diagram indicate the conditions under which two phases are in equilibrium. If a change in conditions results in the **crossing of a boundary line**, a substance undergoes the corresponding change in phase (Figure 3.30).

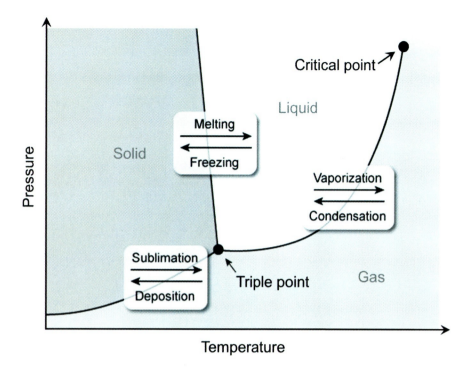

Figure 3.30 Components and features of a phase diagram.

Because the boiling point of a liquid occurs when its vapor pressure is equal to the ambient pressure, the boundary line between the liquid and the gas (ie, vapor) phase on a phase diagram is also a **vapor pressure curve**, indicating the vapor pressure of the liquid across a range of temperatures. In a similar way, the vapor pressure of a solid must equal the ambient pressure for sublimation to take place. Therefore, the boundary line between the solid and the gas phase on a phase diagram is the vapor pressure curve for the solid.

The **triple point** is the point at which all three boundary lines intersect, and it indicates the temperature and pressure at which all three phases simultaneously exist in equilibrium (Figure 3.31). A substance cannot exist in liquid form at a pressure below that of its triple point.

Figure 3.31 All three phases of water in equilibrium at its triple point (0.006 atm and 0.01 °C).

The **critical temperature** of a substance is the maximum temperature at which a substance can exist as a liquid. Above the critical temperature, a substance cannot be converted to the liquid phase under any pressure, no matter how great. The pressure required to convert a substance to its liquid phase at the critical temperature is called the **critical pressure**.

On a phase diagram, the critical point indicates the critical temperature and pressure of a substance. Any substance having a temperature and pressure above its critical point exists as a **supercritical fluid**, having properties between those of a liquid and a gas.

Intensive Properties Influencing the Shape of a Phase Diagram

Concept 3.4.02 discusses how the effect of external pressure on the melting point of a substance depends on the relative densities of the solid and liquid phases of a substance, with higher pressures favoring the denser of the two phases. When the solid phase is denser than the liquid phase (as is the case for most substances), the **melting curve** (ie, the solid-liquid phase boundary) shows a positive slope on the phase diagram. For substances with a liquid phase that is denser than its solid phase (eg, water), the slope of the melting curve is negative, as shown by Figure 3.32.

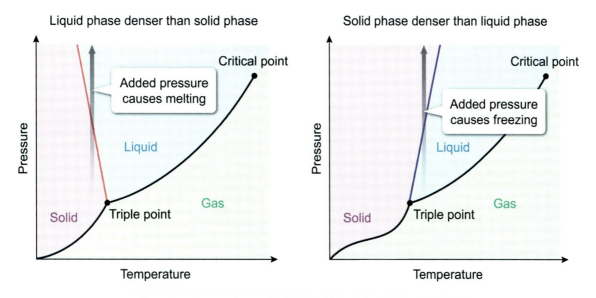

Higher pressure always favors the densest phase of matter.

Figure 3.32 Shapes of phase diagrams based on the density of the solid phase versus the liquid phase.

Another factor influencing the appearance of a phase diagram is the strength of a substance's intermolecular forces. Substances with stronger intermolecular forces have higher melting and boiling points and can also exist in the liquid phase at lower pressures. As such, substances with stronger intermolecular forces have phase diagrams that are both lower and farther to the right than the phase diagrams of substances with weaker intermolecular forces.

Effects of a Dissolved Solute on Phase Diagrams

As explained in Concepts 3.4.02 and 3.4.03, a dissolved solute in a liquid has the dual effects of lowering the melting point and vapor pressure. On a phase diagram, these changes appear as a shift of the melting curve to the left and a shift of the vapor pressure curves (ie, the liquid-gas and solid-gas phase boundaries) down, as Figure 3.33 illustrates.

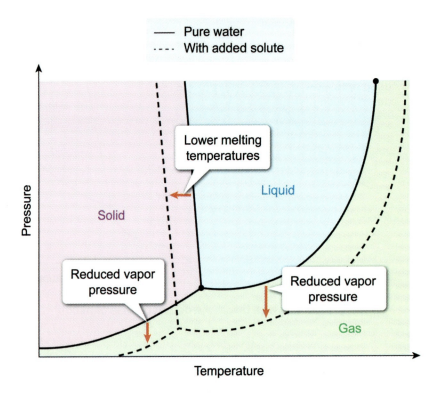

Figure 3.33 Effects of a dissolved solute on a phase diagram.

This downward shift in the vapor pressure curves has the effect of increasing the boiling and sublimation temperatures across all ambient pressures (Figure 3.34).

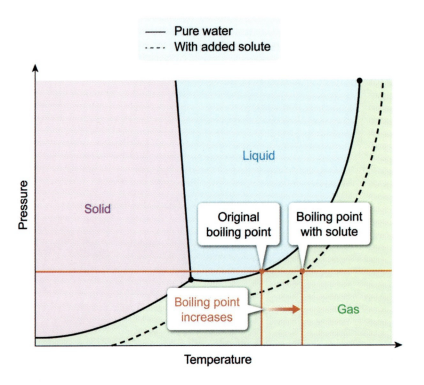

Figure 3.34 Boiling point elevation due to the addition of a dissolved solute.

Concept Check 3.11

Consider the given phase diagram of an unknown substance:

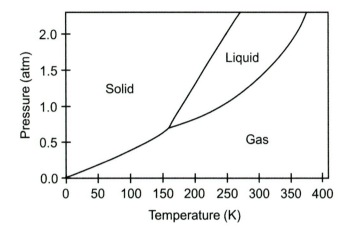

1. What is the estimated normal boiling point of the substance?
2. What are the temperature and pressure at the triple point?
3. How does the density of the solid phase compare to the density of the liquid phase for the substance?
4. Would this substance be expected to have stronger or weaker intermolecular forces than water?

Solution

Note: The appendix contains the answer.

3.4.05 Heating Curves

A **heating curve** graphically represents how the temperature of a substance changes as heat is absorbed at a constant rate. The plateaus (ie, the horizontal flat segments) of a heating curve represent phase changes (eg, melting, boiling). Along these segments, added heat does not result in a temperature increase. Instead, heat is used to overcome the attractive intermolecular forces holding the melting or boiling molecules together, as discussed in Concepts 3.4.02 and 3.4.03. After the substance has fully changed phase, any additional heat contributes to increasing the temperature. Figure 3.35 illustrates these features.

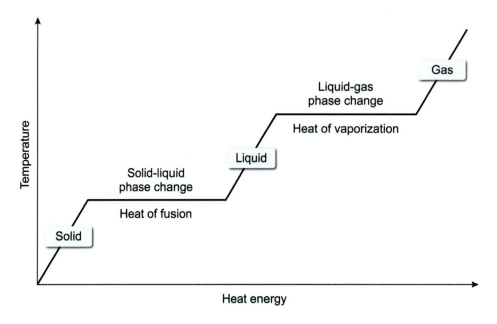

Figure 3.35 General structure of a heating curve.

As seen in Figure 3.35, heat added to (or removed from) a substance can either change the temperature or change the phase, but not both at the same time. Whenever a substance experiences a change in temperature but does not undergo a phase change, the corresponding heat transfer (q) is directly proportional to the mass of the substance (m), its specific heat capacity for a given phase (C_p), and the change in temperature (ΔT):

$$q = mC_p\Delta T$$

Because the latent heat for a phase change (eg, ΔH_{fusion}, $\Delta H_{vaporization}$) is given as thermal energy per mole, the amount of heat transferred during a change in phase can be calculated using the equation:

$$q = n\Delta H$$

where n is the number of moles of the substance undergoing the phase change, and ΔH is the latent heat of the particular phase change that occurs.

For transitions involving both a temperature change and a phase change, the calculation must be broken into multiple parts, with each part being a different stage in the transition represented by its corresponding region of the heating curve, as shown in Figure 3.36.

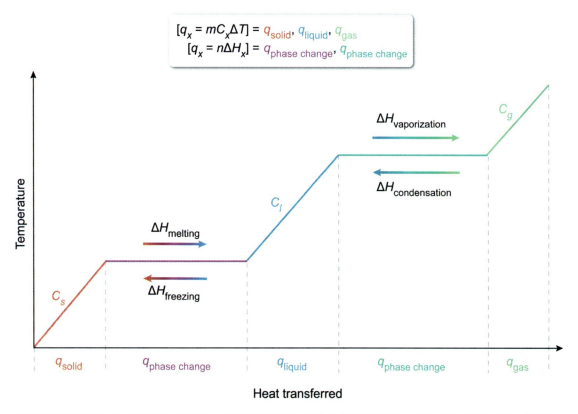

Figure 3.36 Calculations associated with different regions of a heating curve.

Consider a 25.0 g sample of ice at −35 °C heated to steam at 125 °C. The calculation of the total heat transfer for this process can be carried out in three steps:

Step 1. Draw a heating curve to illustrate the different parts of the heating process (Figure 3.37).

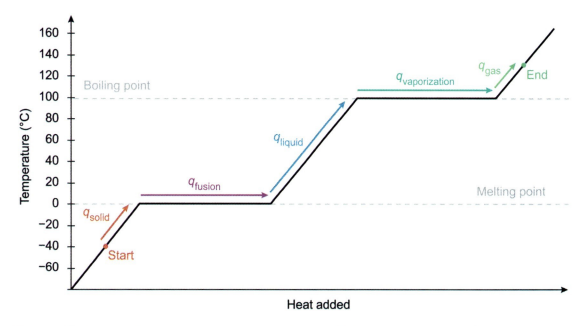

Figure 3.37 Heat transfers associated with the transition of H_2O from −35 °C ice to 125 °C steam.

Step 2. Use thermodynamic data for the substance being heated to solve for the heat transfer associated with each region of the heating curve using the appropriate equations (see Figure 3.36 for reference). The relevant thermodynamic data of water in this example is given in Table 3.4.

Table 3.4 Selected thermodynamic quantities for water.

Thermodynamic property	Value
Specific heat for $H_2O(s)$	2.092 J/g°C
ΔH_{fusion}	6.008 kJ/mol
Specific heat for $H_2O(l)$	4.184 J/g°C
$\Delta H_{vaporization}$	40.67 kJ/mol
Specific heat of $H_2O(g)$	1.841 J/g°C

$$q_{solid} = mC_s\Delta T \implies q_{solid} = (25.0 \text{ g}) \cdot (2.092 \text{ J/g} \cdot °C) \cdot [0\,°C - (-35\,°C)] = 1{,}830 \text{ J}$$

$$q_{fusion} = n\Delta H_{fusion} \implies q_{fusion} = \left[25.0 \text{ g H}_2\text{O} \times \left(\frac{1 \text{ mol H}_2\text{O}}{18.02 \text{ g H}_2\text{O}}\right)\right] \cdot (6{,}008 \text{ J/mol}) = 8{,}340 \text{ J}$$

$$q_{liquid} = mC_l\Delta T \implies q_{liquid} = (25.0 \text{ g}) \cdot (4.184 \text{ J/g} \cdot °C) \cdot (100\,°C - 0\,°C) = 10{,}500 \text{ J}$$

$$q_{vaporization} = n\Delta H_{vaporization} \implies q_{vaporization} = \left[25.0 \text{ g H}_2\text{O} \times \left(\frac{1 \text{ mol H}_2\text{O}}{18.02 \text{ g H}_2\text{O}}\right)\right] \cdot (40{,}670 \text{ J/mol}) = 56{,}400 \text{ J}$$

$$q_{gas} = mC_g\Delta T \implies q_{gas} = (25.0 \text{ g}) \cdot (1.841 \text{ J/g°C}) \cdot (125\,°C - 100\,°C) = 1{,}150 \text{ J}$$

Step 3. Sum the calculated q values for each part of the heating process to get the total heat transferred.

$$q_{total} = q_{solid} + q_{fusion} + q_{liquid} + q_{vaporization} + q_{gas}$$

$$q_{total} = 1{,}830 \text{ J} + 8{,}340 \text{ J} + 10{,}500 \text{ J} + 56{,}400 \text{ J} + 1{,}150 \text{ J} = 78{,}200 \text{ J}$$

Concept Check 3.12

A sample consisting of 1.12 kg of molten iron at 1,650 °C is cooled to a solid at room temperature (22 °C). How much heat (in kJ) is transferred during this process? (Selected thermodynamic values for iron are given in the table shown.)

Thermodynamic property	Value
Specific heat for Fe(s)	0.450 J/g·°C
Melting Point	1,538 °C
ΔH_{fusion}	13.8 kJ/mol
Specific heat for Fe(l)	0.902 J/g·°C

Solution

Note: The appendix contains the answer.

Lesson 3.5

Heat Transfer in Chemical Changes

Introduction

Stoichiometry is concerned with relating the amounts of substances involved in chemical reactions (Lesson 2.5). In addition to the rearrangement of atoms and chemical bonds that occurs during a chemical reaction, an exchange of energy also occurs. In chemical reactions, heat can behave as a reactant and be absorbed from the surrounding environment to form higher-energy products. Alternatively, heat can behave as a product and be released into the surrounding environment as higher-energy reactants are converted into lower-energy products (Figure 3.38).

Heat as a reactant

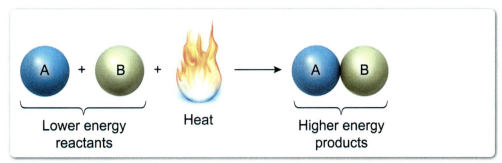

Heat as a product

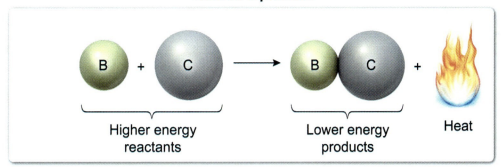

Figure 3.38 Heat as a reactant or product in chemical reactions.

When a reaction takes place under constant pressure (which is usually the case), the heat transfer that occurs is equal to the enthalpy change of the reaction. This lesson explores the concepts and calculations associated with the enthalpy changes that occur during such chemical reactions.

3.5.01 Enthalpy of a Reaction

As Concept 3.2.03 explains, enthalpy H is defined as the sum of a system's internal energy (U) and the product of its pressure (P) and volume (V). The change in enthalpy ΔH resulting from a chemical reaction can be expressed mathematically as:

$$\Delta H_{\text{rxn}} = H_{\text{products}} - H_{\text{reactants}}$$

Under constant pressure, the total amount of heat q_{rxn} released or consumed by a chemical reaction is equal to ΔH_{rxn}:

$$\Delta H_{rxn} = q_{rxn} \quad \text{(at constant pressure)}$$

In thermodynamic analyses of chemical reactions, the **enthalpy of reaction ΔH_{rxn}** (ie, the heat of a reaction) is a useful means to determine how the energy of reactants change as they are converted into products because heat transfers during a reaction can be easily measured.

The sign of ΔH_{rxn} indicates whether a reaction is **endothermic** or **exothermic**. In endothermic reactions, heat is absorbed by the reactants to form higher-energy products, resulting in a positive ΔH_{rxn}. In exothermic reactions, heat is released by the reactants as they form lower-energy products, resulting in a negative ΔH_{rxn}. The energy changes that occur during a reaction can be represented on a diagram that plots the enthalpy H (y-axis) versus reaction progress (x-axis), as Figure 3.39 illustrates.

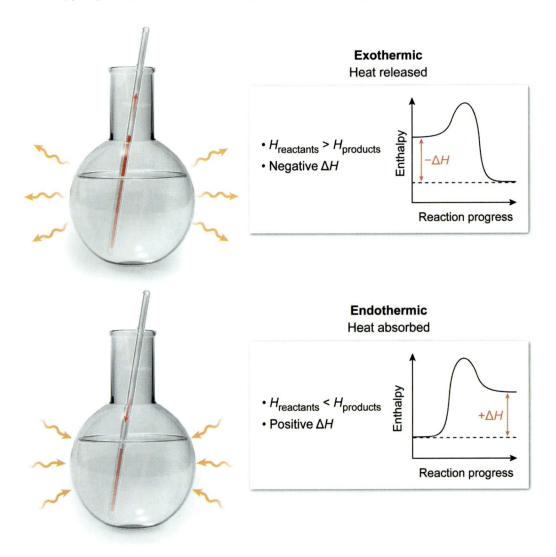

Figure 3.39 Energy diagrams for endothermic and exothermic reactions.

Values of ΔH_{rxn} are stated as the amount of heat transferred per mole of reaction (ie, the total heat absorbed or released by a balanced reaction that occurs 6.022×10^{23} times). Although the *per mole* part of the units is sometimes omitted as an abbreviation in cited values, the molar basis is implied by the reaction. Evaluating ΔH_{rxn} under standard conditions (ie, 1 atm pressure and 1 M concentration for all species in solution) yields the **standard enthalpy change** $\Delta H°_{rxn}$ of a reaction, which is denoted by the addition of the degree symbol.

For example, consider the following chemical equation:

$$CH_4(g) + 2\,O_2(g) \rightarrow CO_2(g) + 2\,H_2O(g) \quad \Delta H°_{rxn} = -802.5 \text{ kJ/mol}_{rxn}$$

The equation indicates that when 1 mole of CH_4 reacts with 2 moles of O_2 to form 1 mole of CO_2 and 2 moles of H_2O, 802.5 kJ of heat is produced. Thus, $\Delta H°_{rxn}$ can be related to the molar amounts of any product or reactant in this reaction, and those relationships can be used as conversion factors. Accordingly, the enthalpy change corresponding to the reaction of 2.0 moles of CH_4 can be determined by using the relationship 1 mol CH_4 = −802.5 kJ as a conversion factor:

$$2.0 \text{ mol CH}_4 \cdot \left(\frac{-802.5 \text{ kJ}}{1 \text{ mol CH}_4}\right) = -1.6 \times 10^3 \text{ kJ}$$

Conversely, the number of moles of CH_4 that must react to produce 400 kJ of heat can be calculated using an inverted version of the same conversion factor:

$$-400 \text{ kJ} \cdot \left(\frac{1 \text{ mol CH}_4}{-802.5 \text{ kJ}}\right) = 0.5 \text{ mol CH}_4$$

Note that the phrase "produce 400 kJ of heat" translates into an enthalpy change of −400 kJ. Heat as a product of a reaction (ie, heat released by the system) is a loss of energy for the system, which is indicated by the minus sign. Careful analysis of the wording in thermochemistry questions is necessary to ensure proper sign conventions are followed.

The direction of a reaction also impacts $\Delta H°_{rxn}$. Reversing a reaction's direction changes the sign of $\Delta H°_{rxn}$. In other words, if a forward reaction is exothermic, the reverse process is endothermic (Figure 3.40).

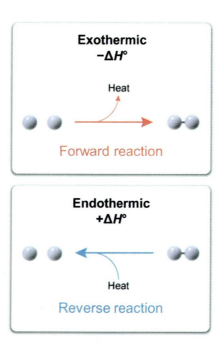

Figure 3.40 The enthalpy change for a reaction has the same magnitude but changes sign (eg, negative to positive) when the reaction is reversed.

Chapter 3: Thermodynamics

> **Concept Check 3.13**
>
> Consider the following reaction for the formation of MgO:
>
> $$2\ Mg(s) + O_2(g) \rightarrow 2\ MgO(s) \quad \Delta H°_{rxn} = -1{,}203.6\ kJ$$
>
> What is the enthalpy change for the *decomposition* of 10.2 g MgO?
>
> **Solution**
>
> *Note: The appendix contains the answer.*

3.5.02 Hess' Law

Because enthalpy is a state function, the enthalpy change depends only upon the initial and final states of the system. Applied to chemical reactions, this principle is known as Hess's law, which states that if a reaction can be performed in more than one step, the enthalpy change of the overall reaction is equal to the sum of the enthalpy changes from each step (Figure 3.41).

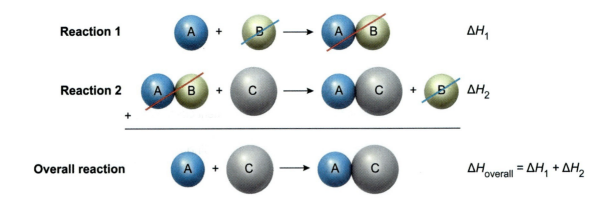

Figure 3.41 When two or more reactions combine to give an overall reaction, the enthalpy change of the overall reaction is equal to the sum of the enthalpy changes of the combined reactions (Hess's law).

In certain cases, directly measuring the enthalpy change of a particular reaction can prove challenging. If such a reaction can be broken down into steps for which the enthalpy changes are known, the overall enthalpy change of the reaction can be easily determined using Hess's law. For example, the reaction of C(s, graphite) with H₂(g) to form C₂H₄(g) is not easily carried out in the lab.

Desired overall reaction: $\quad 2\ C(s) + 2\ H_2(g) \rightarrow C_2H_4(g) \quad \Delta H°_{rxn} = ?$

However, the following reactions and their enthalpy changes are well known:

Reaction 1 $\quad C_2H_4(g) + 3\ O_2(g) \rightarrow 2\ CO_2(g) + 2\ H_2O(l) \quad \Delta H°_{rxn} = -1{,}411.1\ kJ$

Reaction 2 $\quad 2\ H_2(g) + O_2(g) \rightarrow 2\ H_2O(l) \quad \Delta H°_{rxn} = -571.7\ kJ$

Reaction 3 $\quad C(s, \text{graphite}) + O_2(g) \rightarrow CO_2(g) \quad \Delta H°_{rxn} = -393.5\ kJ$

Although these reactions do not add up to the desired overall reaction as written, Figure 3.42 shows how they can be modified to do so.

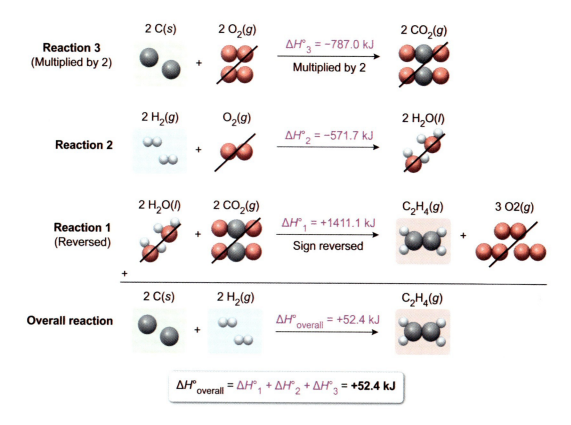

Figure 3.42 Determination of $\Delta H°_{rxn}$ for the formation of C_2H_4 from its constituent elements by Hess's law.

As the given example illustrates, a reaction equation (and its corresponding $\Delta H°$) can be modified in one or both of the following ways:

- A reaction can be reversed, which changes the sign of $\Delta H°$.
- A reaction can be multiplied by a whole number or fraction, which also multiplies the value of $\Delta H°$ by that same value.

When determining the enthalpy change of a reaction using Hess's law, the initial focus should be on substances from the reaction that appear in only one of the given equations. By modifying the given equations so that each of these substances matches the target reaction, the given equations (modified as needed) are certain to add up to the target reaction, which also means that their respective $\Delta H°$ values add up to the $\Delta H°_{rxn}$ of the target reaction.

Although Hess's law applies specifically to the enthalpy change of reactions, its principles can be applied to any of the state functions (eg, ΔG, ΔS) associated with reactions.

Concept Check 3.14

Reaction 1	$2\,S(s) + 3\,O_2(g) \rightarrow 2\,SO_3(g)$	$\Delta H°_{rxn} = -791.4$ kJ
Reaction 2	$2\,SO_2(g) + O_2(g) \rightarrow 2\,SO_3(g)$	$\Delta H°_{rxn} = -197.8$ kJ

Using the given equations and their corresponding $\Delta H°_{rxn}$ values, calculate the enthalpy change for the following reaction:

$$SO_2(g) \rightarrow S(s) + O_2(g) \qquad \Delta H°_{rxn} = ?$$

Solution

Note: The appendix contains the answer.

3.5.03 Heats of Formation

The **standard enthalpy of formation $\Delta H°_f$** (also called the standard heat of formation) of a substance is the enthalpy change that accompanies the formation of exactly 1 mole of a substance from its constituent elements in their most stable states under standard thermodynamic conditions. An element already in its most stable state under standard conditions has a $\Delta H°_f$ of zero.

The heats of formation for a vast number of substances have been experimentally determined. Using these known values and the principles of Hess's law, the enthalpy change of various reactions can be calculated. For example, consider the reaction of titanium(IV) chloride with water:

$$TiCl_4(l) + 2\,H_2O(l) \rightarrow TiO_2(s) + 4\,HCl(aq)$$

The heats of formation for each of the substances in this reaction are:

$$Ti(s) + 2\,Cl_2(g) \rightarrow TiCl_4(l) \qquad \Delta H°_f = -804.2 \text{ kJ}$$

$$H_2(g) + \tfrac{1}{2}\,O_2(g) \rightarrow H_2O(l) \qquad \Delta H°_f = -285.8 \text{ kJ}$$

$$Ti(s) + O_2(g) \rightarrow TiO_2(s) \qquad \Delta H°_f = -944.7 \text{ kJ}$$

$$\tfrac{1}{2}\,H_2(g) + \tfrac{1}{2}\,Cl_2(g) \rightarrow HCl(aq) \qquad \Delta H°_f = -167.2 \text{ kJ}$$

The heats of formation and their corresponding reaction equations must be modified and combined as shown in Figure 3.43 to give the $\Delta H°_{rxn}$ for the overall process.

Reactant ΔH°f equations

Reversed equations multiplied by the reactant coefficient	$TiCl_4(l) \rightarrow Ti(s) + 2\ Cl_2(g)$	$-\Delta H°_f = +804.2\ kJ$
	$2 \cdot \left[H_2O(l) \rightarrow H_2(g) + \frac{1}{2}O_2(g) \right]$	$2 \cdot (-\Delta H°_f) = 2 \cdot (+285.8\ kJ)$

Product ΔH°f equations

Each equation multiplied by the product coefficient	$Ti(s) + O_2(g) \rightarrow TiO_2(s)$	$\Delta H°_f = -944.7\ kJ$
+	$4 \cdot \left[\frac{1}{2}H_2(g) + \frac{1}{2}Cl_2(g) \rightarrow HCl(aq) \right]$	$4 \cdot \Delta H°_f = 4 \cdot (-167.2\ kJ)$

Net reaction	$TiCl_4(l) + 2\ H_2O(l) \rightarrow TiO_2(s) + 4\ HCl(aq)$	$\Delta H°_{rxn} = -238.7\ kJ$

Figure 3.43 Using heats of formation to solve for the enthalpy change of the reaction of titanium(IV) chloride with water.

As shown by this example, two patterns emerge when solving for $\Delta H°_{rxn}$ using heats of formation:

- $\Delta H°_f$ of every reactant must have its sign changed.
- Every $\Delta H°_f$ must be multiplied by its corresponding stoichiometric coefficient.

Mathematically, these patterns are expressed in the following equation:

$$\Delta H°_{rxn} = \sum n\Delta H°_f\ (\text{products}) - \sum m\Delta H°_f\ (\text{reactants})$$

where n and m refer to the coefficients of the balanced chemical equation for which the $\Delta H°_{rxn}$ is being determined. Because of the way that $\Delta H°_f$ and state changes are defined, the signs of the enthalpy values are automatically given the correct treatment in the equation, and therefore enthalpy of formation values can be directly substituted into this equation without modification.

Concept Check 3.15

Consider the reaction of $N_2O(g)$ with $NO_2(g)$ to form NO:

$$N_2O(g) + NO_2(g) \rightarrow 3\,NO(g)$$

Using the heats of formation in the given table, determine the standard enthalpy change for this reaction.

Substance	$\Delta H°_f$ (kJ/mol)
$N_2O(g)$	81.6
$NO_2(g)$	33.84
$NO(g)$	90.37

Solution

Note: The appendix contains the answer.

3.5.04 Bond Dissociation Energy

Bond enthalpy (ie, bond dissociation energy) is the amount of energy needed to break 1 mole of a bond between two atoms in the gas phase. Although the bond enthalpy for each type of bond (eg, a C–H bond) varies slightly from molecule to molecule, average bond enthalpies for different types of bonds are useful for estimating the enthalpy change of reactions. Table 3.5 gives the average bond enthalpies for some common types of bonds.

Table 3.5 Average Bond Enthalpies for Selected Common Bond Types.

Bond type	Bond enthalpy (kJ/mol)	Bond type	Bond enthalpy (kJ/mol)
C–H	413	H–H	436
C–C	348	H–F	567
C=C	614	H–Cl	431
C–N	293	H–Br	366
C–O	358	H–I	299
C=O	799	O–H	463
C–F	485	O–O	146
C–Cl	328	O=O	495
C–Br	276	O–F	190
C–I	240	O–Cl	203
N–H	391	Cl–Cl	242
N–N	163	Br–Cl	218
N–O	201	Br–Br	193
N–F	272	I–Cl	208
N–Cl	200	I–Br	175
N–Br	243	I–I	151

Breaking a bond requires energy and is always an endothermic process ($\Delta H°_{\text{bond breaking}} > 0$) whereas forming a bond is always an exothermic process ($\Delta H°_{\text{bond formation}} < 0$). Consequently, the enthalpies associated with breaking or forming a given type of bond are equal in magnitude but opposite in sign:

$$\Delta H°_{\text{bond formation}} = -\Delta H°_{\text{bond breaking}}$$

A chemical reaction can be viewed as a combined series of steps in which some bonds within the reactants are broken as new bonds are formed to make the products. Applying the principle of Hess's law, the overall enthalpy change of a reaction is the sum of the enthalpy changes for all bonds broken and formed (Figure 3.44).

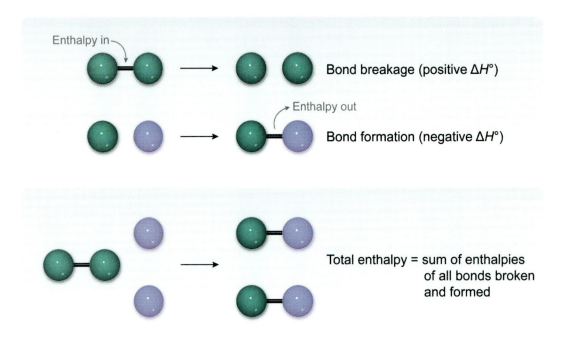

Figure 3.44 Total enthalpy change of a reaction in terms of the enthalpies of bonds broken and formed.

Because bond enthalpies are typically given as the energy required to *break* a bond, the sign of these values must be changed (eg, from positive to negative) to represent the enthalpy change of any bond *formed* in the products. Therefore, the overall enthalpy change of a reaction ($\Delta H°_{rxn}$) can be solved by subtracting the sum of the bond enthalpies of the bonds formed from the sum of the bond enthalpies of the bonds broken:

$$\Delta H°_{rxn} = \sum \binom{\text{bond enthalpies}}{\text{of bonds broken}} - \sum \binom{\text{bond enthalpies}}{\text{of bonds formed}}$$

Calculations using this equation can be approached in two ways. The first approach is to focus only on those bonds that are broken and formed in the reaction and to ignore those bonds that remain unchanged. The second approach is to assume all bonds within the reactants are broken and all bonds within the products are formed (Figure 3.45).

Reaction: CH₃(OH) + H—Cl → CH₃Cl + H—O—H

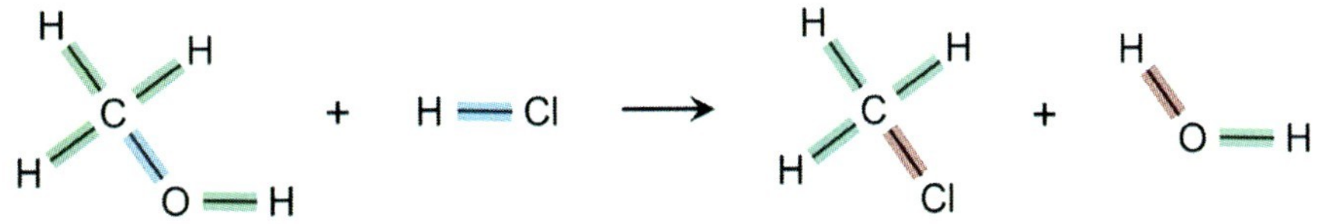

Approach 1:

$$\Delta H°_{rxn} = (\text{C–O} + \text{H–Cl}) - (\text{C–Cl} + \text{O–H})$$

Bonds broken — Bonds formed

$$\Delta H°_{rxn} = \left(358\,\frac{kJ}{mol} + 431\,\frac{kJ}{mol}\right) - \left(328\,\frac{kJ}{mol} + 463\,\frac{kJ}{mol}\right) = -2\,\frac{kJ}{mol}$$

Approach 2:

Reactant bonds (broken and unchanged) — Product bonds (formed and unchanged)

$$\Delta H°_{rxn} = [(\text{C–O}) + (\text{H–Cl}) + 3(\cancel{\text{C–H}}) + (\cancel{\text{O–H}})] - [(\text{C–Cl}) + (\text{O–H}) + 3(\cancel{\text{C–H}}) + (\cancel{\text{O–H}})]$$

$$\Delta H°_{rxn} = 358\,\frac{kJ}{mol} + 431\,\frac{kJ}{mol} + 1239\,\frac{kJ}{mol} + 463\,\frac{kJ}{mol}$$

$$- \left(328\,\frac{kJ}{mol} + 463\,\frac{kJ}{mol} + 1239\,\frac{kJ}{mol} + 463\,\frac{kJ}{mol}\right) = -2\,\frac{kJ}{mol} \quad \text{Unchanged bonds cancel}$$

Figure 3.45 Two different approaches to using bond enthalpies to solve for $\Delta H°_{rxn}$.

The advantage of the second approach is that it requires less analysis (eg, determining precisely which bonds are broken and formed), and as a result, mistakes are usually less likely. The disadvantage is that it can take extra time to tally all the bonds of each type. However, both approaches are valid ways to solve for $\Delta H°_{rxn}$ using bond enthalpies.

Concept Check 3.16

Using the table of average bond enthalpies provided, calculate $\Delta H°_{rxn}$ for the reaction shown:

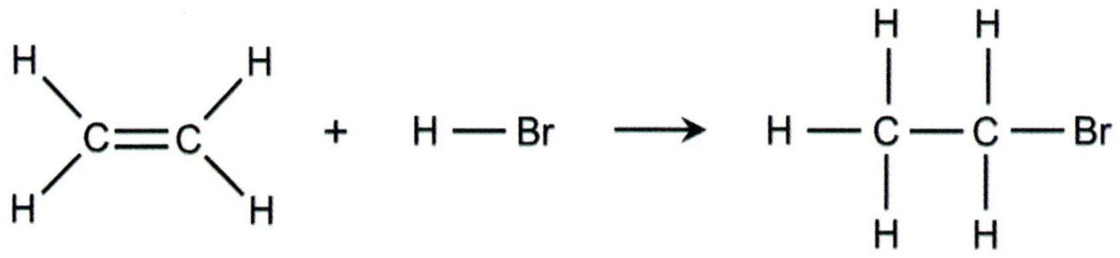

$$H_2C=CH_2 + H-Br \longrightarrow H_3C-CH_2-Br$$

Bond type	Bond enthalpy (kJ/mol)
C–H	413
C–C	348
C=C	614
C–Br	276
H–Br	366

Solution
Note: The appendix contains the answer.

Confusion often arises when comparing the method of solving for $\Delta H°_{rxn}$ using bond enthalpies to the method of solving for $\Delta H°_{rxn}$ using heats of formation. At first glance, the two equations that correspond to these methods seem to say opposite things:

Using heats of formation: $\Delta H°_{rxn} = \sum n\Delta H°_f \text{ (products)} - \sum m\Delta H°_f \text{ (reactants)}$

Using bond enthalpies: $\Delta H°_{rxn} = \sum \binom{\text{bond enthalpies of}}{\text{reactant bonds broken}} - \sum \binom{\text{bond enthalpies of}}{\text{product bonds formed}}$

However, heat of formation (ie, the enthalpy of *forming* a substance) and bond enthalpy (ie, the enthalpy of *breaking* the bond of a substance) are essentially opposite processes. In terms of products and reactants, these two methods necessarily have opposite sign conventions. Remembering the definitions of heat of formation and bond enthalpy helps in discerning the order of products and reactants in each equation.

Lesson 3.6

Calorimetry

3.6.01 Experimental Measurement of Heat Transfer

Calorimetry is the measurement of the amount of heat transferred within a system during a reaction or physical process. Such measurements require a determination of how much heat is gained or lost by the substances in the system. However, various substances respond differently (ie, have different changes in temperature) when exposed to the same amount of heat.

These differences in the thermal properties of substances are quantified by the heat capacity C of a sample, which is the amount of heat q required to cause the entire sample to have a change in temperature ΔT of 1 °C:

$$C = \frac{q}{\Delta T}$$

Because the amount of heat depends upon the amount of mass m in a sample (ie, more mass absorbs/releases more heat), comparisons between two different samples are made by expressing C per unit of mass as a specific heat capacity C_p, which is the amount of heat q_x required to cause *1 gram* of a substance x to have $\Delta T = 1$ °C under constant pressure:

$$C_p = \frac{q_x}{m \Delta T}$$

Consequently, two objects that have different masses but are made of the same substance each have a different overall heat capacity C but an identical specific heat capacity C_p. In most cases, C_p is used when making comparisons that consider the composition of a sample whereas C is used when a sample or apparatus is being evaluated as a single object apart from its composition.

If the specific heat capacity of a substance x is known, the amount of heat q_x gained or released by the mass of the sample during a corresponding change in temperature can be calculated by:

$$q_x = m C_p \Delta T$$

By convention, a negative q indicates that heat is released from a substance, and a positive q indicates that heat is absorbed by a substance. When the sample is a dilute aqueous solution, the C_p of the solution can often be assumed to be equivalent to the specific heat capacity of water because the mixture is mostly water. However, the mass of the *solution* is the sum of the masses of both the solute (ie, the dissolved substance) and the water.

$$q_\text{solution} = (m_\text{solute} + m_{H_2O}) C_{p,H_2O} \Delta T$$

If the heat gained or lost by the system is due to a chemical reaction (ie, heat is produced or consumed by a chemical change), the process has a specific **molar enthalpy** ΔH_rxn, which is the change in enthalpy per mole of the reacting species. As such, the heat q_rxn transferred by the reaction is found by multiplying ΔH_rxn by the number of moles n of the reaction:

$$q_\text{rxn} = n \Delta H_\text{rxn}$$

When two samples of matter that are initially at different temperatures come into thermal contact, the samples undergo heat transfer until they reach thermal equilibrium at some new, final temperature.

Transferred heat can be measured based on the initial and final temperatures ($T_{initial}$ and T_{final}) of a sample (or a chemical reaction) that is placed into a known amount of water inside the insulated chamber of a **calorimeter** (ie, an apparatus for measuring heat). An example of one type of calorimeter is shown in Figure 3.46.

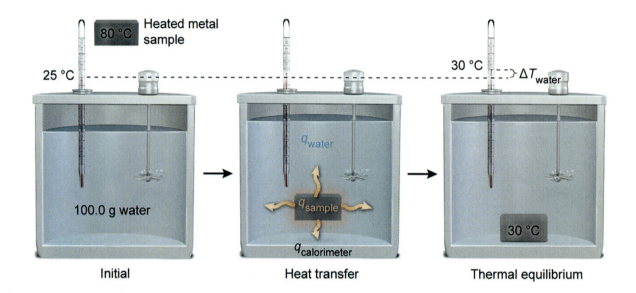

Figure 3.46 Analysis of a heated sample using a calorimeter.

According to the first law of thermodynamics, the transferred heat energy is conserved and the heat lost from the sample (indicated by a negative sign) must be equal to the total amount of heat gained by the calorimeter and the water (or solution) inside the calorimeter:

$$-q_{sample} = q_{water} + q_{calorimeter}$$

In an ideal system, no heat would be absorbed by the calorimeter itself (ie, $q_{calorimeter} = 0$), and all the heat released from the sample would be retained by the water. In rough measurements of heat transfer, it is sometimes assumed that $q_{calorimeter} \approx 0$ to simplify calculations.

$$-q_{sample} = q_{water} + 0 = q_{water}$$

However, in real systems some heat is always absorbed by the calorimeter itself. So, for more accurate measurements, it is necessary to determine the heat capacity of the calorimeter $C_{calorimeter}$ by performing an experiment involving a known amount of heat. Based on the known amount of heat transferred from a standard sample and the associated change in temperature, $C_{calorimeter}$ can be determined prior to performing a measurement of an unknown sample.

Because a calorimeter is a composite apparatus with a mass and composition that are not readily known, it is more convenient to determine the heat capacity of the entire apparatus (treated as a single object). Once $C_{calorimeter}$ is known, the value can be used in all measurements involving that calorimeter assembly. Accordingly, the heat absorbed by the calorimeter is simply:

$$q_{calorimeter} = C_{calorimeter} \Delta T$$

Concept Check 3.17

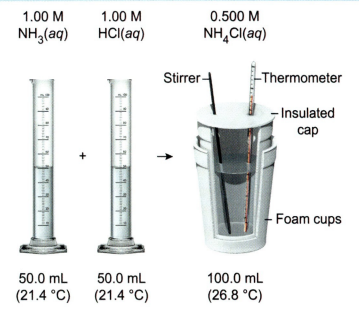

A student mixes 50.0 mL of 1.00 M NH3(aq) and 50.0 mL of 1.00 M HCl(aq) in a coffee cup calorimeter to produce 100.0 mL of 0.500 M NH4Cl(aq) in an exothermic reaction. The temperatures of the solutions before mixing are measured to be 21.4 °C, and the temperature of the mixture after the reaction is measured to be 26.8 °C. What is the change in enthalpy of the reaction ΔH_{rxn}? (Assume that the heat absorbed by the calorimeter is negligible and that the solution has a mass of 100 g with a specific heat capacity of 4.184 J/(g·°C).)

Solution

Note: The appendix contains the answer.

Chapter 3: Thermodynamics

Lesson 3.7

Gibbs Free Energy

Introduction

As outlined in Concept 3.2.03, endothermic processes (eg, phase changes, reactions) have a positive change in **enthalpy** ΔH and require a continuous input of thermal energy (ie, heat) to occur. In contrast, exothermic processes have a negative ΔH (ie, release heat) and occur without needing a supply of thermal energy. As such, the ΔH of a process is a contributing factor in determining if that process can occur spontaneously (ie, proceed unaided after being initiated). However, ΔH alone does not determine spontaneity.

According to the **second law of thermodynamics**, any spontaneous process results in an increase in **entropy** (ie, disorder and the dispersal of energy), as discussed in Lesson 3.3. Consequently, the associated change in entropy ΔS for a process is also a significant factor in determining if the process is spontaneous. As such, both ΔH and ΔS must be considered. The net effect of ΔH and ΔS at a given temperature can be evaluated by the change in **Gibbs free energy** ΔG, which indicates how much energy is made available in the system to perform work. Therefore, ΔG determines if a process is spontaneous.

This lesson examines the relationship between ΔH, ΔS, and ΔG at constant temperature and discusses some important associated concepts.

3.7.01 Gibbs Free Energy as a Thermodynamic Quantity

The change in **Gibbs free energy** ΔG is a thermodynamic potential that measures the maximum amount of energy available to accomplish some process or chemical conversion (ie, useful work) in a system at a constant temperature and pressure. On a reaction energy diagram (ie, a graph of energy versus the reaction progress), ΔG is equal to the difference between the energy levels of the products and reactants:

$$\Delta G = G_{products} - G_{reactants}$$

ΔG is the result of the changes in both the enthalpy ΔH and entropy ΔS of a system at a given temperature (measured in Kelvin), as described by the equation:

$$\Delta G = \Delta H - T\Delta S$$

This relationship can be evaluated under any set of conditions but for making comparisons between different systems, measurements are often made under the standard state in which all reactants and products have concentrations of 1.0 mol/L and all gases have partial pressures of 1.0 atm. Unlike the standard conditions used for gas laws, no temperature is specified for the standard state in thermodynamics; however, 298 K (25 °C) is often used for convenience when reporting data. Measurements made in the standard state are indicated by the inclusion of the degree symbol beside the thermodynamic variables (ie, $\Delta G°$, $\Delta H°$, $\Delta S°$).

If a reaction is carried out in the standard state, ΔG and $\Delta G°$ are equal, but under any other conditions, ΔG differs from $\Delta G°$. As such, $\Delta G°$ has only one value for a given temperature, but the value of ΔG varies depending on the specific concentrations of the species in solution. The relationship between ΔG and $\Delta G°$ is discussed in Concept 5.3.03.

Although Δ*H* describes the flow of heat in a reaction and Δ*S* describes the disorder of the products relative to the reactants, Δ*G* does not have a direct physical correlation. Instead, Δ*G* describes the thermodynamic favorability of a process or reaction (ie, its likelihood to occur without additional intervention). A reaction with a negative Δ*G* (ie, Δ*G* < 0) is thermodynamically favorable and will proceed spontaneously after initiation whereas a positive Δ*G* (ie, Δ*G* > 0) indicates a reaction that is thermodynamically unfavorable and nonspontaneous (ie, will not proceed without a continuous input of energy), as illustrated in Figure 3.47.

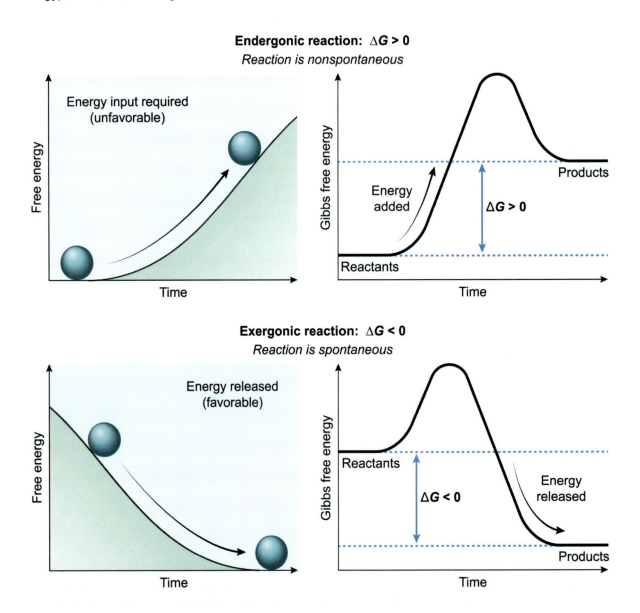

Figure 3.47 A comparison of thermodynamically favorable and unfavorable reactions.

Reactions with Δ*G* < 0 are **exergonic** (ie, involve a net release of energy) because the energy of the reactants is greater than the energy of the products, and the excess energy is liberated. Conversely, reactions with Δ*G* > 0 are **endergonic** (ie, involve a net intake of energy) because the energy of the reactants is less than the energy of the products. However, a reaction that is thermodynamically favorable and exergonic may still require a small activation energy to initiate the reaction.

For example, methane gas reacts with oxygen gas in a favorable and highly exergonic combustion reaction, but mixing the two gases does not cause an immediate reaction because a small amount of energy (eg, a spark) is needed to start the process (see Figure 3.48).

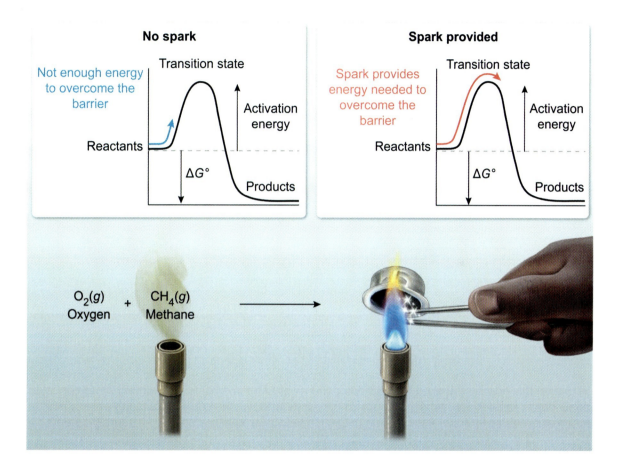

Figure 3.48 Example of a thermodynamically favorable exergonic reaction that requires an activation energy.

On a reaction energy diagram, the activation energy is seen as an energy peak between the energy levels of the reactants and products (Concept 4.1.02). The activation energy peak is the energy barrier required to reach the transition state and initiate the reaction. In an exergonic reaction, once the activation energy is supplied, the amount of energy released by the reaction ($-\Delta G$) eliminates any further need for energy input and sustains the reaction until the reactants are fully converted to products.

The thermodynamic quantities ΔG, ΔH, and ΔS are state functions, which means they describe differences between the reactants and the products in their current state. Therefore, these values are independent of the chemical pathway that a reaction takes to get from reactants to products (Concept 3.1.01). As such, ΔG indicates if a reaction is spontaneous but does not say anything about the kinetics of a reaction (ie, how fast a reaction happens for a given reaction pathway).

Because thermodynamic quantities are independent of the reaction pathway, catalysts added to increase the reaction rate do not change the amounts of products produced or the relative energies of the reactants and products (ie, catalysts do not alter the ΔG, ΔH, or ΔS of a reaction). Reaction rates (ie, chemical kinetics) and the action of catalysts are discussed in detail in Chapter 4.

Concept Check 3.18

$$2\ NO(g) + O_2(g) \rightarrow 2\ NO_2(g)$$

$$\Delta H° = -116.2\ \text{kJ/mol}$$

$$\Delta S° = -146.4\ \text{J/mol·K}$$

Based on the given thermodynamic parameters for the reaction of $NO(g)$ and $O_2(g)$, what is the value of $\Delta G°$ at room temperature (25 °C)?

Solution

Note: The appendix contains the answer.

3.7.02 Enthalpic and Entropic Contributions to Gibbs Free Energy

As noted in Concept 3.7.01, ΔG is the result of the changes in both the enthalpy ΔH and entropy ΔS of a system at a given temperature (measured in Kelvin), as described by the equation:

$$\Delta G = \Delta H - T\Delta S$$

According to this equation, the ΔH and ΔS components of ΔG can provide complimentary or opposing contributions to favorability. An exothermic reaction ($-\Delta H$) assists favorability because excess heat is released whereas an endothermic reaction ($+\Delta H$) hinders favorability by needing a source of heat.

The sign of ΔS also influences favorability, and this effect is amplified by temperature. Increasing entropy ($+\Delta S$) assists favorability because the $-T\Delta S$ term stays negative whereas decreasing entropy ($-\Delta S$) hinders favorability because the term turns positive (ie, negative signs cancel):

$$-T(-\Delta S) = +T\Delta S.$$

Although quantitative calculations of ΔG can be performed by the equation if values for ΔH, ΔS, and T are known, a qualitative assessment of the favorability of a process or reaction can be made without rigorous calculations of ΔG by considering the four possible cases that can occur:

- $-\Delta H$ and $+\Delta S$ always results in a process that is spontaneous ($-\Delta G$) at any temperature.
- $+\Delta H$ and $-\Delta S$ always results in a process that is nonspontaneous ($+\Delta G$) at any temperature.
- $-\Delta H$ and $-\Delta S$ results in a process that is spontaneous ($-\Delta G$) at low temperatures and nonspontaneous ($+\Delta G$) at high temperatures (ie, the favorability decreases as temperature increases).
- $+\Delta H$ and $+\Delta S$ results in a process that is nonspontaneous ($+\Delta G$) at low temperatures and spontaneous ($-\Delta G$) at high temperatures (ie, the favorability increases as temperature increases).

An example of the third case (ie, a reaction with $-\Delta H$ and $-\Delta S$) is illustrated by the graph in Figure 3.49.

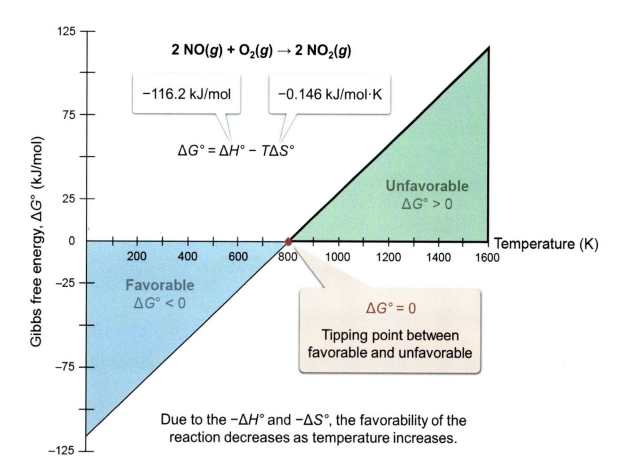

Figure 3.49 Graph of $\Delta G°$ for the reaction of NO(g) and O₂(g) at various temperatures with $-\Delta H°$ and $-\Delta S°$.

Figure 3.49 illustrates that it is possible for an exothermic reaction (ie, a reaction with $-\Delta H$) to still be unfavorable and endergonic (ie, have $+\Delta G$), depending on the temperature and contributions from ΔS. Consequently, not all exothermic reactions are exergonic, and not all endothermic reactions are endergonic (ie, $-\Delta H$ does not always result in $-\Delta G$, and $+\Delta H$ does not always result in $+\Delta G$). Contributions from entropy amplified by temperature can play a significant role in the favorability of a reaction.

 Concept Check 3.19

$$2\ KClO_4(s) \rightarrow 2\ KClO_3(s) + O_2(g)$$

The decomposition of potassium perchlorate to potassium chlorate and oxygen gas is thermodynamically unfavorable at 100 °C but favorable at 200 °C. Are the signs of $\Delta H°$ and $\Delta S°$ positive or negative for this reaction?

Solution

Note: The appendix contains the answer.

Chapter 3: Thermodynamics

Lesson 4.1

Reaction Mechanisms

Introduction

When a molecule undergoes a chemical reaction, each step of the reaction occurs through either a **unimolecular process** or through an **intermolecular process**. In unimolecular processes (eg, decomposition, rearrangement), only one molecule is involved. In contrast, all intermolecular processes involve at least two molecules, and **bimolecular processes** are the most common.

According to **molecular collision theory**, a reaction can occur between molecules only when the molecules collide with one another in the correct orientation and with the necessary amount of energy. Although a reaction equation describes a chemical conversion, the equation may only represent the outcome (ie, the products) of the collisions between reactants. A reaction equation does not necessarily indicate the sequence of molecular collisions that resulted in the products.

For example, consider the following reaction:

$$X_2 + 2\,Y \rightarrow X + XY_2$$

Although the equation indicates that one molecule of X_2 reacts with two atoms of Y, this does not necessarily mean that the reaction proceeds by a single collision of these three particles. For example, the reaction could proceed in more than one step (ie, more than one collision, each involving only two particles), as detailed here:

Step 1: $X_2 + Y \rightarrow X_2Y$

Step 2: $X_2Y + Y \rightarrow X + XY_2$

Net: $X_2 + 2\,Y \rightarrow X + XY_2$

In Step 1 of this scenario, the first collision forms an X_2Y intermediate, which is then converted by a collision with a second Y atom in Step 2. Because the intermediate is consumed in Step 2, the net reaction is the same as the original equation. However, without experimental evidence, it is not possible to say whether the reaction occurs by one collision (one step) or by more than one collision (two or more steps).

As such, the **mechanism** (ie, the exact sequence of unimolecular processes and molecular collisions that result in a chemical conversion described by a chemical equation) must be determined experimentally. Mechanisms include the reactants, intermediates, products, and catalysts involved in each step.

The reaction pathway within a mechanism fundamentally influences the energy profile, thermodynamic favorability, and rate of a reaction. These aspects are examined in this lesson.

4.1.01 Elementary Reactions and Net Reactions

Consider the hypothetical mechanism presented in the introduction:

Step 1:	$X_2 + Y \rightarrow X_2Y$		Elementary reaction
Step 2:	$X_2Y + Y \rightarrow X + XY_2$		Elementary reaction
Net:	$X_2 + 2Y \rightarrow X + XY_2$		(Overall result)

The overall (net) reaction is the result of the two simpler reactions (Steps 1 and 2). In contrast, Steps 1 and 2 each represent a single molecular collision. As such, the reactions of Steps 1 and 2 are said to be elementary reactions. Reactions that are the *result* of two or more elementary reactions are said to be **net reactions**. Accordingly, elementary reactions can represent molecular collisions, but net reactions cannot.

The most common molecular collisions are bimolecular collisions (ie, collisions involving two molecules) because the probability of two molecules colliding under conditions favorable for a reaction is reasonably high. Termolecular collisions (ie, simultaneous collisions involving three molecules) are rare because the probability of three (or more) molecules meeting at the same place is low under normal conditions.

Although unimolecular processes are not collision events (ie, only one molecule), these processes are also considered elementary reactions because unimolecular conversions are also not the net result of other reactions and cannot be restated in those terms.

It is not possible to tell with certainty just by looking at a reaction equation whether the reaction is an elementary reaction or not. As a result, reactions must be experimentally studied to determine the mechanism associated with the described molecular conversions.

4.1.02 Energy Considerations

The collision of two molecules does not necessarily guarantee that a reaction will occur. To react, the molecules must collide with an orientation that facilitates the formation of new bonds, and the molecules must also have enough energy to initiate and complete a reaction. Molecular collisions that occur with either an incorrect orientation or insufficient energy are unproductive (ie, no reaction takes place), as shown in Figure 4.1.

$$AB + C \rightarrow A + BC$$

Productive collision (products form with proper orientation and energy)

Unproductive collision (products do not form)

Figure 4.1 Productive and unproductive molecular collisions.

The minimum energy needed for two molecules to initiate a reaction is called the **activation energy** E_a. At a given temperature, some molecules move more slowly (ie, lower kinetic energy), whereas other molecules move more quickly (ie, higher kinetic energy). This difference causes the molecules to display a wide range of kinetic energies, represented graphically by a Maxwell-Boltzmann distribution.

However, molecules that collide with less than the minimum E_a cannot react to form products because the E_a represents an **energy barrier** to the conversion, as illustrated by the reaction energy diagram in Figure 4.2.

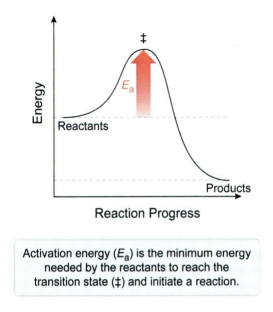

Activation energy (E_a) is the minimum energy needed by the reactants to reach the transition state (‡) and initiate a reaction.

Figure 4.2 Energy diagram showing the activation energy of a reaction.

Molecules that collide with the correct orientation and the necessary E_a can reach the **transition state** (sometimes called an **activated complex**) in which new bonds are formed as old bonds are broken. Although specific transition states (denoted by the ‡ symbol) are unique to each reaction, a simplified model of this transient (ie, short-lived), high-energy state is shown in Figure 4.3.

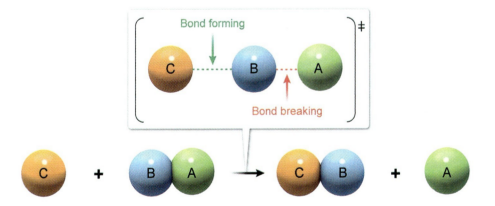

Figure 4.3 Conceptual model of a transition state in a reaction.

4.1.03 Rate-Determining Step

A **reaction rate** indicates how fast a chemical reaction occurs by measuring the change in the concentration of reactants or products per unit of time. In a chemical conversion that involves a mechanism with more than one elementary reaction step, each elementary reaction step in the mechanism has its own rate that depends on the activation energy E_a of that step. As a result, the overall reaction can only go as fast as the slowest step allows.

The effect of E_a on the rate can be seen by examining a reaction energy diagram for the mechanism, which plots the energy of the chemical species (y-axis) versus the reaction progress (x-axis). Each elementary reaction step of the mechanism produces its own high-energy transition state, seen as a peak on the diagram, and the resulting intermediates formed are seen as valleys between the peaks. An annotated hypothetical energy diagram for a two-step reaction mechanism is shown in Figure 4.4.

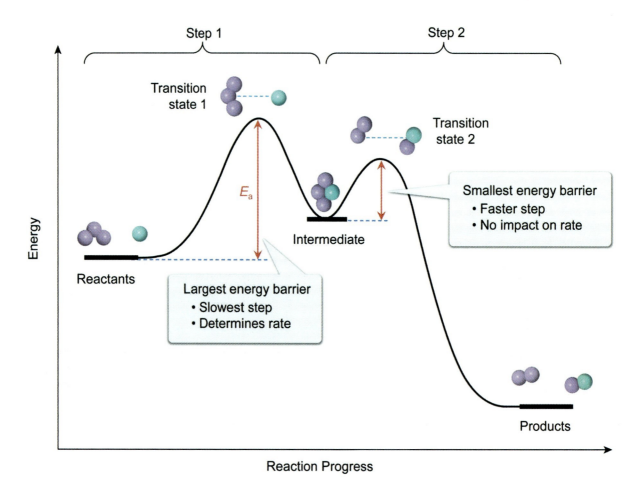

Figure 4.4 Energy diagram for a reaction mechanism involving two elementary reaction steps.

Elementary reaction steps with a higher E_a occur more slowly than steps with a lower E_a. In a reaction mechanism sequence, the slowest step (ie, the step with the highest E_a) is the **rate-determining step** for the overall (net) reaction because any subsequent, faster steps (ie, those with a lower E_a) can only occur after the slower step has completed.

In the example in Figure 4.4, Step 1 is the rate-determining step because it has the highest E_a. Consequently, the rate of the overall reaction is determined by the rate of the elementary reaction of Step 1.

Chapter 4: Kinetics

> ☑ **Concept Check 4.1**
>
> Suppose that a three-step reaction mechanism has the energy profile shown in the diagram. Which step is the rate-determining step of the overall reaction?
>
>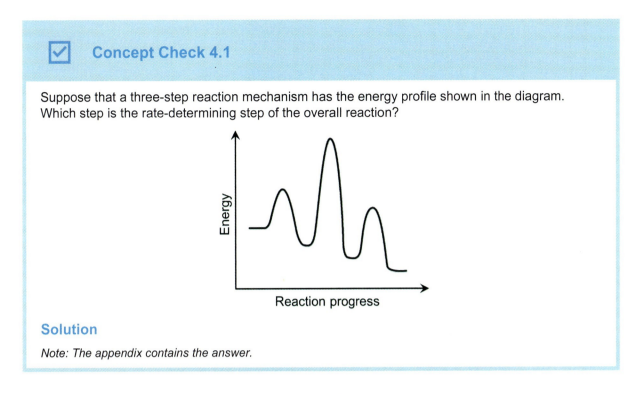
>
> **Solution**
> *Note: The appendix contains the answer.*

4.1.04 Thermodynamic versus Kinetic Control

As discussed in Concept 3.7.02, the thermodynamic favorability of a reaction describes if the reaction can proceed without a continuous input of energy; however, the favorability of a reaction does not indicate anything about the *rate* (ie, kinetics) of the reaction. The rate depends on the activation energy E_a for the reaction and the kinetic energy of the reactant molecules, as described in Concept 4.1.03.

If the temperature is too low, the reaction will not occur even if it is thermodynamically favorable because the reactant molecules do not have enough energy to overcome the E_a barrier. Favorable reactions that require a large initial energy input to occur at a measurable rate (ie, reactions with a high E_a) are under **kinetic control**, as illustrated by the general scheme in Figure 4.5.

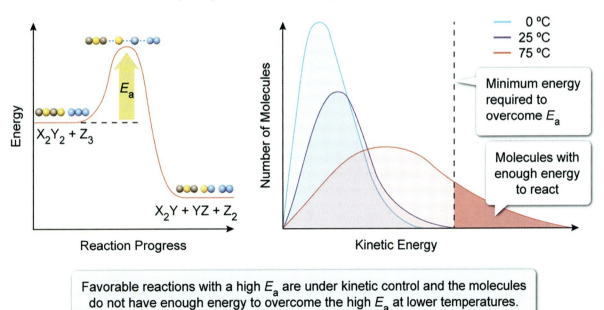

Favorable reactions with a high E_a are under kinetic control and the molecules do not have enough energy to overcome the high E_a at lower temperatures.

Figure 4.5 General depiction of a reaction under kinetic control.

A commonly cited example of kinetic control in nature is that of diamond. Although it is thermodynamically favorable for diamond to revert to graphite, the E_a for the process is so great that the conversion does not proceed at an observable rate (Figure 4.6).

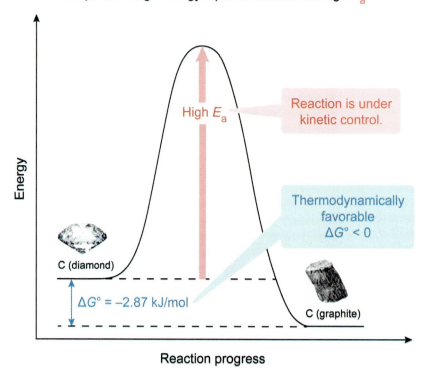

Figure 4.6 Kinetic control of the conversion of diamond into graphite.

If the temperature of a system is high enough that the E_a barrier is no longer an obstacle, the system then shifts to form the most thermodynamically stable products (ie, the products with the lowest energy). Under such conditions, the system is said to be under **thermodynamic control**.

When a reaction can produce more than one product, how much of each product is formed often depends on thermodynamic and kinetic considerations. During product formation, the compound with the lowest E_a forms fastest, making it the **kinetic product**. In contrast, the compound with the most stable (lowest energy) structure is the **thermodynamic product**, but because it has a higher E_a, it forms more slowly.

Consequently, experimental conditions can be used to bias the products of a reaction toward a desired outcome. At lower temperatures (ie, under kinetic control), the kinetic product is the majority product produced. Conversely, at higher temperatures (ie, under thermodynamic control), the additional energy enables the reaction to overcome the larger energy barrier to form the thermodynamic product as the majority product.

A general reaction energy diagram for a system that can form both a kinetic product and a thermodynamic product is shown in Figure 4.7.

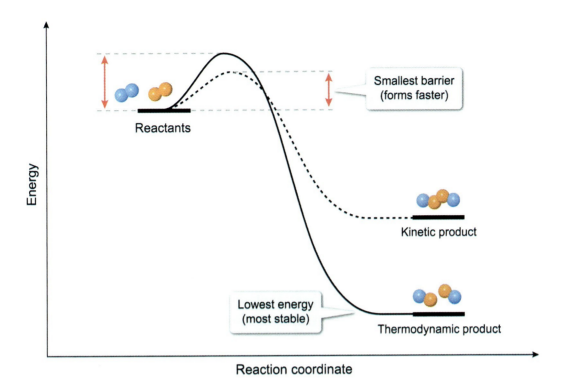

Figure 4.7 Reaction energy diagram for a reaction with a kinetic and a thermodynamic product.

Lesson 4.2

Reaction Rates

Introduction

Reaction rates indicate how quickly a chemical conversion happens by describing the change in the concentration of reactants or products per unit of time. An **instantaneous rate** indicates how rapidly the molar concentration of a species [X] changes at a given point in time t, whereas an **average rate** measures the change in the molar concentration Δ[X] over a time interval Δt. The difference between these two types of rates is shown on the graphs in Figure 4.8.

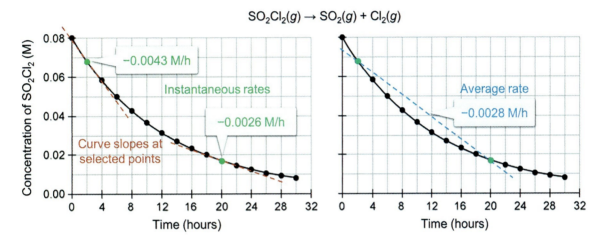

Figure 4.8 Selected instantaneous and average rates evaluated from experimental data collected for the decomposition of $SO_2Cl_2(g)$.

Determining an instantaneous rate requires either a graph of experimental data collected at many points in time (as shown in Figure 4.8) or a **rate law** (ie, an equation of the relationship between the reaction rate and the molar concentration of the relevant reacting species) evaluated from the experimental data. Using the graphical method, the instantaneous rate is found by evaluating the slope of the curve at the times of interest. Alternatively, using a rate law yields the instantaneous rate at a molar concentration of interest.

In contrast, an average rate for a reaction can be determined from two data points:

$$\text{Average rate} = \frac{\text{Change in concentration of X}}{\text{Change in time}} = \frac{[X]_{\text{final}} - [X]_{\text{initial}}}{t_{\text{final}} - t_{\text{initial}}} = \frac{\Delta[X]}{\Delta t}$$

In this lesson, reaction rates are examined with respect to the molar concentrations of different species in a chemical reaction, and the effect of temperature on the rate is considered. This lesson also explores the evaluation of rate laws from experimental data and reaction mechanisms.

4.2.01 Reaction Rates and Rate Measurements

For the general reaction:

$$aA + bB \rightarrow cC + dD$$

the rate of the reaction can be expressed in terms of the change in concentration of any of the chemical species. As such, four valid rate expressions can be written:

$$\text{Rate of consumption of A} = -\frac{\Delta[A]}{\Delta t}$$

$$\text{Rate of consumption of B} = -\frac{\Delta[B]}{\Delta t}$$

$$\text{Rate of formation of C} = \frac{\Delta[C]}{\Delta t}$$

$$\text{Rate of formation of D} = \frac{\Delta[D]}{\Delta t}$$

The rates for the consumption or production of individual species are related to each other and to the overall reaction rate by the stoichiometric coefficients of the reaction:

$$\text{Reaction rate} = -\frac{1}{a} \cdot \frac{\Delta[A]}{\Delta t} = -\frac{1}{b} \cdot \frac{\Delta[B]}{\Delta t} = \frac{1}{c} \cdot \frac{\Delta[C]}{\Delta t} = \frac{1}{d} \cdot \frac{\Delta[D]}{\Delta t}$$

The reactants include a negative sign because the concentrations *decrease* as the reaction proceeds, whereas the concentrations of products *increase* and have positive rate values.

For example, consider the decomposition of $N_2O_5(g)$:

$$2\ N_2O_5(g) \rightleftarrows 4\ NO_2(g) + O_2(g)$$
colorless　　　　　brown

To determine the rate of the reaction, the concentration of $N_2O_5(g)$ is measured at regular intervals during the reaction, yielding the results shown in Table 4.1.

Table 4.1 Kinetic data for the decomposition of $N_2O_5(g)$.

Time(s)	[$N_2O_5(g)$]
0	0.0200
100	0.0169
200	0.0142
300	0.0120
400	0.0101
500	0.0086
600	0.0072
700	0.0061

Using this data, both the average rate of $N_2O_5(g)$ decomposition and the average rate of $NO_2(g)$ formation can be found. For example, substituting the values for the $N_2O_5(g)$ decomposition during the first 300 s into the rate expression gives:

$$\text{Average rate of } N_2O_5 \text{ decomposition} = -\frac{\Delta[N_2O_5(g)]}{\Delta t} = -\frac{[N_2O_5(g)]_{\text{final}} - [N_2O_5(g)]_{\text{initial}}}{t_{\text{final}} - t_{\text{initial}}}$$

$$-\frac{\Delta[N_2O_5(g)]}{\Delta t} = -\frac{(0.0120 \text{ M} - 0.0200 \text{ M})}{(300 \text{ s} - 0 \text{ s})} = -\frac{(-0.0080 \text{ M})}{300 \text{ s}}$$

$$-\frac{\Delta[N_2O_5(g)]}{\Delta t} = 2.7 \times 10^{-5} \text{ M/s}$$

Based on the stoichiometric coefficients of the balanced reaction equation, the rate of $N_2O_5(g)$ decomposition is related to the rate of $NO_2(g)$ formation as:

$$\frac{1}{2} \cdot \left(-\frac{\Delta[N_2O_5(g)]}{\Delta t}\right) = \frac{1}{4} \cdot \frac{\Delta[NO_2(g)]}{\Delta t}$$

Solving for the rate of $NO_2(g)$ formation yields:

$$\text{Average rate of } NO_2 \text{ formation} = \frac{\Delta[NO_2(g)]}{\Delta t} = \frac{4}{2} \cdot \left(-\frac{\Delta[N_2O_5(g)]}{\Delta t}\right)$$

$$\frac{\Delta[NO_2(g)]}{\Delta t} = \frac{4}{2} \cdot \left(2.7 \times 10^{-5} \frac{\text{M}}{\text{s}}\right) = 5.4 \times 10^{-5} \text{ M/s}$$

A summary of the reaction with calculations is shown in Figure 4.9.

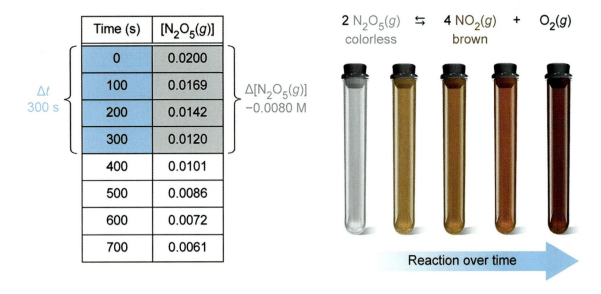

Average rate of NO$_2$ formation $= \dfrac{\Delta[NO_2(g)]}{\Delta t} = \dfrac{4}{2} \cdot \left(-\dfrac{\Delta[N_2O_5(g)]}{\Delta t}\right) = 5.4 \times 10^{-5}$ M/s

Figure 4.9 Evaluation of the average rate of N$_2$O$_5$(g) decomposition and the average rate of NO$_2$(g) formation during the first 300 s of the reaction.

✓ Concept Check 4.2

If the average rate of NO$_2$(g) formation during the first 300 s of the reaction shown is 5.4×10^{-5} M/s, what is the average rate of O$_2$(g) formation during the same time?

$$2\ N_2O_5(g) \rightleftharpoons 4\ NO_2(g) + O_2(g)$$
colorless   brown

Solution
Note: The appendix contains the answer.

The rate of a reaction can also be used as a conversion factor to determine the change in the molar concentration of a reactant or product [X] over a given amount of time. Accordingly, the change in concentration is given by:

$$\text{\sout{Time}} \times \frac{\Delta[X]}{\sout{\Delta t}} = \Delta[X]$$

If the initial concentration of a reactant or product is known, the concentration of that species after the elapsed reaction time can be found by:

$$\Delta[X] = [X]_{final} - [X]_{initial}$$

$$[X]_{final} = [X]_{initial} + \Delta[X]$$

Alternatively, the time required to produce a specific change in the molar concentration of a species at a given rate can be found by using the inverted rate as a conversion factor:

$$\sout{\Delta(\text{Molarity X})} \times \frac{\Delta t}{\sout{\Delta[X]}} = \Delta t$$

> **Concept Check 4.3**
>
> Suppose that a sample of $N_2O_5(g)$ with an initial concentration of 0.0200 M decomposes in a 0.50 L vessel at an average rate of -2.7×10^{-5} M/s during the first 300 s, as shown. How many moles of $N_2O_5(g)$ remain in the vessel after the 300 s have elapsed?
>
> $$2\ N_2O_5(g) \rightleftharpoons 4\ NO_2(g) + O_2(g)$$
> colorless   brown
>
> **Solution**
>
> Note: The appendix contains the answer.

4.2.02 Rate Laws

A rate law relates the rate of a reaction to the molar concentration of each reactant that participates in the rate-determining step. A rate law is expressed generally by the equation:

$$\text{rate} = k[A]^m[B]^n$$

where k is a constant and m and n are the **reaction orders** of the respective reactants A and B. The values of m and n are unrelated to the stoichiometric coefficients of the reactants and must be determined experimentally, except when the reaction is an elementary reaction.

One method used to experimentally determine the values of m and n is to measure the initial rate of the reaction in a series of reaction trials in which the concentration of one reactant changes as the other is held constant. By comparing the concentration changes to any resulting changes in the initial rates, the reaction order with respect to each reactant can be found.

For example, consider the following reaction between unspecified reactants X_2 and Y:

$$X_2 + 2\,Y \longrightarrow 2\,XY$$

An initial expression of the rate law for the reaction can be written as:

$$\text{rate} = k[X_2]^m[Y]^n$$

To determine the values of *m* and *n* for the rate law, suppose that the initial rate of the reaction was measured in three reaction trials using different initial concentrations of X_2 and Y, yielding the results shown in Table 4.2.

Table 4.2 Data for the kinetics of the reaction between X_2 and Y.

Trial	[X₂]	[Y]	Initial rate (M/s)
1	1.00	1.00	4.00
2	2.00	1.00	8.00
3	1.00	2.00	4.00

Comparing Trials 1 and 2, the data shows that doubling $[X_2]$ doubles the reaction rate, which indicates that *m* = 1 for this reaction. This result can be verified mathematically by evaluating a **p**roportion of rates from these trials:

$$\frac{\text{rate}_2}{\text{rate}_1} = \frac{k[X_2]_2^m[Y]_2^n}{k[X_2]_1^m[Y]_1^n}$$

Substituting the experimental values from Table 4.2 (with units canceled out) yields:

$$\frac{8.00 \text{ M/s}}{4.00 \text{ M/s}} = \frac{k(2.00 \text{ M})^m(1.00 \text{ M})^n}{k(1.00 \text{ M})^m(1.00 \text{ M})^n}$$

Canceling common terms and solving for *m* gives:

$$2.00 = \frac{k(2.00)^m \cancel{(1.00)^n}}{k(1.00)^m \cancel{(1.00)^n}} = \frac{(2.00)^m}{(1.00)^m} = \left(\frac{2.00}{1.00}\right)^m = (2.00)^m$$

$$2.00 = (2.00)^m \quad \Longrightarrow \quad m = 1$$

Similarly, the data from Trials 1 and 3 shows that doubling [Y] has no effect on the reaction rate, which indicates that *n* = 0 for this reaction. Setting up a ratio of the data from these trials to verify this value gives:

$$\frac{\text{rate}_3}{\text{rate}_1} = \frac{k[X_2]_3^m[Y]_3^n}{k[X_2]_1^m[Y]_1^n}$$

$$\frac{4.00 \text{ M/s}}{4.00 \text{ M/s}} = \frac{k(1.00 \text{ M})^m(2.00 \text{ M})^n}{k(1.00 \text{ M})^m(1.00 \text{ M})^n}$$

$$1.00 = \frac{\cancel{k(1.00)^m}(2.00)^n}{\cancel{k(1.00)^m}(1.00)^n} = \frac{(2.00)^n}{(1.00)^n} = \left(\frac{2.00}{1.00}\right)^n = (2.00)^n$$

$$1.00 = (2.00)^n \quad \Longrightarrow \quad n = 0$$

Substituting the determined values of *m* and *n* into the general rate law expression yields the specific rate law for the reaction of X_2 and Y:

$$\text{rate} = k[X_2]^m[Y]^n = k[X_2]^1[Y]^0 = k[X_2]$$

Therefore, the reaction is first-order with respect to $[X_2]$ and zero-order with respect to [Y], and the reaction is first-order overall.

$$\text{Overall reaction order} = m + n = (1 + 0) = 1$$

The values of *m* and *n* do not match the stoichiometric coefficients of the reactants because the reaction is not an elementary reaction, and *m* and *n* are determined by the slowest *elementary step* of the reaction mechanism (see Concept 4.1.03). The absence of [Y] from the rate law indicates that Y is not involved in the rate-determining step of the reaction mechanism and underscores why *m* and *n* must be determined experimentally. The mechanism for the reaction between X_2 and Y is shown in Figure 4.10.

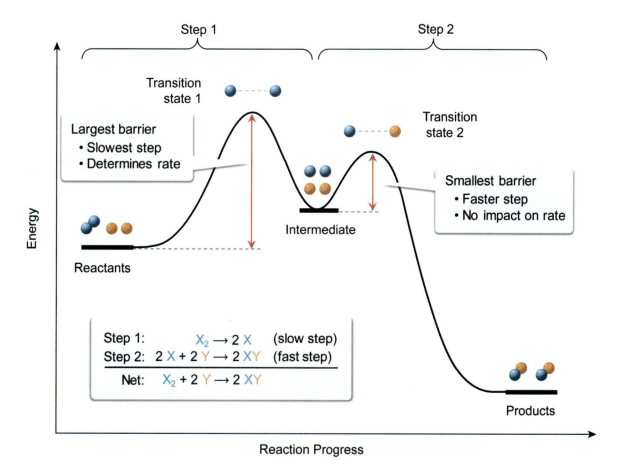

Figure 4.10 Mechanism for the reaction between X_2 and Y.

Because Y does not participate in the rate-determining (slowest) step of the reaction mechanism, [Y] is not relevant to the rate of the overall reaction and therefore is not represented in the rate law. Consequently, the rate law of the first elementary reaction (ie, the first step) is also the rate law for the net reaction.

Concept Check 4.4

The precipitation reaction between aqueous mercury(II) chloride and oxalate ions is represented by the equation shown. The kinetics data from three experimental trials of the reaction are given in the following table.

$$2\ HgCl_2(aq) + C_2O_4^{2-}(aq) \rightarrow 2\ Cl^-(aq) + 2\ CO_2(g) + Hg_2Cl_2(s)$$

Trial	[HgCl$_2$]	[C$_2$O$_4^{2-}$]	Initial rate (M/s)
1	0.070	0.16	2.3×10^{-7}
2	0.070	0.32	9.2×10^{-7}
3	0.035	0.32	4.6×10^{-7}

What is the rate law and overall order for this reaction?

Solution

Note: The appendix contains the answer.

4.2.03 Temperature Dependence of the Reaction Rate

The rate of a chemical reaction depends on the product of the relevant reactant concentrations (each raised to an experimentally determined power) and an experimentally determined rate constant k, which is expressed mathematically as a rate law:

$$\text{rate} = k[A]^m[B]^n$$

where A and B are the reacting species, and n and m are the experimentally determined exponents.

Conceptually, the rate constant k measures how frequently reactants collide in a manner that results in a reaction (ie, a productive collision). Temperature is a measure of the average kinetic energy of the molecules in a system. As temperature increases, the average kinetic energy (and therefore the average speed) of the molecules increases, resulting in an increased collision frequency.

More collisions result in more reactions and a faster reaction rate. Consequently, the rate constant and reaction rate both increase with rising temperature, as depicted in Figure 4.11.

Rate = k[reactants]

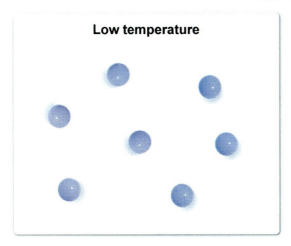
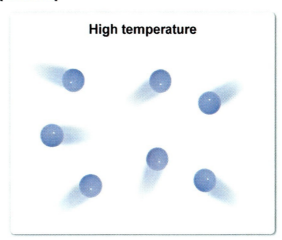

Low temperature
- Molecules move slower
- Collisions are less frequent
- Rate constant k is smaller
- Reaction rate is slower

High temperature
- Molecules move faster
- Collisions are more frequent
- Rate constant k is larger
- Reaction rate is faster

Figure 4.11 Comparison of the molecular motion, rate constant, and reaction rate of a system at low temperature and at high temperature.

The extent to which k changes with temperature is described mathematically by the **Arrhenius equation**:

$$k = A e^{-\frac{E_a}{RT}}$$

where A is a constant that represents the collision frequency of reactants, and the exponential term $e^{-E_a/RT}$ corresponds to the fraction of collisions with enough kinetic energy to overcome the activation energy Ea at a given temperature T. In the exponential term, the gas constant R is expressed in terms of energy (R = 8.314 J·mol^{-1}·K^{-1}).

Solutions to the Arrhenius equation are beyond the scope of the exam. However, because the rate of a reaction is directly proportional to its rate constant k, knowing the Arrhenius equation is useful for understanding how its variables influence the rate of reaction. A summary of the effects resulting from changes to the different variables is given in Table 4.3.

Chapter 4: Kinetics

Table 4.3 Summary of effects from changes to individual variables in the Arrhenius equation.

Change	Reaction rate	Rate constant, k	Activation energy, E_a	Temperature T
Increase (↑) T	↑	↑	Constant	↑
Decrease (↓) T	↓	↓	Constant	↓
Increase (↑) k	↑	↑	Constant	↑
Decrease (↓) k	↓	↓	Constant	↓
Increase (↑) E_a	↓	↓	↑ (uncatalyzed)	Constant
Decrease (↓) E_a	↑	↑	↓ (catalyzed)	Constant

Raising the temperature of a reaction increases the energy of the molecules and the number of productive collisions between them, which increases both the rate and the rate constant of a reaction. In the Arrhenius equation, this result is signaled by a smaller negative value of the exponential term (ie, a smaller negative exponent corresponds to a larger value of k). The exponential term decreases because T is in the denominator, and dividing E_a by a larger T yields a smaller number. If the temperature decreases, the opposite results are obtained.

In most circumstances, the E_a of a reaction is constant and is an intrinsic property of the reaction system. However, the E_a can sometimes be lowered using a catalyst (ie, a substance that increases the reaction rate but is not consumed by the reaction). The lower E_a causes an increase in k and the reaction rate. Catalysts are discussed in greater detail in Lesson 4.3.

✓ **Concept Check 4.5**

For a reaction with $E_a = x$ initially performed at a temperature $T = y$, what is the net effect on the rate of the reaction if adding a catalyst results in $E_a = 0.25x$ but the temperature of the reaction is also decreased to $T = 0.50y$?

Solution

Note: The appendix contains the answer.

Lesson 4.3

Chemical Catalysis

4.3.01 Types and Mechanisms of Catalysis

In contrast to the **thermodynamics** of a reaction, which indicates its favorability, the **kinetics** of a reaction describes how quickly a reaction reaches completion under a given set of conditions. A reaction may be thermodynamically favorable but proceeds so slowly that it does not reach completion in a reasonable amount of time.

The rate of a reaction depends on the activation energy E_a of the slowest elementary step. A larger E_a makes initiating the reaction more difficult and causes a reaction to proceed more slowly. A **catalyst** is a substance that increases the **rate** (kinetics) of a reaction by decreasing the E_a required for the reaction to generate products at a given temperature.

However, a catalyst does not affect the *amount* of products formed nor the reaction's thermodynamics (ie, the energy difference between reactants and products), as illustrated in Figure 4.12.

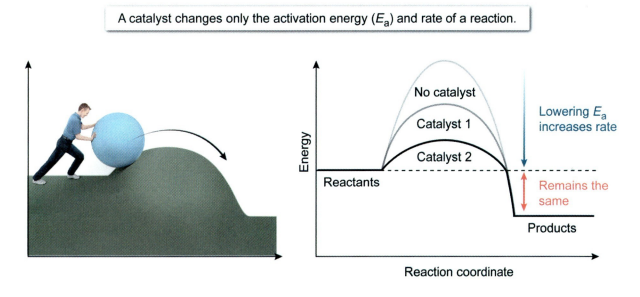

Figure 4.12 Catalysts influence the kinetics but not the thermodynamics of a reaction.

During a catalyzed reaction, the catalyst often reacts with one of the reactants and forms an **intermediate**. This intermediate then reacts in a subsequent step to complete the formation of a product, and the catalyst is regenerated; consequently, the catalyst is not consumed in the overall reaction. This characteristic feature of a catalyzed reaction is represented by the generalized scheme in Figure 4.13.

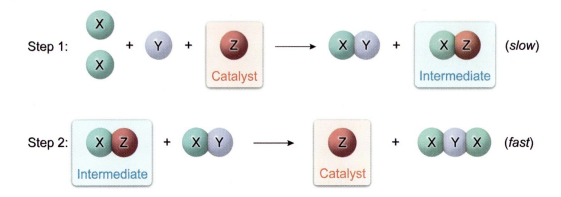

A catalyst forms an intermediate with a reactant in one step and is regenerated in a later step (**not consumed** in the reaction).

Figure 4.13 Action and regeneration of a catalyst in a reaction mechanism.

When viewed from the perspective of a reaction mechanism, catalysts function by lowering the energy of the transition state responsible for forming products. The catalyst accomplishes this by providing an alternate reaction pathway with a different, lower-energy transition state, as represented in Figure 4.14. The catalyzed mechanism often includes more steps than the uncatalyzed mechanism.

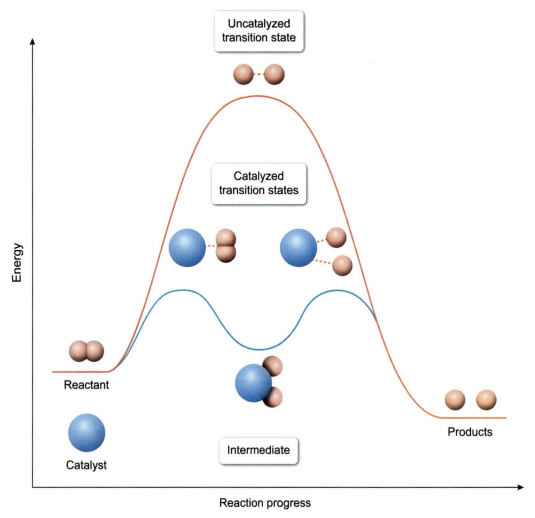

Figure 4.14 Alternate reaction pathway (mechanism) provided by a catalyst in a decomposition reaction.

The uncatalyzed decomposition of the reactant molecule depicted in Figure 4.14 proceeds via a mechanism that has a single, high-energy transition state. The addition of the catalyst provides an alternative reaction pathway with a mechanism that has two lower-energy transition states. As a result, the catalyzed reaction occurs at a significantly faster rate.

Although many variations in catalysts exist, two types of catalysts commonly seen in chemical systems are worth noting: surface catalysts and enzymes.

Surface Catalysts

A surface catalyst is an example of a heterogeneous catalyst (ie, one that exists in a different phase than the reactants). Surface catalysts work by providing active surfaces that facilitate interactions between reactants and lower the energy of the transition state necessary to form products.

Reactions mediated by surface catalysts typically have four main steps:

1. **Adsorption**. The reactants adsorb (ie, cling to the surface) on the catalyst.
2. **Activation**. The catalyst material primes the reaction via surface interactions.
3. **Reaction**. The activated reactants proceed through a transition state and form products.
4. **Desorption**. The products diffuse and release from the catalyst surface.

These four steps in a reaction involving a surface catalyst are illustrated in Figure 4.15.

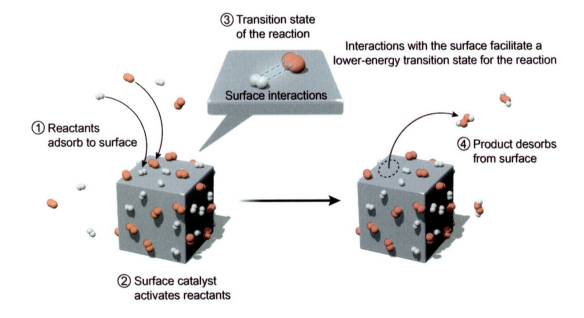

Figure 4.15 Reaction mediated by a surface catalyst.

Because reactions with a surface catalyst involve the available surfaces on the catalyst material, the rate of such reactions depends on the total amount of surface area present. As such, increasing the surface area of the catalyst (eg, grinding a larger catalyst sample into a greater number of smaller granules) increases the rate of the reaction because more surface area is made available on which the reaction can occur.

Enzymes

Enzymes are biological catalysts. These biomolecules are usually **proteins** (ie, very long amino acid chains) that fold into particular shapes (often represented in diagrams as lumpy blobs). The shape of the folded enzyme protein provides pockets called **binding sites** where other biomolecules can be bound into a complex that facilitates a reaction.

In addition to providing a lower-energy reaction pathway for reactants, such binding sites also enhance reactivity by fixing reactants closer together and in specific orientations that increase the probability of a productive interaction, as illustrated in Figure 4.16.

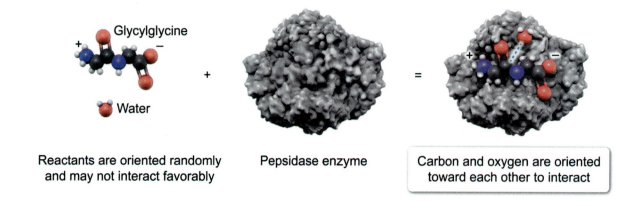

Figure 4.16 Peptidase enzyme acts by binding the biomolecule glycylglycine and a water molecule into a lower-energy transition state with an orientation more favorable for a reaction.

Lesson 5.1

Equilibrium in Reversible Reactions

5.1.01 Dynamic Equilibrium

The chemical reactions presented in previous chapters have been considered **irreversible**, meaning the reactions proceed in only one direction (ie, the products formed cannot react to regenerate the initial reactants), as indicated by a single reaction arrow (→). Consequently, irreversible reactions go to completion.

However, most reactions proceed with some degree of reversibility. A **reversible** reaction can proceed in both the forward direction (toward products) and reverse direction (toward reactants), as indicated by double reaction arrows (⇄). For example, consider the decomposition of $N_2O_4(g)$ to form $NO_2(g)$:

Forward reaction: $N_2O_4(g) \rightarrow 2\ NO_2(g)$

After some $NO_2(g)$ is formed, it can react with itself to regenerate $N_2O_4(g)$:

Reverse reaction: $2\ NO_2(g) \rightarrow N_2O_4(g)$

As such, the decomposition of $N_2O_4(g)$ can then be written as a reversible reaction:

$$N_2O_4(g) \rightleftarrows 2\ NO_2(g)$$

Reversible reactions do not go to completion because as the products are formed, some amount of the products react together to reform the reactants. The extent to which a reaction can regenerate reactants from products (and vice versa) is quantified by the equilibrium constant K_{eq}, which is discussed in Concept 5.3.01.

As a reversible reaction proceeds over time in a closed system (ie, no addition or removal of reactants or products), the reaction eventually reaches a state of **dynamic equilibrium** in which the rate of the forward reaction is equal to the rate of the reverse reaction (Figure 5.1). Note that this differs from a system in **static equilibrium**, where the opposing reactions have completely stopped.

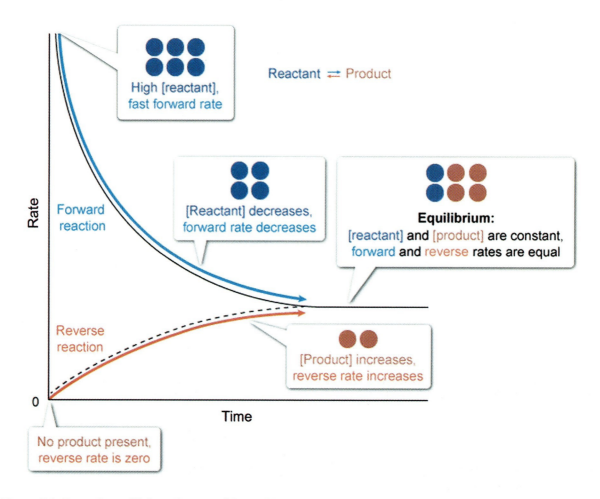

Figure 5.1 Dynamic equilibrium of a reversible reaction.

When a reaction is in a state of dynamic equilibrium, both reactions occur continuously but their effects cancel each other, so no net change in reactant or product concentration is observed (ie, the concentrations are constant). Although constant, the equilibrium concentrations are not necessarily equal because equilibrium refers to a state of equal reaction *rates* (ie, changes in concentration over time) and not to a state of equal concentrations. This difference is demonstrated in Figure 5.2, where product and reactant concentrations are shown as a function of time. Equilibrium concentrations are seen as horizontal segments in the graph.

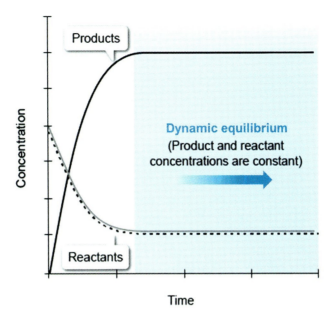

Figure 5.2 Product and reactant concentrations of a reversible reaction as a function of time.

Dynamic equilibria can be categorized as one of two types: homogeneous and heterogeneous. **Homogeneous equilibria** involve a single phase, such as a reaction involving only gases:

$$SiH_4(g) + 2\ Cl_2(g) \rightleftarrows SiCl_4(g) + 2\ H_2(g)$$

In contrast, **heterogeneous equilibria** involve more than one phase, such as a reaction involving both solid and aqueous phases:

$$2\ Cr^{3+}(aq) + Cd(s) \rightleftarrows Pb^{2+}(aq) + 2\ Cr^{2+}(aq)$$

In this chapter, both homogeneous and heterogeneous equilibria are explored in detail.

Chapter 5: Chemical Equilibrium

Lesson 5.2

Law of Mass Action

5.2.01 Introduction to the Law of Mass Action

Consider the general reversible reaction in which chemical species A and B react to form products C and D.

$$aA + bB \rightleftarrows cC + dD$$

The **law of mass action** states that for any reaction at equilibrium, the ratio of the molar concentrations (indicated by square brackets) of products to reactants, each raised to the power of their respective stoichiometric coefficient, is always constant (ie, does not change) for a given temperature:

$$\frac{[\text{Products}]^{\text{coefficients}}}{[\text{Reactants}]^{\text{coefficients}}} = \frac{[C]^c[D]^d}{[A]^a[B]^b} = constant$$

This ratio is known as the **mass action ratio**. Substituting the molar concentrations at equilibrium into the mass action ratio yields a quantity called the **equilibrium constant K_{eq}** (sometimes written as K_c, where the subscript c indicates concentration):

$$\frac{[C]^c[D]^d}{[A]^a[B]^b} = K_c = K_{eq}$$

As will be seen in Concept 5.3.01, the numerical value of K_{eq} is generally expressed without units (ie, K_{eq} is a dimensionless quantity). The value of K_{eq} is based on a unitless quantity called an *activity*, which is a ratio of the measured concentration to the standard state concentration (defined as 1.0 M). For example, the activity of reactant A with a measured concentration of X mol/L is given by:

$$\text{Activity of A} = \frac{[A]_{\text{measured}}}{[A]_{\text{standard state}}} = \frac{X \:\cancel{\text{mol/L}}}{1.0 \:\cancel{\text{mol/L}}} = X \:(unitless)$$

As such, the numerical values of the molar concentrations are often assumed to be equivalent to the activities when performing calculations. This approximation does not significantly alter the numerical value of K_{eq} but instead conveniently eliminates all units. Note that the activities of pure solids and pure liquids are defined as 1. Consequently, the concentrations of pure solids and pure liquids are not included in the mass action ratio or K_{eq} expression.

Alternatively, the equilibrium molar concentrations in the mass action ratio can be expressed in terms of partial pressures (P) when dealing with gases. In such cases, K_{eq} is often referred to as K_p, where the subscript indicates partial pressures:

$$\frac{(P_C)^c(P_D)^d}{(P_A)^a(P_B)^b} = K_p$$

In general, K_p and K_c are not equal because the partial pressures of the reacting species are not equal to the molar concentrations. However, the exam typically refers to any equilibrium constant simply as K_{eq}.

When applied to the dynamic equilibrium of an elementary reaction (ie, a reaction that occurs in a single step, as outlined in Lesson 4.1), the law of mass action also states that the *rate* of the reaction is proportional to the molar concentrations of each reactant raised to the power of its reaction order, as discussed in Concept 4.2.02. For example, suppose the following hypothetical reaction proceeds via a single elementary step:

$$2\,A + B \rightleftharpoons A_2B$$

The rate of the forward reaction ($r_{forward}$) and reverse reaction ($r_{reverse}$) can be written as:

$$r_{foward} = k_{forward}[A]^2[B] \quad \text{and} \quad r_{reverse} = k_{reverse}[A_2B]$$

where $k_{forward}$ and $k_{reverse}$ are the rate constants for the forward and reverse reactions, respectively. Recall from Lesson 5.1 that when $r_{forward}$ is equal to $r_{reverse}$, the system is at equilibrium. As such, the rate equations can be set equal to each other:

$$k_{forward}[A]^2[B] = k_{reverse}[A_2B]$$

Rearranging the expression to separate the rate constants from the molar concentrations yields:

$$\frac{k_{forward}}{k_{reverse}} = \frac{[A_2B]}{[A]^2[B]}$$

Because both $k_{forward}$ and $k_{reverse}$ are constants for a given temperature, the ratio is also constant and equal to K_{eq}:

$$\frac{k_{forward}}{k_{reverse}} = K_{eq}$$

Therefore, the law of mass action establishes a fundamental connection between chemical equilibrium and chemical kinetics.

Concept Check 5.1

For the reaction below, what is the expression for K_{eq}?

$$SnO_2(s) + 2\,CO(g) \rightleftharpoons Sn(s) + 2\,CO_2(g)$$

Solution

Note: The appendix contains the answer.

Lesson 5.3

Reaction Quotient

Introduction

As stated in Lesson 5.2, the law of mass action defines the equilibrium position of a reversible reaction by the **equilibrium constant** K_{eq}:

$$aA + bB \rightleftarrows cC + dD \qquad K_{eq} = \frac{[C]^c[D]^d}{[A]^a[B]^b}$$

However, a reaction does not reach a state of dynamic equilibrium instantaneously; instead, equilibrium is achieved over a period of time depending on the rates of the forward and reverse reactions. For a reaction not yet at equilibrium, the ratio of products to reactants is not numerically equal to K_{eq} but is instead equal to the reaction quotient Q, which is the value of the mass action ratio of the unequilibrated system at a given moment in time:

$$Q = \frac{[C]^c[D]^d}{[A]^a[B]^b}$$

Although the expression for Q is identical to that of K_{eq}, the difference between the two values at a given temperature is that the value of Q depends on the *current state* of the system at a particular time (ie, Q can have multiple values). In contrast, K_{eq} has only one value, and it represents the relative amounts of reactants and products present at *equilibrium*. As such, Q indicates how far a reaction is from its equilibrium position and which way the reaction must shift to reach equilibrium.

In this lesson, various applications of Q and K_{eq} are explored in detail, including the use of K_{eq} in equilibrium calculations. This lesson also examines the relationship between thermodynamics and equilibrium.

5.3.01 Equilibrium Constant

Every reaction that is reversible moves toward a state of equilibrium, regardless of the starting conditions or complexity of the reaction. Consequently, every reversible reaction has an equilibrium constant K_{eq} associated with it for a given temperature.

Interpreting the Magnitude of K_{eq}

The magnitude of the equilibrium constant reflects the relative amounts of reactants and products present in a reaction mixture at equilibrium. For example, the K_{eq} for the decomposition of NO(g) at 298 K is very large:

$$2\,NO(g) \rightleftarrows N_2(g) + O_2(g) \qquad K_{eq} = \frac{[N_2][O_2]}{[NO]^2} = 1 \times 10^{10}$$

A large value of K_{eq} signifies that the numerator (ie, products) of the mass action ratio must be much greater than the denominator (ie, reactants). Therefore, the concentrations of $N_2(g)$ and $O_2(g)$ must be much greater than the concentration of NO(g), indicating that the forward reaction is favored over the reverse reaction.

Alternatively, the K_{eq} for the decomposition of $COCl_2(g)$ at 373 K is very small:

$$COCl_2(g) \rightleftharpoons CO(g) + Cl_2(g) \qquad K_{eq} = \frac{[CO][Cl_2]}{[COCl_2]} = 4 \times 10^{-9}$$

Because the value of K_{eq} is small, the concentration of $COCl_2(g)$ must be much greater than the concentrations of $CO(g)$ and $Cl_2(g)$ (ie, the denominator is greater than the numerator) and the reverse reaction is favored.

In general, the magnitude of K_{eq} can be interpreted as follows:

- Scenario 1: If $K_{eq} \ll 1$ (a small number), the equilibrium position favors the reverse reaction (ie, more reactants than products are present in the equilibrium mixture).
- Scenario 2: If $K_{eq} = 1$, neither the forward or reverse reaction is favored (ie, significant amounts of both reactants and products are present in the equilibrium mixture).
- Scenario 3: If $K_{eq} \gg 1$ (a large number), the equilibrium position favors the forward reaction (ie, more products than reactants are present in the equilibrium mixture).

Figure 5.3 summarizes the relationship between the magnitude of K_{eq} and the relative concentrations of reactants and products.

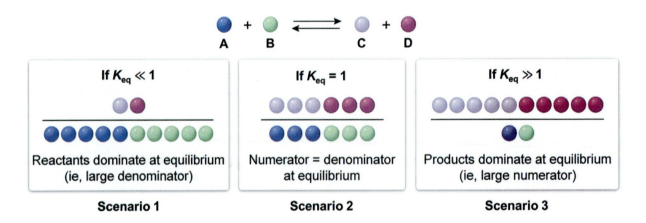

Figure 5.3 The relationship between K_{eq} and the composition of a mixture at equilibrium.

Concept Check 5.2

When $CO(g)$ is mixed with $Cl_2(g)$, the following reaction takes place:

$$CO(g) + Cl_2(g) \rightleftharpoons COCl_2(g) \qquad K_{eq} = 4.5 \times 10^9$$

At equilibrium, which species is present in the greatest amount?

Solution

Note: The appendix contains the answer.

Chapter 5: Chemical Equilibrium

Calculations Involving K_{eq}

If the equilibrium concentrations of all reactants and products are known for a reaction, the numerical value of K_{eq} for the reaction can easily be calculated using the mass action ratio (see Lesson 5.2). For example, consider the following reaction:

$$HCN(aq) + H_2O(l) \rightleftarrows H_3O^+(aq) + CN^-(aq)$$

At equilibrium, the concentrations of HCN(aq), H_3O^+(aq), and CN^-(aq) are 0.15 M, 8.6 × 10^{-6} M, and 8.6 × 10^{-6} M, respectively. Substituting these equilibrium concentrations into the mass action ratio yields the unitless value of K_{eq}:

$$K_{eq} = \frac{[H_3O^+][CN^-]}{[HCN]} = \frac{(8.6 \times 10^{-6})(8.6 \times 10^{-6})}{(0.15)} = 4.9 \times 10^{-10}$$

The equilibrium concentrations of all species in an equilibrium mixture are often unknown. However, if the equilibrium constant and initial conditions for a reaction are known, the equilibrium concentrations can be calculated. For example, suppose in the previous example that the equilibrium concentrations of HCN(aq), H_3O^+(aq), and CN^-(aq) are unknown but the initial concentration of HCN(aq) is 0.25 M and K_{eq} is known. As the reaction shifts toward equilibrium, HCN(aq) is consumed as H_3O^+(aq) and CN^-(aq) are formed.

At equilibrium, $[H_3O^+]$ and $[CN^-]$ can be defined as +x, where the plus sign indicates formation. As such, the equilibrium concentration of HCN(aq) must be equal to $[HCN]_{initial} - x$ because a corresponding amount (−x) of HCN(aq) is consumed in a 1:1 mole ratio during the reaction. This relationship can be demonstrated using an **ICE** (**I**nitial, **C**hange, **E**quilibrium) table, as shown in Figure 5.4.

Reaction	HCN	+ H₂O	$\xrightleftarrows{K_{eq}}$	H₃O⁺	+ CN⁻
Initial	0.25 M	Pure liquids not included		0 M	0 M
Change	−x consumed	—		+x produced	+x produced
Equilibrium	0.25 − x	—		x	x

Figure 5.4 Determining equilibrium concentrations using an ICE table.

Substituting the equilibrium concentrations (stated in terms of x) into the K_{eq} expression gives:

$$K_{eq} = 4.9 \times 10^{-10} = \frac{(x)(x)}{(0.25\ M - x)} = \frac{x^2}{(0.25\ M - x)}$$

Solving for x requires the use of the quadratic formula but because K_{eq} is such a small number (ie, the reverse reaction is favored), it can be assumed that x is negligible compared to the initial 0.25 M concentration (ie, 0.25 M − x ≈ 0.25 M). Note that this estimation can only be used if the initial concentration is at least 100 times greater than K_{eq}:

$$4.9 \times 10^{-10} \approx \frac{x^2}{(0.25 \text{ M})}$$

Solving for x yields:

$$x = \sqrt{(4.9 \times 10^{-10})(0.25 \text{ M})} = 1.1 \times 10^{-5} \text{ M}$$

Therefore, the equilibrium concentrations are:

$$[HCN] = (0.25 \text{ M} - x) = 0.25 \text{ M} - (1.1 \times 10^{-5} \text{ M}) = 0.25 \text{ M}$$

$$[H^+] = x = 1.1 \times 10^{-5} \text{ M}$$

$$[CN^-] = x = 1.1 \times 10^{-5} \text{ M}$$

 Concept Check 5.3

Consider the following reaction at 1,400 K:

$$2 \text{ H}_2\text{S}(g) \rightleftarrows 2 \text{ H}_2(g) + \text{S}_2(g) \qquad K_{eq} = 2.2 \times 10^{-4}$$

At equilibrium, the concentrations of $H_2(g)$ and $S_2(g)$ are 4.5×10^{-4} M and 1.0×10^{-3} M, respectively. What is the equilibrium concentration of $H_2S(g)$?

Solution

Note: The appendix contains the answer.

Predicting Reaction Shifts

Recall from this lesson's introduction that when a system is not at equilibrium, the product to reactant molar ratio is the reaction quotient Q. Comparing the value of Q to K_{eq} indicates the direction a reaction will proceed to achieve a state of equilibrium. Three possible scenarios for Q exist during a reaction:

- **Scenario 1**: If $Q < K_{eq}$, the concentration of products is too small and the concentration of reactants is too large. Therefore, the reaction is driven toward products (ie, the forward direction is favored).
- **Scenario 2**: If $Q = K_{eq}$, the reaction is at equilibrium and no net reaction takes place.
- **Scenario 3**: If $Q > K_{eq}$, the concentration of products is too large and the concentration of reactants is too small. Therefore, the reaction is driven toward reactants (ie, the reverse direction is favored).

The relationship between Q and K_{eq} is summarized in Figure 5.5.

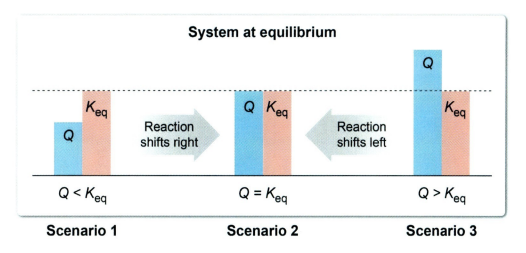

Figure 5.5 Predicting reaction shifts by comparing the reaction quotient Q to the equilibrium constant K_{eq}.

Concept Check 5.4

Nitrosyl chloride (NOCl) decomposes at 226 °C according to the following reaction:

$$2\ NOCl(g) \rightleftarrows Cl_2(g) + 2\ NO(g) \qquad K_{eq} = 4.2 \times 10^{-5}$$

If a vessel contains 2.0×10^{-6} M of NOCl(g), 1.0×10^{-2} M of $Cl_2(g)$, and 6.0×10^{-2} M of NO(g), in which direction must the reaction shift for the system to achieve equilibrium?

Solution

Note: The appendix contains the answer.

5.3.02 Effect of Catalysts

In Lesson 4.3, a catalyst is described as a substance that increases the rate of a reaction by lowering the activation energy required for the reaction to proceed. Not only does a catalyst increase the rate of the forward reaction, it also increases the rate of the reverse reaction by the same amount. Consequently, a dynamic equilibrium is established more quickly in the presence of a catalyst.

Because the forward and reverse rates increase equally, the equilibrium concentrations do not change. As such, the equilibrium position (ie, K_{eq}) is not changed due to the presence of a catalyst, as Figure 5.6 illustrates.

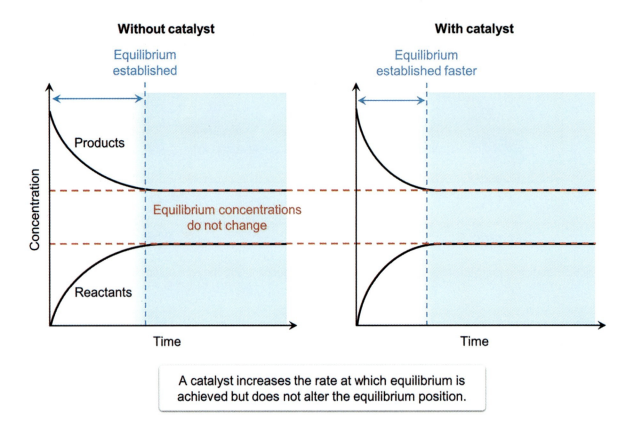

Figure 5.6 A catalyst does not affect the equilibrium position (K_{eq}) of a reaction.

5.3.03 Gibbs Free Energy and the Equilibrium Constant

The change in Gibbs free energy ΔG (see Lesson 3.7) can be described as a measure of a reaction's driving force to reach a state of equilibrium. A negative ΔG indicates a driving force in the forward direction (ie, the reaction spontaneously moves toward products) whereas a positive ΔG indicates a driving force in the reverse direction (ie, the reaction spontaneously moves toward reactants).

For a system that is not yet at equilibrium, the spontaneity of a reaction in either the forward or reverse direction depends on the composition of the mixture as described by the reaction quotient Q. As such, a standard state is defined to establish a reference for measuring ΔG. Under the **standard state**, all reactants and products are initially at concentrations of 1.0 M, and any gases have partial pressures of 1.0 atm. Unlike the standard conditions defined for the gas laws (Concept 6.3.02), the standard state in thermodynamics does *not* define a standard temperature; however, 298 K (25 °C) is often used in data tables.

Measurements of ΔG made under standard conditions are indicated with the addition of the degree symbol, as in $\Delta G°$. Using $\Delta G°$ as a reference, the *actual* ΔG observed at any point in a reaction can be quantitatively evaluated by the equation:

$$\Delta G = \Delta G° + RT \ln Q$$

where $\Delta G°$ is the change in Gibbs free energy under standard conditions, R is the gas constant (8.314 J/K·mol), and T is the absolute (Kelvin) temperature. In this equation, the $RT\ln Q$ term adjusts ΔG (relative to $\Delta G°$) in proportion to the deviations from the standard state as measured by changes in Q. If a reaction is carried out under standard conditions, $Q = 1$ and $RT\ln Q = 0$ with $\Delta G = \Delta G°$; however, under any other conditions ΔG will differ from $\Delta G°$, as illustrated in Figure 5.7. Note that for a given temperature, $\Delta G°$ has only one value but the value of ΔG varies depending on Q.

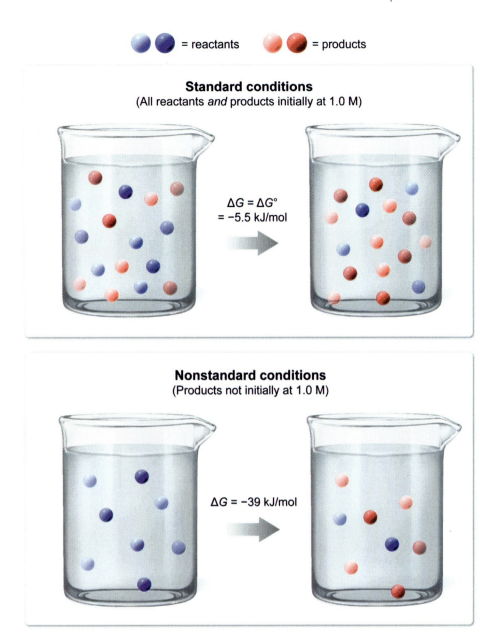

Figure 5.7 Standard and observed Gibbs free energy changes.

✓ **Concept Check 5.5**

If a sealed vessel contains 1.0 atm of $N_2(g)$, 2.0 atm of $H_2(g)$, and 0.50 atm of $NH_3(g)$ at 1,000 K, is the reaction spontaneous in the forward direction?

$$N_2(g) + 3\,H_2(g) \rightleftarrows 2\,NH_3(g) \qquad \Delta G° = -1.6 \times 10^4 \text{ J/mol}$$

Solution

Note: The appendix contains the answer.

When a state of equilibrium has been achieved, $Q = K_{eq}$ and $\Delta G = 0$ because the driving forces for both the forward and reverse directions are equal. Consequently, for a reaction at equilibrium, the previously stated equation for ΔG can be written as:

$$0 = \Delta G° + RT \ln K_{eq}$$

This variation of the equation establishes a relationship between K_{eq} and the spontaneity of a reaction under *standard state conditions*. As such, if K_{eq} is known for a reaction, the equation can be rearranged to solve for $\Delta G°$:

$$\Delta G° = -RT \ln K_{eq}$$

Alternatively, if $\Delta G°$ is known, the equation can be rewritten to calculate K_{eq}:

$$K_{eq} = e^{-\frac{\Delta G°}{RT}}$$

When $K_{eq} < 1$, $\Delta G° > 0$ and the reverse reaction is spontaneous (ie, reactants are favored). In contrast, when $K_{eq} > 1$, $\Delta G° < 0$ and the forward reaction is spontaneous (ie, products are favored). Figure 5.8 summarizes the relationship between K_{eq} and $\Delta G°$.

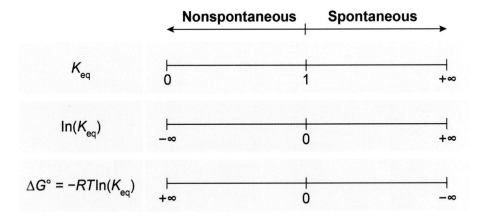

Figure 5.8 The relationship between K_{eq} and the spontaneity of a reaction under standard conditions.

A reaction that has not yet reached equilibrium spontaneously moves in the direction that minimizes ΔG.

One method of predicting the direction a reaction will spontaneously shift to minimize ΔG under any set of conditions is to compare the values of Q and K_{eq}. Rewriting the equation for ΔG in terms of K_{eq} and Q yields:

$$\Delta G = \Delta G° + RT \ln Q = -RT \ln K_{eq} + RT \ln Q$$

Combining the logarithmic terms results in a simplified equation:

$$\Delta G = RT \ln\left(\frac{Q}{K_{eq}}\right)$$

This simplified equation can be used to qualitatively predict the direction a reaction shifts to reach equilibrium. There are three possible scenarios to consider:

- **Scenario 1**: When $Q < K_{eq}$, $\frac{Q}{K_{eq}} < 1$ and $\ln\left(\frac{Q}{K_{eq}}\right)$ is negative. Therefore, ΔG is negative and the reaction spontaneously moves in the forward direction.
- **Scenario 2**: When $Q = K_{eq}$, $\frac{Q}{K_{eq}} = 1$ and $\ln\left(\frac{Q}{K_{eq}}\right) = 0$. Therefore, $\Delta G = 0$ and the reaction is at equilibrium (ie, no further movement in either direction).
- **Scenario 3**: When $Q > K_{eq}$, $\frac{Q}{K_{eq}} > 1$ and $\ln\left(\frac{Q}{K_{eq}}\right)$ is positive. Therefore, ΔG is positive (ie, the forward reaction is nonspontaneous) and the reaction spontaneously moves in the reverse direction.

In Figure 5.9, a graph of Gibbs free energy is shown as a function of reaction progress (as reflected in the value of Q). The graph demonstrates that a reaction not yet at equilibrium spontaneously shifts in any direction needed to minimize ΔG and reach equilibrium.

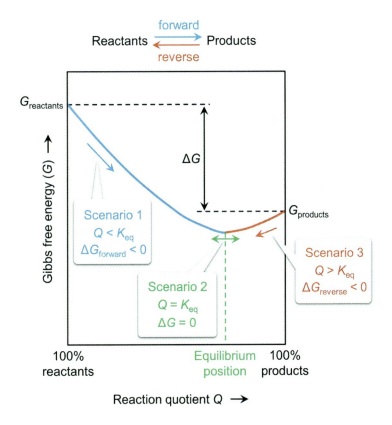

Figure 5.9 Gibbs free energy G as a function of the reaction quotient Q.

Concept Check 5.6

Consider the reaction shown.

$$PH_3BCl_3(s) \rightleftarrows PH_3(g) + BCl_3(g) \qquad K_{eq} = 1.6$$

If the partial pressures of $PH_3(g)$ and $BCl_3(g)$ are 0.5 atm and 2.1 atm, respectively, is ΔG positive, negative, or equal to zero for the forward reaction?

Solution

Note: The appendix contains the answer.

Lesson 5.4

Le Châtelier's Principle

Introduction

As discussed in Lesson 5.3, the reaction quotient Q is the value of the mass action ratio for a reversible reaction at a given moment in time.

$$a\text{A} + b\text{B} \rightleftarrows c\text{C} + d\text{D} \qquad Q = \frac{[\text{C}]^c[\text{D}]^d}{[\text{A}]^a[\text{B}]^b}$$

When the forward and reverse reactions take place at the same rate in a chemical system, a dynamic equilibrium is reached due to the balance between the opposing reactions. The concentrations of the reactants and products present at equilibrium yield a specific value of Q called the equilibrium constant K_{eq} (ie, $Q = K_{eq}$).

However, if a change in the conditions of the chemical system affects the rates of the forward and reverse reactions differently, the system is no longer at equilibrium because the rates of the forward and reverse reactions are no longer equal (ie, $Q \neq K_{eq}$). According to **Le Châtelier's principle**, this "stress" or disturbance causes the system to shift in the direction (either toward products or reactants) that restores a state of equilibrium, as demonstrated in Figure 5.10.

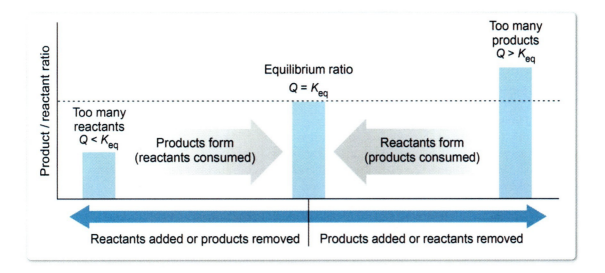

Figure 5.10 Le Châtelier's principle predicts the direction a system's equilibrium position shifts after a disturbance.

In this lesson, Le Châtelier's principle is applied to assess how a system responds to the following stresses:

- Changes in reactant or product concentrations
- Changes in volume and pressure
- Changes in temperature

5.4.01 Effect of Concentration Changes

The equilibrium position of a chemical system can be disturbed by altering the concentrations (or partial pressures) of one or more of the products or reactants. Under constant-temperature conditions, this disturbance changes the value of Q (ie, Q no longer equals K_{eq}). For example, suppose an equilibrium is established between the general aqueous species A, B, and C (note that this example also applies if these species are gaseous):

$$A(aq) + B(aq) \rightleftarrows C(aq) \qquad K_{eq} = \frac{[C]}{[A][B]}$$

If some amount of species C is *removed* from the system (ie, concentration decreases) or an additional amount of species A or B is *added* to the system (ie, concentration increases), the value of Q *decreases*, resulting in $Q < K_{eq}$. To counteract this disturbance, the equilibrium position shifts toward the product side of the reaction (ie, the forward reaction is favored) to regenerate species C or deplete species A or B until $Q = K_{eq}$. As such, removing products or adding reactants shifts the equilibrium position toward products, as illustrated in Figure 5.11.

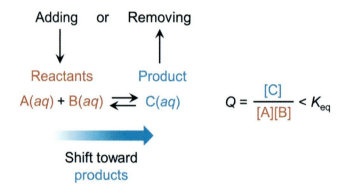

Figure 5.11 Effect of removing products from or adding reactants to a mixture at equilibrium.

Conversely, if some amount of species C is *added* to the system or a portion of species A or B is *removed* from the system, the value of Q *increases*, resulting in $Q > K_{eq}$. As such, the equilibrium position shifts toward the reactant side of the reaction (ie, the reverse reaction is favored) to deplete species C or regenerate species A or B until $Q = K_{eq}$. Therefore, removing products or adding reactants shifts the equilibrium position toward reactants, as summarized in Figure 5.12.

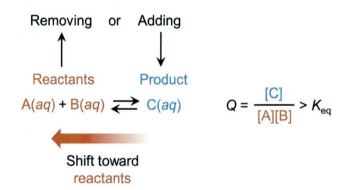

Figure 5.12 Effect of adding products to or removing reactants from a mixture at equilibrium.

Another way to alter the reactant and product concentrations is through dilution (eg, adding solvent) or concentration (eg, evaporating solvent). Diluting an aqueous solution at equilibrium by adding water can alter Q (ie, $Q \neq K_{eq}$) and cause the equilibrium position to shift toward whichever side has a greater number (ie, more moles) of species in solution. Conversely, concentrating the solution by evaporating water has the opposite effect (see Figure 5.13). However, if a reaction has *equal* numbers of species in solution on each side of the equation, diluting or concentrating the solution does not affect the equilibrium position (ie, Q is not affected).

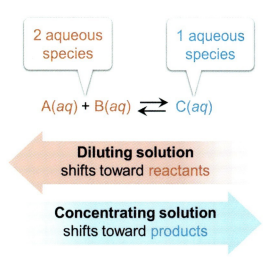

Figure 5.13 Effect of diluting or concentrating a mixture at equilibrium.

Concept Check 5.7

Consider the following reaction at equilibrium:

$$2\ HgCl_2(aq) + C_2O_4^{2-}(aq) \rightleftharpoons 2\ Cl^-(aq) + 2\ CO_2(g) + Hg_2Cl_2(s)$$

Which of the following changes to the system will cause the equilibrium to shift to the left toward the reactants? (Select all that apply.)

I. Adding water to the mixture.
II. Adding more $CO_2(g)$ to the mixture.
III. Removing $HgCl_2(aq)$ from the mixture.

Solution

Note: The appendix contains the answer.

5.4.02 Effect of Pressure Changes

Because gases are highly compressible compared to liquids and solids, the equilibrium position of a system containing at least one gaseous species can be disturbed by changes in the system's volume and pressure at a constant temperature. For example, consider the following gas-phase system held in a cylinder fitted with a movable piston (ie, the volume is adjustable):

$$A(g) + B(g) \rightleftarrows C(g) \qquad K_p = \frac{P_C}{P_A P_B}$$

Pressing down on the cylinder's piston decreases the volume of the system, which increases the system's total pressure (Concept 6.1.01). This pressure increase also increases the partial pressures of each gaseous species present in the mixture. Because two gaseous species are in the denominator of the Q expression and only one is in the numerator, the increase in pressure increases the denominator more than it does the numerator. As a result, the value of Q decreases (ie, $Q < K_p$) and the system responds by shifting toward products until $Q = K_p$.

In general, an *increase* in pressure due to a *decrease* in volume shifts a reaction in the direction that *decreases* the number of moles of gas molecules, as illustrated in Figure 5.14.

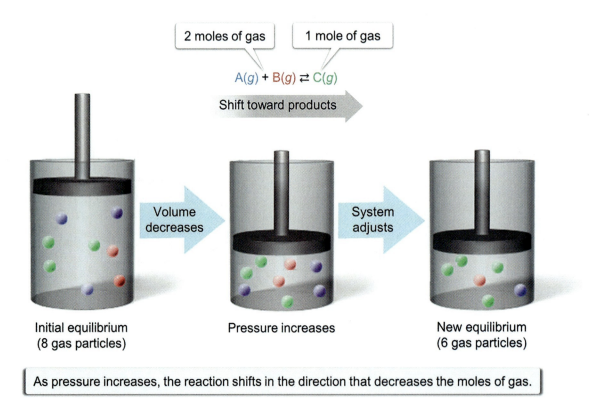

As pressure increases, the reaction shifts in the direction that decreases the moles of gas.

Figure 5.14 Effect of increasing pressure by decreasing the volume of a mixture at equilibrium.

Conversely, pulling the cylinder's piston upward increases the volume of the system and decreases the system's total pressure (ie, the partial pressures of each gaseous species also decrease). Consequently, the value of Q increases (ie, $Q > K_p$) and the system responds by shifting toward reactants until $Q = K_p$. In general, a *decrease* in pressure due to an *increase* in volume shifts a reaction in the direction that *increases* the number of moles of gas molecules (Figure 5.15). Note that if an equal number of moles of gas are reacted and formed during a reaction, the mixture at equilibrium is unaffected by a volume change.

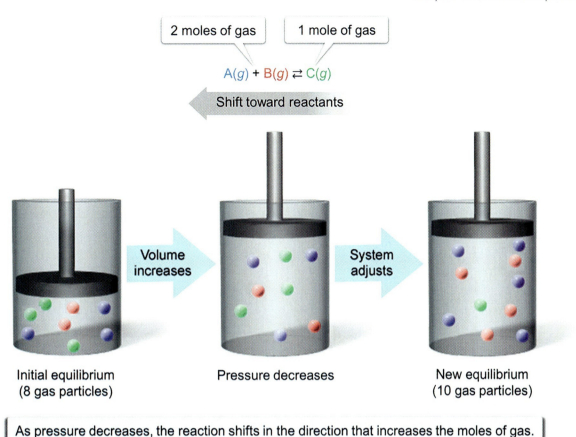

Figure 5.15 Effect of decreasing pressure by increasing the volume of a mixture at equilibrium.

Another way to increase the pressure of the system is by adding an inert (ie, nonreactive) gas. If an inert gas is added under constant volume (ie, the size of the container is fixed), the total pressure of the system increases; however, the partial pressures of each gas present in the mixture do not change because the number of moles of each gas per unit volume stays constant. Consequently, adding an inert gas has no effect on the equilibrium position of a system kept at constant volume. Alternatively, if an inert gas is added to a system under constant pressure (ie, no fixed volume), the equilibrium position shifts in response to the volume change, as discussed previously.

 Concept Check 5.8

The reaction shown is performed in a sealed reaction vessel:

$$UO_2(s) + 4\,HF(g) \rightleftarrows UF_4(g) + 2\,H_2O(g)$$

If the reaction is at equilibrium, how will decreasing the volume of the reaction vessel affect the partial pressure of $HF(g)$?

Solution

Note: The appendix contains the answer.

5.4.03 Effect of Temperature Changes

Recall from Concepts 5.4.01 and 5.4.02 that when a system at equilibrium is disturbed by a change in concentration or pressure at a constant temperature, the composition of the reaction mixture changes (ie, the value of Q changes) but the value of K_{eq} remains the same. As long as the temperature remains constant, Le Châtelier's principle states that the system responds to the stress by shifting in a direction that alters the value of Q so that it again becomes equal to K_{eq}.

In contrast, changing the temperature of a system at equilibrium does not *immediately* change the concentrations of the products and reactants; that is, the value of Q is the same before and after the temperature change. However, the value of K_{eq} is temperature-dependent and does change. Altering the temperature changes the energy, strength, and frequency of molecular collisions, which changes the rates of the forward and reverse reactions. Consequently, a new equilibrium position (ie, a new K_{eq}) is established. As a result, the product-to-reactant ratio Q adjusts over time until it is equal to the new K_{eq}.

Because K_{eq} no longer equals Q immediately after a temperature change, Le Châtelier's principle states that the system shifts in a direction that allows it to reach a new state of equilibrium. The direction of the shift after a temperature change depends on whether the reaction is exothermic ($\Delta H < 0$) or endothermic ($\Delta H > 0$).

For exothermic reactions, the heat released is treated as a hypothetical *product*:

$$A + B \rightleftharpoons C + \text{heat} \qquad \Delta H < 0$$

Adding heat (ie, increasing the temperature) causes the reaction to shift toward reactants to consume the excess heat, which *decreases* K_{eq}. Alternatively, removing heat (ie, decreasing the temperature) shifts the reaction toward products to produce more heat, which *increases* K_{eq}. Figure 5.16 summarizes the direction an exothermic reaction shifts in response to a temperature change, as predicted by Le Châtelier's principle.

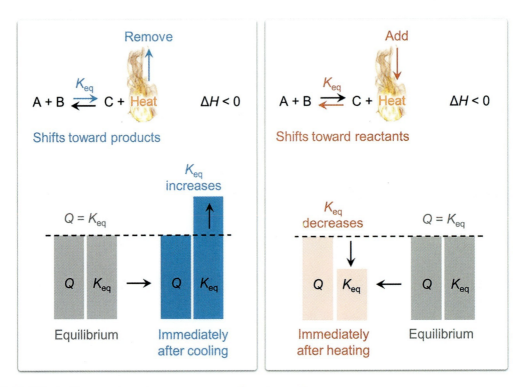

Figure 5.16 Effect of temperature changes on an exothermic reaction.

For endothermic reactions, the heat absorbed is treated as a hypothetical *reactant*:

$$A + B + \text{heat} \rightleftarrows C \qquad \Delta H > 0$$

Adding heat to an endothermic reaction shifts the reaction toward products, and removing heat shifts the reaction toward reactants (ie, the opposite of the responses seen from exothermic reactions). Therefore, increasing the temperature of an endothermic reaction results in an increased value of Keq, and decreasing the temperature results in a decreased value of Keq, as summarized in Figure 5.17.

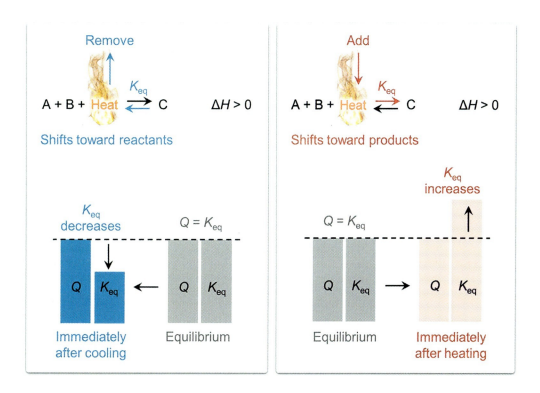

Figure 5.17 Effect of temperature changes on an endothermic reaction.

✓ Concept Check 5.9

Hydrogen gas is produced industrially using the following reaction:

$$CO(g) + H_2O(g) \rightleftarrows CO_2(g) + H_2(g) \qquad \Delta H° = -41 \text{ kJ}$$

To increase the equilibrium yield of $H_2(g)$, should the reaction be carried out at higher or lower temperatures?

Solution

Note: The appendix contains the answer.

Lesson 6.1

Pressure, Volume, and Temperature Relationships

Introduction

The unique behavior of particles in the gas phase is due to gas particles (ie, atoms or molecules) generally being spaced far apart and moving very quickly. Many of the factors ordinarily considered when analyzing the behavior of liquids and solids can be ignored in all but the most extreme circumstances.

Under these assumptions, the behavior of gases can be simplified into the following tenets of the **kinetic molecular theory of gases** (illustrated in Figure 6.1):

- Gases consist of large numbers of particles in continuous, random motion.
- The combined molecular volume of all particles of a gas sample is negligible when compared to the total volume of the container containing the gas sample.
- Attractive and repulsive forces between gas particles are negligible (ie, assumed to be absent).
- Energy can be transferred between particles during collisions but, as long as the temperature remains constant, the average kinetic energy of the particles remains constant (ie, the collisions are elastic).
- The average kinetic energy of the particles is proportional to the absolute temperature (ie, the Kelvin scale). At any given temperature, the particles of all gases have the same average kinetic energy.
- Pressure results from the collisions that gas particles make with the walls of a container. Some equivalent values for pressure used to convert between pressure units are:

$$1 \text{ atm} = 760.0 \text{ mmHg} = 760.0 \text{ torr} = 14.7 \text{ psi} = 101{,}325 \text{ Pa} = 1.01325 \text{ bar}$$

Chapter 6: Gas Laws

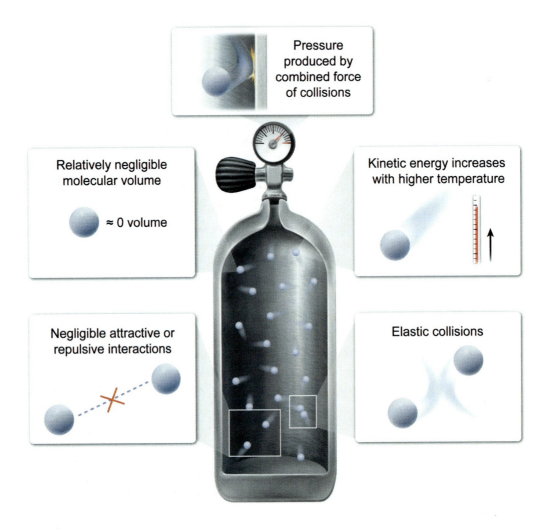

Figure 6.1 Principles and assumptions of the kinetic molecular theory of gases.

These principles are the foundation of the temperature, volume, and pressure relationships of gases expressed by Boyle's law, Charles's law, Gay-Lussac's law, and the combined gas law, which are covered in this lesson.

6.1.01 Boyle's Law

As the kinetic molecular theory of gases explains, pressure is the result of gas particles colliding with the walls of their container. When temperature is held constant, the same number of gas particles compressed into a smaller volume results in more collisions with the walls of the container, therefore increasing pressure (Figure 6.2). Conversely, increasing the volume of the container results in fewer collisions, thereby decreasing pressure.

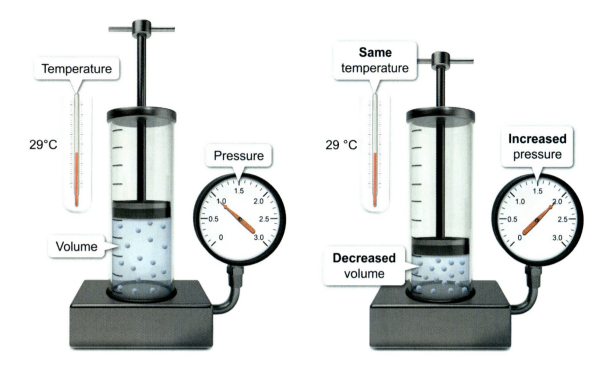

Figure 6.2 At constant temperature and number of moles, the pressure of an ideal gas is inversely proportional to its volume.

Boyle's law describes this inverse relationship between volume *V* and pressure *P* for a fixed number of molecules of an ideal gas at constant temperature according to the following equations:

$$P \propto \frac{1}{V}$$

$$P_1 V_1 = P_2 V_2$$

where P_1 and V_1 are the initial pressure and volume of the system, and P_2 and V_2 are the final pressure and volume of the system, respectively. These equations show that pressure increases as volume decreases (and vice versa).

> **Concept Check 6.1**
>
> A sample of a gas occupying a volume of 1.25 L is compressed into a volume of 255 mL at a constant temperature. If the pressure of the sample is originally 1.10 atm, what is the pressure of the compressed gas?
>
> **Solution**
> *Note: The appendix contains the answer.*

6.1.02 Guy-Lussac's Law

Since the average kinetic energy of gas particles is proportional to the absolute temperature (measured in kelvin), the average kinetic energy increases (ie, the particles move faster) as the temperature increases.

As seen in Figure 6.3, particles moving faster due to an increase in temperature results in more collisions with the surface of the container, which increases the pressure if the volume is held constant.

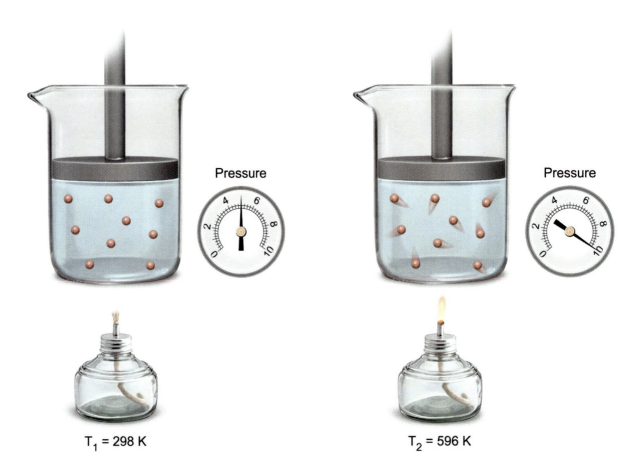

Figure 6.3 At a constant volume and number of moles, the pressure of a gas is directly proportional to its absolute temperature.

For a closed, rigid container with a constant volume, **Gay-Lussac's law** states that the pressure exerted by a gas inside the container is directly proportional to the absolute temperature of the system, as described by the following equations:

$$P \propto T$$

$$\frac{P_1}{T_1} = \frac{P_2}{T_2}$$

where P_1 and T_1 are the initial pressure and temperature of the system, and P_2 and T_2 are the final pressure and temperature of the system, respectively. (Note that this relationship is only valid when the temperature values are in kelvin.) These equations show that if the volume of the system remains constant, the pressure exerted by the gas increases as the temperature increases (and vice versa).

> **☑ Concept Check 6.2**
>
> A can of compressed air accidentally left out on a hot, sunny day reaches an internal pressure of 10.5 atm. If the pressure inside the can is 135 psi when stored at room temperature (22 °C), what temperature (in °C) does the can reach when left outside in the sun?
>
> **Solution**
>
> *Note: The appendix contains the answer.*

6.1.03 Charles' Law

When the temperature of a gas increases in a closed, *flexible* container, any increase in pressure is only momentary because the container immediately expands to keep the pressure on the inside equal to the pressure on the outside. Under such conditions of (essentially) constant pressure and amount of gas, a direct relationship exists between the volume of a gas and its absolute temperature, as illustrated in Figure 6.4.

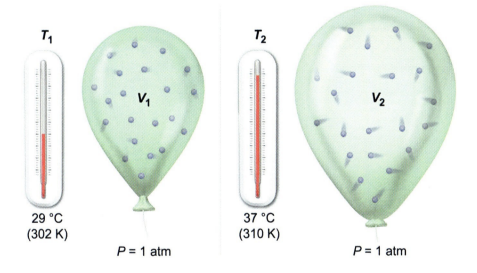

Figure 6.4 At constant pressure and number of moles, the volume of a gas is directly proportional to its absolute temperature.

This relationship, known as **Charles' law**, is expressed by the following equations:

$$V \propto T$$

$$\frac{V_1}{T_1} = \frac{V_2}{T_2}$$

where V_1 and T_1 are the initial volume and temperature of the system, and V_2 and T_2 are the final volume and temperature of the system, respectively. (Once again, temperatures must be in kelvin for this relationship to hold true.) These equations show that if the pressure remains constant, the temperature increases as the volume of gas also increases (and vice versa).

> **Concept Check 6.3**
>
> A helium-filled balloon originally at room temperature (22 °C) has a volume of 750 mL. If the balloon is dipped into a vat of liquid nitrogen (−196 °C), what is the new volume of the balloon?
>
> **Solution**
>
> Note: The appendix contains the answer.

6.1.04 Combined Gas Law

The relationships described in Boyle's law, Charles's law, and Gay-Lussac's law can be synthesized into one unified relationship known as the **combined gas law**, which is expressed by the following equation:

$$\frac{P_1 V_1}{T_1} = \frac{P_2 V_2}{T_2}$$

where P_1, V_1, and T_1 are the initial pressure, volume, and absolute temperature of a closed system of gases, and P_2, V_2, and T_2 are the final pressure, volume, and absolute temperature of the same closed system of gases, respectively. This relationship is valid for any closed system of gases for which the assumptions of the kinetic molecular theory of gases hold true.

As an example, suppose a child holds a 1.25 L helium balloon on a hot summer day (32 °C) in which the pressure is 745 mmHg. The balloon is then released into the sky, eventually reaching an altitude where the pressure is 482 mmHg and the temperature is 11 °C (see Figure 6.5). What would the volume of the balloon be at this altitude?

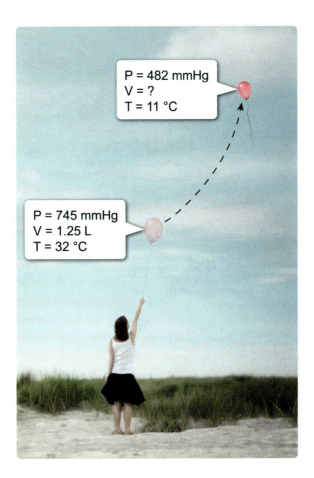

Figure 6.5 A helium balloon is released into the upper atmosphere, where pressure and temperature are lower than at ground level.

A useful approach to this problem is as follows:

Step 1. List out all the variables and identify what quantity is unknown:

$V_1 = 1.25$ L $\qquad V_2 = ?$

$P_1 = 745$ mmHg $\qquad P_2 = 482$ mmHg

$T_1 = 32\ °C$ $\qquad T_2 = 11\ °C$

Step 2. Rearrange the equation to solve for the missing variable:

$$V_2 = \frac{P_1 V_1 T_2}{T_1 P_2}$$

Step 3. Check pressure and volume units to see if they match (for proper cancellation), and make sure all temperatures are in kelvin. Carry out any conversions as necessary:

$V_1 = 1.25$ L $\qquad\qquad V_2 = ?$

$P_1 = 745$ mmHg $\qquad\qquad P_2 = 482$ mmHg

$T_1 = 32\ °C + 273 = 305$ K $\qquad\qquad T_2 = 11\ °C + 273 = 284$ K

Step 4. Plug the quantities into the rearranged equation and solve:

$$V_2 = \frac{P_1 V_1 T_2}{T_1 P_2} = \frac{(745\ \text{mmHg})(1.25\ \text{L})(284\ \text{K})}{(305\ \text{K})(482\ \text{mmHg})} = 1.80\ \text{L}$$

It is helpful to recognize that the combined gas law readily reduces to one of the simpler laws when one of the conditions is constant. When the values of two variables are the same on both sides of the equation, the variables cancel from the equation:

- Constant pressure ($P_1 = P_2$): Charles's law

$$\frac{\cancel{P_1} V_1}{T_1} = \frac{\cancel{P_2} V_2}{T_2} \quad \Rightarrow \quad \frac{V_1}{T_1} = \frac{V_2}{T_2}$$

- Constant volume ($V_1 = V_2$): Gay-Lussac's law

$$\frac{P_1 \cancel{V_1}}{T_1} = \frac{P_2 \cancel{V_2}}{T_2} \quad \Rightarrow \quad \frac{P_1}{T_1} = \frac{P_2}{T_2}$$

- Constant temperature ($T_1 = T_2$): Boyle's law

$$\frac{P_1 V_1}{\cancel{T_1}} = \frac{P_2 V_2}{\cancel{T_2}} \quad \Rightarrow \quad P_1 V_1 = P_2 V_2$$

 Concept Check 6.4

An anesthesiologist administers nitrous oxide to a young patient prior to surgery. A volume of 75 mL of gas exits the pump at a pressure of 1.5 atm and 18 °C. If the gas expands to a volume of 120 mL inside the lungs of the patient (whose body temperature is 37 °C), what is the pressure of the gas inside the patient's lungs?

Solution

Note: The appendix contains the answer.

Lesson 6.2

Pressure and Volume Relationships to Molar Amount

Introduction

Up to this point, the gas laws discussed have only addressed cases of closed systems containing one type of gas; however, most systems in nature are **open systems** in which matter and energy are exchanged between the system and its surroundings. Furthermore, gaseous systems are often complex mixtures of many substances. Consider the human respiratory system as an example (Figure 6.6).

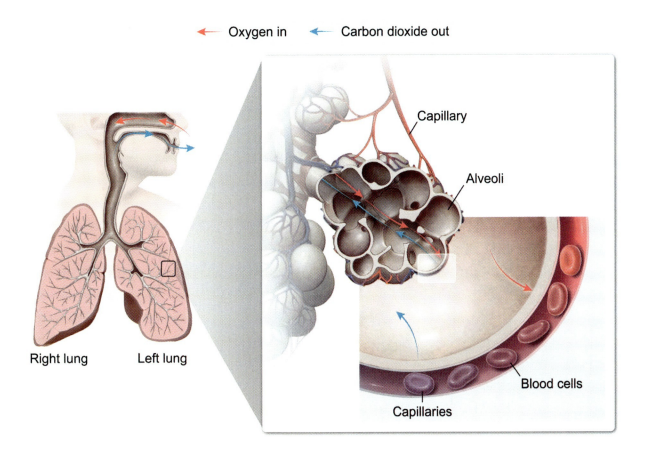

Figure 6.6 Human respiration is a complex open system involving multiple gases.

Carbon dioxide, oxygen, and other gases are exchanged between the capillaries and the alveoli in the lungs and, from a macroscopic perspective, between the lungs and exterior air. In open systems such as these, the molar amounts of the different substances must be considered when analyzing the system. The gas laws presented in this lesson address such cases.

6.2.01 Avogadro's Law

As more gas particles are added to an open system, the number of collisions with the walls of the container increases and begins to raise the pressure inside the container. However, if the container is expandable, the pressure remains effectively constant because the added collisions push the container to expand and increase its volume until the pressure inside the container equals the external pressure. The identities of the gas particles within the container do not matter, only the total quantity of gas.

These principles are expressed by **Avogadro's law**, which states that at a constant temperature and pressure, the volume (V) of a gas is directly proportional to the number of moles (n) of the gas. This is mathematically stated by the following relationship:

$$V \propto n$$

When evaluating a change in which gas is added to (or removed from) an open system under constant temperature and pressure (or when comparing two different systems under the same temperature and pressure), Avogadro's law can be applied as:

$$\frac{V_1}{n_1} = \frac{V_2}{n_2}$$

where V_1 and n_1 indicate the volume and number of moles for the initial state of the system, and V_2 and n_2 represent the volume and number of moles for the final state of the system, respectively (Figure 6.7).

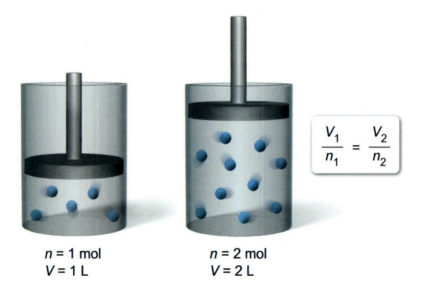

Figure 6.7 The volume occupied by a gas (at constant temperature and pressure) is directly proportional to the number of moles.

Chapter 6: Gas Laws

 Concept Check 6.5

A balloon containing 0.0831 mol of helium has a total volume of 1.46 L. If the balloon is inflated to a total volume of 4.83 L, how many moles of helium are added to the balloon?

Solution

Note: The appendix contains the answer.

6.2.02 Dalton's Law of Partial Pressures

When gas particles are added to a rigid container (ie, constant volume) at a constant temperature, the number of particle collisions with the walls of the container (and consequently the pressure) both increase in direct proportion to the amount of gas present (Figure 6.8). This direct relationship between the moles of gas and its pressure is expressed by the following equation:

$$\frac{P_1}{n_1} = \frac{P_2}{n_2}$$

where P_1 and n_1 indicate the pressure and number of moles for the initial state of the system, and P_2 and n_2 represent the pressure and number of moles for the final state of the system, respectively.

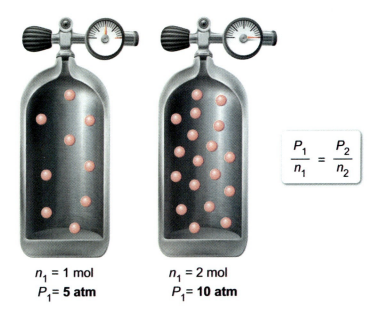

Figure 6.8 For two systems with the same volume and temperature, the pressure in each system is directly proportional to the number of moles of gas in that system.

As with Avogadro's law, this proportional relationship between pressure and moles of gas can be used to compare two different closed systems at the same volume and temperature *or* two different states of one open system to which gas is added or removed. However, this relationship can also be used to compare one gas (ie, "gas A") from a mixture to the total amount of gas in the mixture, which leads to the following equation:

$$\frac{P_A}{n_A} = \frac{P_{total}}{n_{total}} \implies \frac{P_A}{P_{total}} = \frac{n_A}{n_{total}}$$

where P_A and n_A indicate the partial pressure and number of moles for the selected gas of interest, and P_{total} and n_{total} represent the total pressure and number of moles of gas for the whole system.

These equations reveal that within a mixture of gases, the **partial pressure** of a gas (ie, the fraction of the total pressure contributed by one gas in the mixture) is directly proportional to the **mole fraction** of that gas, as expressed in the following equation:

$$P_A = X_A \cdot P_{total}$$

where X_A is the mole fraction $\left(\frac{n_A}{n_{total}}\right)$ of gas A. Derived from this relationship, Dalton's law of partial pressures states that for a mixture of gases at constant volume and temperature, the total pressure of the system is equal to the sum of the partial pressures of each constituent gas, as illustrated in Figure 6.9. This law is expressed by the following equation:

$$P_{total} = P_A + P_B + P_C + \cdots$$

where P_{total} is the total pressure of the system, and P_A, P_B, and P_C are the partial pressures of the individual gases within the system.

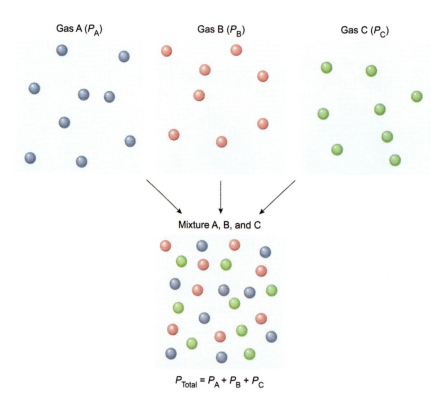

Figure 6.9 The partial pressures of gases A, B, and C add up to the total pressure of the system containing these three gases.

Concept Check 6.6

When a diving tank is charged with 0.74 mol N_2 gas, the pressure in the tank reads 1.68 atm. If the tank is then filled to a total pressure of 10.0 atm by adding oxygen, how many moles of O_2 are added to the tank?

Solution

Note: The appendix contains the answer.

Lesson 6.3

Ideal Gas Model

Introduction

All the gas law relationships discussed thus far assume **ideal gas** behavior. The properties of an ideal gas, which are derived from the assumptions of the **kinetic molecular theory of gases**, are as follows:

- Particles of an ideal gas have no volume; therefore, they are infinitely compressible.
- Particles of an ideal gas have no attractive or repulsive forces between the particles.
- All collisions between ideal gas particles are perfectly elastic (ie, no energy is lost to heat, sound, material deformation).
- Particles of an ideal gas have an average kinetic energy (energy of motion) directly proportional to the temperature of the gas.

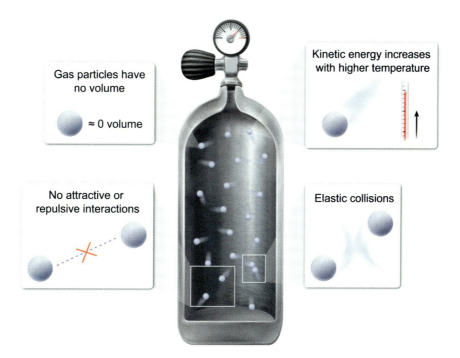

Figure 6.10 Characteristics of an ideal gas.

Although no real gas behaves perfectly like an ideal gas, deviations from ideal gas behavior tend to be minimal under normal conditions. However, under extreme conditions (eg, very high pressures, very low temperatures) the ideal gas model does not accurately predict the behavior of a gas. Such cases are addressed later in this lesson.

6.3.01 The Ideal Gas Law

A single formula for describing the behavior of an ideal gas under a given set of conditions is derived by combining the relationships described by Boyle's law, Charles's law, Gay-Lussac's law, and Avogadro's law.

This **ideal gas law** relates pressure (P), volume (V), moles (n), and absolute temperature (T) with the universal gas constant (R), which is derived from experimental values (Figure 6.11).

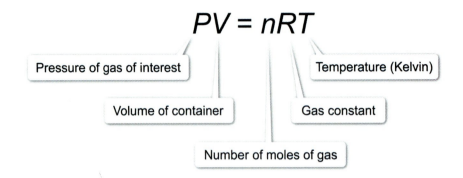

Figure 6.11 The ideal gas law and its variables.

This relationship provides a useful model to approximate the behavior of real gases. However, in calculations, it is important to keep the units of measurement consistent with those of the universal gas constant (R), either by converting the given units to match R or by using a value of R that corresponds with the given units. The value of R may differ depending on the units used to express it:

$$R = 0.08206 \; \frac{L \cdot atm}{K \cdot mol} = 8.314 \; \frac{m^3 \cdot Pa}{K \cdot mol} = 8.314 \; \frac{L \cdot kPa}{K \cdot mol} = 62.36 \; \frac{L \cdot torr}{K \cdot mol}$$

The ideal gas law can be altered from its original form to solve for certain quantities more easily. Substituting the ratio of mass (m) over molar mass (M) in place of moles (n) yields:

$$PV = \left(\frac{m}{M}\right) RT \quad \Rightarrow \quad PVM = mRT$$

This equation can be used to solve for the molar mass of an unknown gas. Further rearrangement allows for the density (D) of a gas to be easily calculated:

$$PM = \left(\frac{m}{V}\right) RT \quad \Rightarrow \quad PM = DRT$$

Other methods using the standard form of the ideal gas law may be used to solve for molar mass or the density of a gas, but those methods require the use of additional steps or assumptions.

 Concept Check 6.7

A closed 750 mL flask is charged with 2.20 g of CO_2 at 25 °C. What pressure (in atm) is present inside the flask?

Solution

Note: The appendix contains the answer.

6.3.02 Molar Volume at STP

Because gases can be evaluated under a range of conditions, a **standard temperature and pressure (STP)** is useful as a reference when comparing different gas samples. Although STP is officially defined as a temperature of 273 K and a pressure of 1 bar (1 bar = 0.987 atm), for convenience, STP is often stated simply as 0 °C (273 K) and exactly 1.00 atm of pressure. Under STP, 1 mol of any gas occupies a volume of 22.4 L (Figure 6.12).

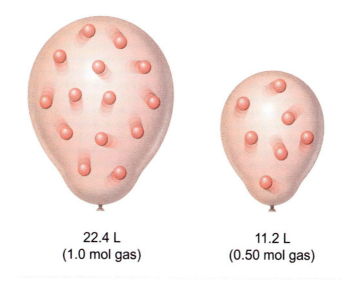

Figure 6.12 Molar volume of a gas at STP.

This molar volume assumes ideal gas behavior and is derived from the ideal gas law by assuming 1.00 mol of a gas at STP:

$$P = 1.00 \text{ atm}$$
$$V = ?$$
$$n = 1.00 \text{ mol}$$
$$R = 0.08206 \, \frac{\text{L} \cdot \text{atm}}{\text{K} \cdot \text{mol}}$$
$$T = 273 \text{ K}$$

Rearranging the ideal gas law to solve for volume and plugging in these values gives the STP molar volume:

$$V = \frac{nRT}{P} = \frac{(1.00 \text{ mol})\left(0.08206 \, \frac{\text{L} \cdot \text{atm}}{\text{K} \cdot \text{mol}}\right)(273 \text{ K})}{1.00 \text{ atm}} = 22.4 \text{ L}$$

Often memorized as a conversion factor (ie, 22.4 L/mol), the molar volume is also a useful reference for other systems at similar pressures and temperatures. This is because relatively small temperature changes have little effect on molar volume. The molar volume at STP can also be a good starting point for making rough estimates of gas volumes under different temperatures and pressures.

> **Concept Check 6.8**
>
> A sample of 132 g NaN₃ in an automobile airbag is ignited following a car crash according to the following reaction:
>
> $$2\,NaN_3(s) \rightarrow 2\,Na(s) + 3\,N_2(g)$$
>
> What is the volume of gas produced inside the airbag at STP?
>
> **Solution**
>
> *Note: The appendix contains the answer.*

6.3.03 Deviations from Ideal Gas Behavior

Although the ideal gas model is effective at modeling the behaviors of gases under most conditions, very high pressures and/or very low temperatures can cause key assumptions made by the ideal gas model to no longer hold true.

At high pressures, a gas can be compressed to a point where the volume of the gas particles themselves occupy a significant portion of the volume of the container. An increased gas particle density also results in a greater likelihood that gas particles interact with each other in an inelastic way because the particles are pushed more closely together (Figure 6.13).

- High temperature
- Low pressure
- Low gas particle density

- Low temperature
- High pressure
- High gas particle density

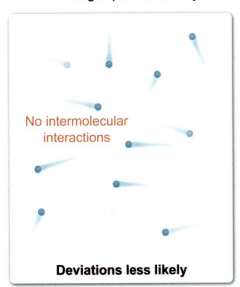
No intermolecular interactions

Deviations less likely

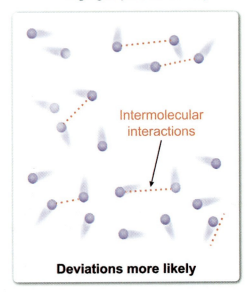
Intermolecular interactions

Deviations more likely

Figure 6.13 Factors contributing to the deviation of a gas from ideal behavior.

Likewise, at low temperatures, particles move slowly enough that attractive and repulsive intermolecular forces between particles can alter the movement of the particles and increase the inelasticity of collisions. In some cases, the gas particles may even begin to condense into a liquid.

One way to evaluate the degree to which a gas deviates from ideal behavior is to calculate the ratio of $\frac{PV}{nRT}$, a parameter also known as **compressibility**. An ideal gas that perfectly follows the ideal gas law has a compressibility value of exactly 1, but gases that deviate from ideal behavior can have different compressibility values. Higher or lower values indicate that the measured volume or pressure of a gas does not exactly match that predicted for an ideal gas (Figure 6.14).

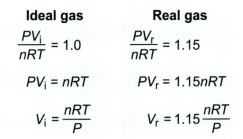

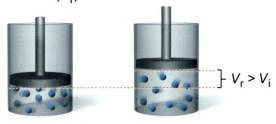

Molecular volume occupies a large fraction of the total volume and adds a small amount of noncompressible volume.

$$V_i < V_r$$

Figure 6.14 Ideal gas volume (V_i) compared to real gas volume (V_r) as derived from the compressibility of 1 mol of gas under the same temperature and pressure.

At the same temperature and pressure, gases with stronger intermolecular forces deviate more from ideal behavior than gases with weaker intermolecular forces. For example, a gas composed of polar NH_3 molecules will occupy a smaller volume than a gas composed of nonpolar N_2 molecules under non-ideal conditions. This is because the strong hydrogen bonding interactions between the NH_3 molecules reduce the elasticity of collisions, inhibiting their movement and bringing them closer together. In contrast, the weaker London dispersion forces between N_2 molecules inhibit their movement to a much lesser degree.

The influence of intermolecular attractions and the nonzero volume of gas molecules are often both contributing factors in the deviation of a gas from ideal behavior, but these factors have competing effects in terms of compressibility (Figure 6.15).

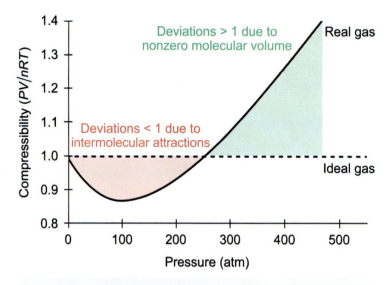

A deviation > 1 at the same temperature and pressure requires a larger real volume compared to the ideal volume.

Figure 6.15 Effects of pressure on compressibility of a real gas vs. an ideal gas at 200 K.

As indicated in Figure 6.15, when intermolecular attractions are the dominant factor in the deviation from ideal gas behavior, the compressibility is less than 1. In contrast, when the contribution from the nonzero molecular volume is the dominant factor, the compressibility is greater than 1.

For cases with significant deviation from ideal gas behavior, the ideal gas law is no longer usable. However, a modified version of the ideal gas law called the **van der Waals equation** accounts (though imperfectly) for the volume of gas particles and the effect of interactions between gas particles. This equation is typically written as:

$$\left(P + \frac{n^2 a}{V^2}\right)(V - nb) = nRT$$

where $\frac{n^2 a}{V^2}$ is a pressure correction term that accounts for the attractive forces between gas particles and nb is a volume correction term that accounts for the volume occupied by the gas particles. The values a and b are van der Waal's constants, which are empirical values different for every gas and must be either experimentally determined or obtained from references.

Although solving the van der Waals equation on the exam is unlikely, a conceptual familiarity with the equation can be useful in understanding the limitations of the ideal gas model.

Concept Check 6.9

Three separate containers containing gas samples of Ne, O_2, and CH_3F are all kept at 220 K and 8.0 atm.

1. Which of these samples would be expected to show the most deviation from ideal behavior?
2. Would the volume of that gas be greater than or less than the volume predicted by the ideal gas law?

Solution

Note: The appendix contains the answer.

END-OF-UNIT MCAT PRACTICE

Congratulations on completing **Unit 3: Thermodynamics, Kinetics, and Gas Laws**.

Now you are ready to dive into MCAT-level practice tests. At UWorld, we believe students will be fully prepared to ace the MCAT when they practice with high-quality questions in a realistic testing environment.

The UWorld Qbank will test you on questions that are fully representative of the AAMC MCAT syllabus. In addition, our MCAT-like questions are accompanied by in-depth explanations with exceptional visual aids that will help you better retain difficult MCAT concepts.

TO START YOUR MCAT PRACTICE, PROCEED AS FOLLOWS:

1) Sign up to purchase the UWorld MCAT Qbank
   IMPORTANT: You already have access if you purchased a bundled subscription.
2) Log in to your UWorld MCAT account
3) Access the MCAT Qbank section
4) Select this unit in the Qbank
5) Create a custom practice test

Unit 4 Solutions and Electrochemistry

Chapter 7 Solutions

7.1 Solutes in Solution

    7.1.01   Solution Formation
    7.1.02   Units of Concentration
    7.1.03   Osmosis

7.2 Solubility of Solutes

    7.2.01   Solubility Measurements
    7.2.02   Solubility Product Constant
    7.2.03   Common-Ion Effect
    7.2.04   Solubility and pH
    7.2.05   Complex Ion Formation
    7.2.06   Gases in Solution (Henry's Law)

Chapter 8 Acid-Base Chemistry

8.1 Definitions of Acids and Bases

    8.1.01   Arrhenius Acids and Bases
    8.1.02   Brønsted-Lowry Acids and Bases
    8.1.03   Lewis Acids and Bases

8.2 Acids and Bases in Aqueous Solution

    8.2.01   Ionization of Water
    8.2.02   The pH and pOH Scales

8.3 Strengths of Acids and Bases

    8.3.01   Relative Strengths of Acids and Bases
    8.3.02   Weak Acid Equilibria
    8.3.03   Polyprotic Acids
    8.3.04   Weak Base Equilibria

8.4 Buffers

    8.4.01   Characteristics and Behavior of Buffer Systems
    8.4.02   Henderson-Hasselbalch Equation
    8.4.03   Formulation of Buffers and Buffering Range

8.5 Hydrolysis of Salts

    8.5.01   Qualitative Analysis of the Acid-Base Properties of Salts
    8.5.02   pH Calculations for Hydrolysis of Salts

8.6 Acid-Base Titrations

    8.6.01   Titration Method
    8.6.02   Indicators
    8.6.03   Interpreting Titration Curves
    8.6.04   Titration Calculations

Chapter 9 Redox Reactions and Electrochemistry

9.1 Review of Redox Reactions

    9.1.01   Important Principles Associated with Redox Reactions

9.2 Redox Titrations

    9.2.01   Titrations Involving Redox Reactions

9.3 Electrochemical Cells

    9.3.01   Cell Components and Their Functions
    9.3.02   Redox Processes at the Electrodes
    9.3.03   Standard Reduction Potentials and Cell Potential
    9.3.04   Gibbs Free Energy of an Electrochemical Cell

9.4 Types of Electrochemical Cells

    9.4.01   Galvanic Cells
    9.4.02   Concentration Cells
    9.4.03   Electrolytic Cells

9.5 Applications of Electrochemical Cells

    9.5.01   Electrolysis
    9.5.02   Batteries

Lesson 7.1

Solutes in Solution

Introduction

A **solution** is a mixture of chemical substances comingled at the molecular level into a single, homogeneous phase (ie, a solid, liquid, or gas state with a uniform composition). The substance that makes up the majority of the mixture (ie, the substance that dissolves the others) is called the **solvent**, and the substances present in smaller amounts (ie, the substances that are dissolved) are called **solutes**. An example of a homogeneous solution is shown in Figure 7.1.

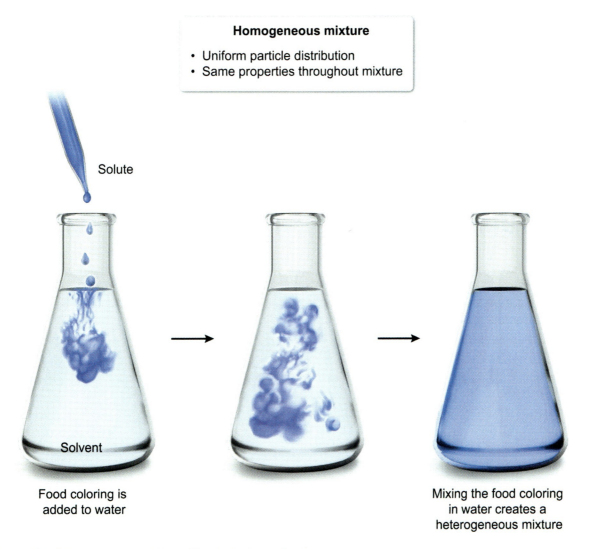

Figure 7.1 A homogeneous mixture of food coloring and water.

Although many solutions are formed by dissolving a solid solute into a liquid solvent (eg, salt dissolved in water), solutions can be formed in a variety of other ways. Some examples are:
- Gas dissolved in another gas (eg, oxygen mixed with atmospheric nitrogen)
- Gas dissolved in a liquid (eg, carbon dioxide in seltzer water)

- Gas dissolved in a solid (eg, hydrogen absorbed by a palladium metal catalyst)
- Liquid dissolved in another liquid (eg, alcohol in water)
- Solid dissolved in a liquid (eg, silver in mercury for a dental filling)
- Solid melted in another solid (eg, metal alloys such as the copper-zinc mixture in brass)

As shown with these examples, the phases of the solute and solvent do not necessarily have to be the same, but the phase of the resulting solution is usually the same as the phase of the solvent because the solvent is the primary component of the mixture.

Forming a solution depends on the ability of the solute and solvent species to interact favorably and form mutual intermolecular attractions. Any such interactions are strongly affected by the conditions of the solvent system (eg, temperature, pressure). As such, a solute's **solubility** (ie, the maximum amount of a solute that can dissolve into a given amount of solvent) is determined by the nature and extent of the intermolecular forces that can form between the solute and solvent species.

Because the amounts of the solutes in a solution can vary, the properties of a solution depend on its composition (ie, how much of each solute is present). Accordingly, the vapor pressure, boiling point, freezing point, and osmotic pressure of a solution are **colligative properties** (ie, depending only on the concentration of solute and not its identity). As such, it is crucial to precisely describe the composition of any solution through a variety of calculations that quantify the relative amounts of solute and solvent in a solution mixture.

This lesson examines the molecular processes involved in solution formation, summarizes the common methods of calculating the composition of a solution, and explores the colligative property of osmosis (ie, diffusion across a semipermeable membrane).

7.1.01 Solution Formation

For one substance to dissolve into another substance and form a solution, the **solute** must be able to form enough complementary intermolecular attractions with the **solvent**. As such, the following simultaneous dissolution steps must occur:

- **Solvent-solvent interactions** are broken. Some of the attractions between solvent molecules must be broken so that new interactions with the solute can begin to form.
- **Solute-solute interactions** are broken. All attractions between solute molecules must be broken so that the solute molecules can be dispersed throughout the solution.
- **Solute-solvent interactions** are formed. New attractions between solute and solvent must be made to carry the solute molecules into the solution.

These dissolution steps are illustrated in Figure 7.2 for an unspecified ionic compound (solute) dissolved into water (solvent).

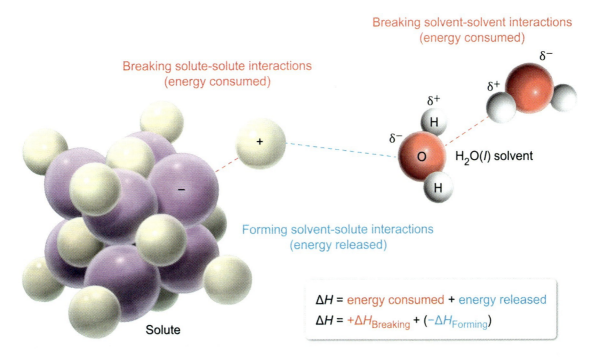

Figure 7.2 Dissolution steps in the process of making of a solution of an ionic compound.

In the example shown in Figure 7.2, the solute-solute interactions (ie, electrostatic attraction between the oppositely charged ions) and the solvent-solvent interactions (ie, dipole-dipole attraction between the partial charges of the polar water molecules) are broken as new solute-solvent interactions (ie, ion-dipole attractions) are formed. In this process, energy must be consumed by the molecules to break the initial attractions ($\Delta H_{breaking} > 0$), but energy is released by the molecules when forming new attractive interactions ($\Delta H_{forming} < 0$).

Consequently, the **enthalpy of dissolution** (ΔH) associated with forming the solution is equal to the net change in energy for the overall process:

$$\Delta H = \Delta H_{breaking} + (-\Delta H_{forming})$$

If the amount of energy released from forming new interactions is greater than the energy consumed to break the existing interactions, then the overall dissolution process is exothermic with a net release of energy ($\Delta H < 0$) that raises the temperature of the solution. Conversely, if more energy is consumed than released, the overall dissolution process is endothermic with a net intake of energy ($\Delta H > 0$) that cools the resulting solution.

As the new solute-solvent interactions form during the dissolution of the solute, each solute molecule becomes solvated (ie, surrounded by a group of attracted solvent molecules called the solvation shell). As a result, the solute particles become dispersed throughout the solution, as shown in Figure 7.3.

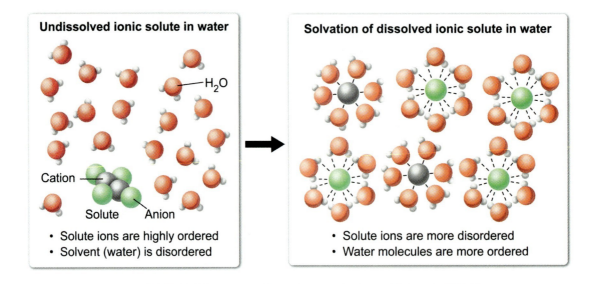

The change in entropy due to dissolution is equal to the net result of the increase in the solute entropy and the decrease in the solvent entropy

Figure 7.3 Solvation of solute ions during solution formation and the resulting effects on entropy.

These dissolution and solvation processes affect the entropy ΔS (ie, the extent of disorder) of the system because the arrangements of the molecules are changed. In many cases, the dissolution process results in a net increase in entropy ($+\Delta S$) because dispersing a highly ordered solid introduces a significant amount of disorder. However, in some cases the solvent molecules may become more ordered (eg, forming solvation shells with solute particles). As such, if the decrease in entropy in the solvent is greater than the increase in entropy in the solute, a net decrease in entropy ($-\Delta S$) can sometimes occur during dissolution.

Concept Check 7.1

Suppose that a solution is prepared by dissolving a sample of KBr(s) in 1.0 L of water. Which of the following aspects of the dissolution process contribute to an increase in entropy? (Select all that apply.)

    I.   Separating the K^+ and Br^- ions in the water.

   II. Dispersing groups of water molecules to accommodate the K^+ and Br^- ions.

   III. Forming a solvation shell of water molecules around the K^+ and Br^- ions.

Solution

Note: The appendix contains the answer.

7.1.02 Units of Concentration

The concentration of a solution is a measurement of the amount of solute (ie, a dissolved compound) present in a given amount of the solution mixture. However, different units can be used to measure the relative amounts of each component in a solution, and the best choice of which units to use depends on the context of the measurements.

When relating the concentration of a solution to a chemical reaction or to the number of solute particles, the concentration is usually best expressed in molarity, which gives the moles of solute per liter of solution:

$$\text{Molarity (M)} = \frac{\text{Moles of solute}}{\text{Solution volume (in liters)}}$$

Using molarity facilitates the determination of the number of moles of solute present in a given solution volume (see Concept 1.6.08). The number of moles of solute can be readily related to other chemical species in a reaction by the stoichiometric coefficients of a balanced reaction equation, as discussed in Concepts 2.5.01 and 2.5.02.

However, in some biological and environmental contexts the bulk amounts of substances in a solution mixture may be more useful than the number of moles of solute. In these cases, expressing solute concentration in terms of the mass of solute per unit volume of solution (eg, g/L, mg/mL) is often more convenient. This mass-to-volume ratio can be used as a conversion factor to determine the mass of solute in a given amount of solution, or the volume of a solution that contains a specified amount of solute mass.

If the amounts of solute and solution are both measured in the same units of mass, the solute concentration can be expressed as a **mass percent (% m/m)**:

$$\text{Mass percent} = \frac{\text{Mass of solute}}{\text{Mass of solution}} \times 100 = \frac{\text{Mass of solute}}{(\text{Mass of solute} + \text{Mass of solvent})} \times 100$$

For very dilute solutions with extremely low concentrations, solute concentrations may be alternatively expressed in **parts per million (ppm)**, which uses a larger one-million-unit basis (ie, 1,000,000 = 10^6) instead of the smaller one-hundred-unit basis used in percentages:

$$\text{ppm} = \frac{\text{Mass of solute}}{\text{Mass of solution}} \times 10^6 = \frac{\text{Mass of solute}}{(\text{Mass of solute} + \text{Mass of solvent})} \times 10^6$$

Because the units in the numerator and denominator of the fraction are the same (or can be converted to an equivalent unit by a conversion factor), the units cancel and ppm (like percent) becomes a **dimensionless quantity**. As such, any mass fraction with a numerator one million times smaller than the denominator is equivalent to ppm. Accordingly, ppm can be expressed as mg/kg when dealing with large amounts of mass:

$$\text{ppm} = \frac{\text{Milligrams of solute}}{\text{Kilogram of solution}}$$

Because water has a density near 1.0 g/mL across a wide temperature range, 1,000 mL (1 L) of water has a mass near 1 kg:

$$1{,}000 \text{ mL } H_2O \times \frac{1.0 \text{ g } H_2O}{1 \text{ mL } H_2O} \times \frac{1 \text{ kg}}{1{,}000 \text{ g}} = 1 \text{ kg } H_2O$$

As a result, a solute concentration of 1 ppm is also approximately equivalent to 1 mg/L in dilute aqueous solutions:

$$1 \text{ ppm} \approx \frac{1 \text{ mg solute}}{1 \text{ kg solution}} \times \frac{1 \text{ kg}}{1{,}000 \text{ g}} \times \frac{1.0 \text{ g } H_2O}{1 \text{ mL } H_2O} \times \frac{1{,}000 \text{ mL}}{1 \text{ L}} \approx \frac{1 \text{ mg solute}}{1 \text{ L solution}}$$

The relationship of 1 ppm = 1 mg/L also applies to gas mixtures contained in a fixed volume. Therefore, ppm can be approximated as:

$$\text{ppm} \approx \frac{\text{Milligrams of solute}}{\text{Liter of solution}}$$

The various units of concentration discussed here represent a few of the ways that the composition of a solution can be quantified. Conversions between these units can be accomplished by applying the appropriate metric conversion factors and the molar mass of the solute (when needed).

> ☑ **Concept Check 7.2**
>
> A gas solution (gas mixture) held in a rigid 0.50 L vessel at 27 °C contains 0.040 g of He(g) blended into 0.14 g N_2(g) with a total pressure of 0.98 atm. What is the concentration of He(g) expressed as molarity and mass percent?
>
> **Solution**
>
> *Note: The appendix contains the answer.*

7.1.03 Osmosis

When two solutions with different solute concentrations are separated by a **semipermeable membrane** (ie, an exclusionary barrier that allows only the solvent or other select chemical species to pass), the solvent from the solution of lower solute concentration diffuses across the membrane into the solution of higher solute concentration until the solution concentrations are equal. This behavior, which is illustrated in Figure 7.4, is called **osmosis**.

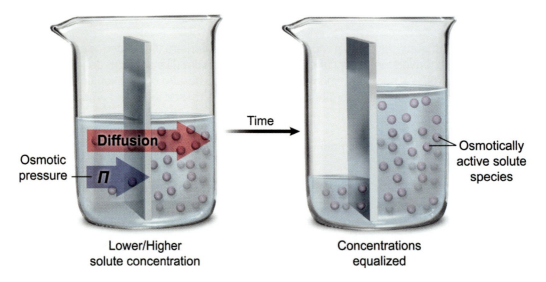

Figure 7.4 Osmosis across a semipermeable membrane.

Dissolved ions or molecules that cannot cross the membrane are **osmotically active** because they contribute to the concentration difference that induces **solvent diffusion** through the membrane. The diffusion of the solvent during osmosis produces an **osmotic pressure** against the membrane, which is calculated for an ideal solution by the relationship:

$$\Pi = iMRT$$

where Π is the osmotic pressure, i is the **van 't Hoff factor** (ie, the number of osmotically active species a molecule yields when dissolved in solution), M is the molarity (ie, molar concentration) of the solute, R is the ideal gas constant (0.0821 L·atm/mol·K), and T is the solution temperature in Kelvin. As such, the osmotic pressure is proportional to the molar concentration of osmotically active solute species (ie, the **osmolarity**), which is given by iM:

$$\text{Osmolarity} = iM = \frac{RT}{\Pi}$$

The molarity M of the bulk solute is not necessarily equivalent to the osmolarity because some solutes can dissociate into ions. For example, NaCl dissociates into Na^+ and Cl^- ions in water, which causes its osmolarity to be twice the NaCl concentration. As such, the van 't Hoff factor i accounts for the number of moles of osmotically active ions or molecules (osmoles) in the solution.

In complex solutions with mixtures of different solutes, determining the effective value of i for the solute mixture can be complicated, and i may be estimated or evaluated directly from the measured solution osmolarity instead.

 Concept Check 7.3

What are the osmolarity and osmotic pressure of 0.10 M solution of $(NH_4)_2SO_4(aq)$ at 27 °C, given that this ionic compound fully dissociates into $NH_4^+(aq)$ and $SO_4^{2-}(aq)$ ions in water?

Solution

Note: The appendix contains the answer.

Lesson 7.2

Solubility of Solutes

Introduction

When a solute is placed in a solvent, the intermolecular attractions formed between the solute and solvent molecules establish an equilibrium between the dissociation of the solute (ie, the forward process) and the precipitation of the solute (ie, the reverse process). The position of this equilibrium, indicated by the equilibrium constant, is related to the maximum concentrations of the solute ions that can dissolve in the solvent.

According to **Le Châtelier's principle**, an equilibrium can be induced to shift in either the forward or the reverse direction by applying a stress to the system, which causes the system to adjust to relieve the stress, as shown in Figure 7.5.

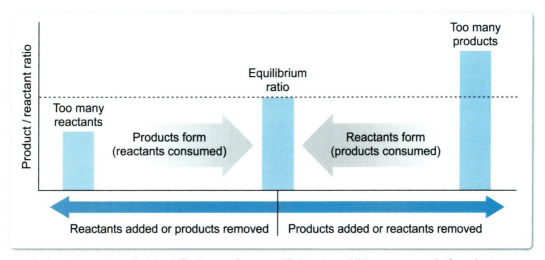

A chemical system that is shifted away from equilibrium by addition or removal of products or reactants will adjust product and reactant levels to establish a new equilibrium.

Figure 7.5 Applying Le Châtelier's principle to influence an equilibrium by adding or removing products.

As such, the solubility equilibrium of a solute-solvent system can be affected by a variety of stresses such as changes in pressure, changes in temperature, and the addition or removal of chemical species within the system. Consequently, these factors can either increase or decrease the solubility of a solute relative to the unstressed system. The most influential type of stress depends on the properties of the solute and solvent.

Applying stress by changing the relative amounts of chemical species in solution is typically achieved by introducing another substance that causes some type of interaction with the dissolved solute. Accordingly, the chemical properties of the solute play a central role in this type of process. This lesson explores solubility equilibria and examines some common ways by which stress can be applied to affect an equilibrium and alter the solubility of a solute.

7.2.01 Solubility Measurements

Solubility refers to the maximum amount of a **solute** (ie, a dissolved compound) that can dissolve in a given amount of a **solvent** (ie, the solution medium) at a given temperature under a given pressure.

Solubilities can be expressed in terms of mass per unit volume (eg, g/L, g/cm^3, mg/mL), moles per unit volume (ie, molar solubility), or as a mass ratio (ie, the grams of solute that will dissolve per 100 g of solvent).

Because the solubility of a solute changes with temperature, reported solubility measurements should always state the temperature at which the measurement was made. Temperatures of 20 or 25 °C are typically used as convenient values for "room temperature" comparisons. Table 7.1 gives an example of a solubility table for some selected ionic compounds dissolved in water.

Table 7.1 Solubility of selected ionic compounds in water at 20 °C.

Compound	Solubility (g/100 g of H_2O)
KBr	65.3
AgCl	0.000192
Na_2SO_4	19.5
$BaSO_4$	0.000245
$CaCO_3$	0.000617
Na_2CO_3	21.5
$PbBr_2$	0.86

Solubility measurements like those in Table 7.1 can be expressed as solute-solvent ratios that are useful for calculations. For example, the solubility of KBr from Table 7.1 can be expressed as:

$$\frac{65.3 \text{ g KBr}}{100 \text{ g H}_2\text{O}}$$

If alternate units are needed for a calculation, unit conversions can be applied to the relevant units in the ratio, as discussed in Concept 7.1.02. For example, the solubility of KBr can be expressed as molar solubility using the density of water at 20 °C (1.0 g/mL) and the molar mass of KBr (119 g/mol):

$$\frac{65.3 \text{ g KBr}}{100 \text{ g H}_2\text{O}} \times \frac{1.0 \text{ g H}_2\text{O}}{1 \text{ mL H}_2\text{O}} = \frac{65.3 \text{ g KBr}}{100 \text{ mL H}_2\text{O}}$$

$$\frac{65.3 \text{ g KBr}}{100 \text{ mL H}_2\text{O}} \times \frac{1000 \text{ mL}}{1 \text{ L}} \times \frac{1 \text{ mol KBr}}{119 \text{ g KBr}} = 5.49 \frac{\text{mol}}{\text{L}}$$

When the amount of solute added to a given amount of solvent results in a solute-solvent ratio less than the solubility limit, the solute fully dissolves and the solution is said to be **unsaturated**. Consequently, additional amounts of solute added to an unsaturated solution will dissolve until the solubility limit is reached. Once the total amount of solute added results in a solute-solvent ratio equal to or greater than the solubility limit, a **saturated solution** is formed and any excess solute will not dissolve.

$$\boxed{\text{Unsaturated}} \quad \boxed{\text{Solubility limit}} \quad \boxed{\text{Saturated}}$$
$$\frac{40.2 \text{ g KBr}}{100 \text{ g H}_2\text{O}} < \frac{65.3 \text{ g KBr}}{100 \text{ g H}_2\text{O}} < \frac{76.5 \text{ g KBr}}{100 \text{ g H}_2\text{O}}$$
$$\text{(No excess)} \quad \text{(No excess)} \quad \text{(Excess)}$$

In some situations, it is necessary to know only if a particular solute has a solubility greater than or equal to some specified solubility threshold. Following this type of scheme permits solutes to be roughly classified as either "soluble" or "insoluble" in a solvent. For example, an ionic compound with a solubility greater than 0.1 g per 100 g of water can be considered "soluble" in water. Using this arbitrary solubility threshold permits the ionic compounds to be categorized, as shown in Table 7.2.

Table 7.2 General categorization of the solubility of ionic compounds in water at 20 °C.

	Compounds containing		Exceptions	
Always soluble	Li^+, Na^+, K^+, Rb^+, Cs^+ $C_2H_3O_2^-$, CH_3COO^- NH_4^+ NO_3^-	(alkali ions) (acetate) (ammonium) (nitrate)	None None None None	Note: "Soluble" in this table means > 0.1 g of the compound will dissolve in 100 g H_2O at 20° C.
Generally soluble	Cl^-, Br^-, I^- F^- SO_4^{2-}		Ag^+, Cu^+, Pb^{2+} Mg^{2+}, Ca^{2+}, Sr^{2+}, Ba^{2+}, Pb^{2+} Ag^+, Ca^{2+}, Sr^{2+}, Ba^{2+}, Pb^{2+}	
Generally insoluble	S^{2-} OH^- PO_4^{3-} CO_3^{2-}		Mg^{2+}, Ca^{2+}, Sr^{2+}, Ba^{2+}, NH_4^+, alkali ions Ca^{2+}, Sr^{2+}, Ba^{2+}, NH_4^+, alkali ions Li^+, Na^+, K^+, Rb^+, Cs^+, NH_4^+ Li^+, Na^+, K^+, Rb^+, Cs^+, NH_4^+	

The solubility guidelines given in Table 7.2 assume that an ionic compound with a solubility less than 0.1 g/100 g H₂O is effectively "insoluble" in water. Such broad classifications can be convenient for relative comparisons when an exact measurement of solubility is not needed; however, even "insoluble" compounds may dissolve to some very small, but measurable, extent.

Although memorizing the information in Table 7.2 is unnecessary, it is useful to remember the ions of the top row because the numerous ionic salts containing these ions are always appreciably soluble in water.

 Concept Check 7.4

Aqueous solutions of KI and Pb(NO₃)₂ are mixed, resulting in the double replacement reaction represented by the equation shown.

$$2\ KI + Pb(NO_3)_2 \rightarrow PbI_2 + 2\ KNO_3$$

During the reaction, a solid precipitate is formed. Which compound is the precipitate?

Solution

Note: The appendix contains the answer.

7.2.02 Solubility Product Constant

For a sample of an ionic compound Z that dissociates (either partially or completely) into ions X and Y when placed in water, an equilibrium is established in which the rate of dissolution into ions (ie, the forward reaction) is equal to the rate of precipitation (ie, the reverse reaction). This equilibrium is represented by:

$$aZ(s) \overset{K_{sp}}{\rightleftarrows} bX + cY$$

where a, b, and c are the stoichiometric coefficients of each species. This equilibrium position is quantified by the **solubility product constant K_{sp}**, which indicates the solubility of a compound at a given temperature under 1 atm of pressure. The general K_{sp} expression is written as:

$$K_{sp} = [X]^b[Y]^c$$

where [X] and [Y] are the molar concentrations of the X and Y ions, each raised to the power of their respective coefficients b and c from the balanced dissolution reaction. The K_{sp} expression does not include a denominator involving [Z] because the reactant is a solid, and the amount of undissolved solid present does not affect the equilibrium. Essentially, the K_{sp} value is a product of the maximum possible concentrations of each ion that can be present in the solution under the given conditions.

As illustrated in Figure 7.6, three cases are possible for the concentrations of [X] and [Y] relative to K_{sp}:

- $[X]^b[Y]^c < K_{sp}$. The solution has not reached the limit of solubility, and more Z(s) will dissolve.
- $[X]^b[Y]^c = K_{sp}$. The solution has reached the limit of solubility and is at equilibrium.
- $[X]^b[Y]^c > K_{sp}$. The solution has exceeded the solubility limit, and no more Z(s) will dissolve (ie, any ions in excess of K_{sp} will precipitate).

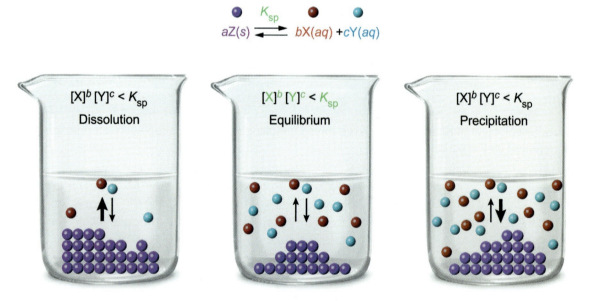

Figure 7.6 Three possible cases for the dissolved ion concentrations [X] and [Y] relative to K_{sp}.

Concept Check 7.5

Magnesium perchlorate salt dissolves in water at 20 °C according to the solubility equilibrium represented here.

$$Mg(ClO_4)_2(s) \rightleftharpoons Mg^{2+}(aq) + 2\ ClO_4^-(aq) \qquad K_{sp} = 43.9$$

Based on the K_{sp} value, is it possible to fully dissolve enough $Mg(ClO_4)_2(s)$ to make a 3 M aqueous solution of the salt at 20 °C?

Solution

Note: The appendix contains the answer.

For compounds that produce the same number of ions per formula unit, K_{sp} values can be directly compared to determine relative solubilities (ie, a larger K_{sp} corresponds to greater solubility, and vice versa). For example, Table 7.3 shows that $MnCO_3$ and CuI both dissociate into two ions per formula unit but $MnCO_3$ is more soluble than CuI, which is verified by the fact that K_{sp} for $MnCO_3$ is larger than K_{sp} for CuI.

Table 7.3 Solubility parameters of selected ionic compounds at 20 °C.

Compound	Solubility equilibrium	K_{sp}	Solubility (g/100 g H_2O)
CuI	$CuI(s) \rightleftharpoons Cu^+(aq) + I^-(aq)$	$[Cu^+][I^-] = 1.1 \times 10^{-12}$	2.0×10^{-5}
$MnCO_3$	$MnCO_3(s) \rightleftharpoons Mn^{2+}(aq) + CO_3^{2-}(aq)$	$[Mn^{2+}][CO_3^{2-}] = 1.8 \times 10^{-11}$	4.9×10^{-5}
$Fe(OH)_2$	$Fe(OH)_2(s) \rightleftharpoons Fe^{2+}(aq) + 2\ OH^-(aq)$	$[Fe^{2+}][OH^-]^2 = 8.2 \times 10^{-16}$	5.3×10^{-5}

Although K_{sp} is an indicator of the solubility of a compound, the K_{sp} value is only a product of the ion concentrations and is not a direct measurement of solubility. Consequently, the K_{sp} values of compounds that produce *different* numbers of ions per dissolved formula unit *cannot* be directly compared to evaluate relative solubilities. For example, Table 7.3 shows that K_{sp} for $Fe(OH)_2$ is smaller than K_{sp} for $MnCO_3$, but $Fe(OH)_2$ has a greater solubility than $MnCO_3$.

The greater solubility of $Fe(OH)_2$ in spite of its smaller K_{sp} is explained by the fact that $Fe(OH)_2$ dissociates into more ions per molecule than does $MnCO_3$. Because each $Fe(OH)_2$ formula unit produces two OH^- ions, $[OH^-]$ is squared when evaluating K_{sp}, which causes the resulting *value* to be smaller than it would be for an equally soluble compound that produces fewer ions. Larger exponents applied to values less than 1 yield smaller numbers.

Chapter 7: Solutions

 Concept Check 7.6

Selected data for the aqueous solubility of two ionic compounds at 20 °C are given in the table shown.

Compound	K_{sp}	Solubility (mol/L)
AgCl	1.8×10^{-10}	?
AgBr	?	7×10^{-7}

Based on the information in the table, what is the K_{sp} value of AgBr, and which of the two compounds is more soluble in water?

Solution

Note: The appendix contains the answer.

The effect of a change in temperature on Ksp depends on whether the dissolution of the solute is exothermic or endothermic. If the dissolution is exothermic (ie, releases heat as a "product"), raising the temperature (ie, adding heat) causes the equilibrium to shift toward the undissolved solute (ie, K_{sp} decreases) to compensate for the heat gain according to Le Châtelier's principle. Conversely, if the dissolution is endothermic (ie, consumes heat as a "reactant"), raising the temperature causes the equilibrium to shift toward the products (ie, K_{sp} increases), as summarized in Figure 7.7.

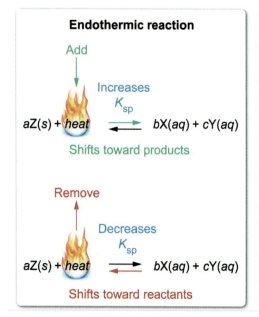
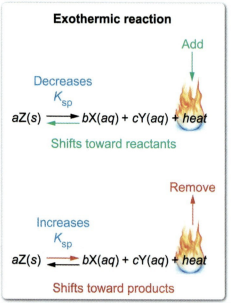

Figure 7.7 The effect of changing temperature on endothermic and exothermic solubility equilibria.

7.2.03 Common-Ion Effect

Le Châtelier's principle says that, if the equilibrium state of a system is disturbed, the system adjusts to counteract the disturbance and establishes a new equilibrium. As such, dissociation equilibria are affected by disruptions to ion concentrations. For example, suppose that a solution contains two ionic compounds (Z and M) that have an ion (X) in common and dissociate according to the following equilibria:

$$a\text{Z}(s) \underset{}{\overset{K_{sp1}}{\rightleftarrows}} b\text{X} + c\text{Y} \implies K_{sp1} = [\text{X}]^b[\text{Y}]^c$$

$$k\text{M}(s) \underset{}{\overset{K_{sp2}}{\rightleftarrows}} m\text{X} + n\text{L} \implies K_{sp2} = [\text{X}]^m[\text{L}]^n$$

where L is a component ion and k, m, and n are the stoichiometric coefficients for the dissolution of M. The common ion X supplied by one compound disturbs the product side of the equilibrium of the other compound, which causes the equilibrium to shift toward the reactants (ie, the undissolved solid).

As a result, both compounds decrease in solubility due to the common-ion effect, which states that if two ionic compounds with the same cation or anion (ie, a common ion) are dissolved together, the amounts of each that dissolve as a mixture are less than the amounts that would dissolve separately. Essentially, the additional amount of the common ion X from the second compound increases [X] beyond that released from only one compound. This causes the ion products to be greater than the respective K_{sp} values of each compound and induces greater precipitation, as illustrated in Figure 7.8.

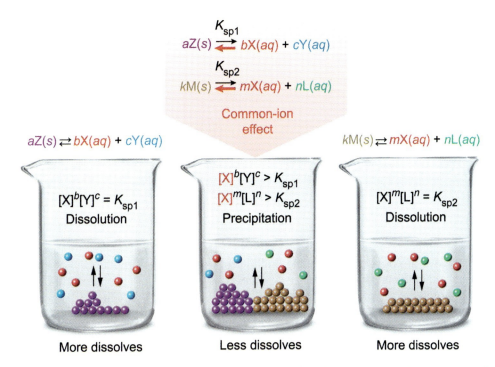

Figure 7.8 The common-ion effect decreases the amount that dissolves from a mixture of two compounds that contain a common ion.

Concept Check 7.7

Three identical samples of a saturated PbSO$_4$(aq) solution are treated with equal volumes of different 0.001 M salt solutions as shown, causing different amounts of PbSO$_4$(s) to precipitate in each flask. Rank the samples in order of how much precipitate will form, from least to most.

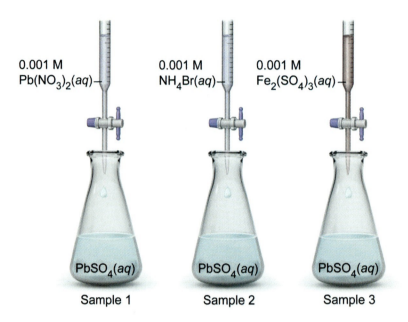

Solution

Note: The appendix contains the answer.

7.2.04 Solubility and pH

Although the common-ion effect is one way Le Châtelier's principle influences the solubility of a compound, other ways are possible depending on the properties of the compound. For example, if one or more of the dissolved ions can act as either an acid or a base, the solubility of the compound can be influenced by changing the pH of the solution.

Consequently, increasing the pH (ie, adding a base such as OH$^-$) increases the solubility of compounds with acidic components and decreases the solubility of compounds with basic components. Decreasing the pH (ie, adding an acid, H$^+$) has the opposite effect. Four possible cases of how changing the pH can affect solubility are summarized in Table 7.4.

Chapter 7: Solutions

Table 7.4 Summary of the effects of pH on the solubility of compounds with acidic or basic components.

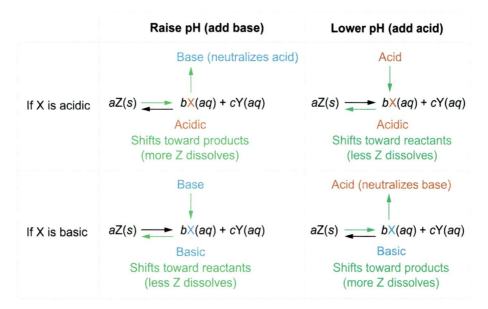

As a specific example of the influence of pH on the solubility of a compound, consider MgNH₄PO₄, which has very low solubility in distilled water (pH 7) and dissociates to a small extent according to the following equilibrium:

$$\text{MgNH}_4\text{PO}_4(s) \underset{}{\overset{K_{sp}}{\rightleftarrows}} \text{Mg}^{2+}(aq) + \text{NH}_4^+(aq) + \text{PO}_4^{3-}(aq) \implies K_{sp} = 2.5 \times 10^{-13}$$

However, the NH₄⁺(aq) ion in the equilibrium is acidic and can react with a base (eg, OH⁻) by donating an H⁺ ion according to the reaction:

$$\text{NH}_4^+(aq) + \text{OH}^-(aq) \rightleftarrows \text{NH}_3(aq) + \text{H}_2\text{O}(l)$$

Adding OH⁻ neutralizes some of the NH₄⁺(aq) and disrupts the equilibrium by decreasing the NH₄⁺(aq) concentration. As a result, the equilibrium shifts toward the products (ie, more MgNH₄PO₄ will dissolve) to replace the neutralized NH₄⁺(aq). Therefore, increasing the pH of the solution increases the amount of MgNH₄PO₄ that dissolves.

 Concept Check 7.8

The oxalate anion ($C_2O_4^{2-}$) acts as a weak base that can accept two H⁺ ions according to the reactions shown.

$$C_2O_4^{2-}(aq) + H^+(aq) \rightleftarrows HC_2O_4^-(aq)$$

$$HC_2O_4^-(aq) + H^+(aq) \rightleftarrows H_2C_2O_4(aq)$$

Lead(II) oxalate (PbC₂O₄) has very low solubility in distilled water. How can a sample of PbC₂O₄ be made more soluble in an aqueous solution at 20 °C?

Solution

Note: The appendix contains the answer.

7.2.05 Complex Ion Formation

As discussed in Concept 2.1.07, a metal cation can be bound to a complex ion by forming coordinate covalent bonds with one or more ligands. Because the solubility equilibrium of most ionic compounds involves the dissociation of a metal cation into solution, forming a soluble complex with the metal ion via the introduction of ligands is another way Le Châtelier's principle can be leveraged to increase the solubility of an ionic compound.

For example, AgBr has very low solubility in water and dissociates only to a small extent to establish the following equilibrium:

$$AgBr(s) \underset{}{\overset{K_{sp}}{\rightleftarrows}} Ag^+(aq) + Br^-(aq) \implies K_{sp} = 5.4 \times 10^{-13}$$

However, the Ag^+ ions readily form a highly soluble complex ion with thiosulfate ($S_2O_3^{2-}$) ligands:

$$Ag^+(aq) + 2\,S_2O_3^{2-}(aq) \underset{}{\overset{K_f}{\rightleftarrows}} [Ag(S_2O_3)_2]^{3-}(aq) \implies K_f = 2.9 \times 10^{13}$$

The formation constant K_f for the complex ion is significantly greater than K_{sp} for AgBr. Consequently, adding the $S_2O_3^{2-}$ ligands to the solution increases the amount of AgBr that dissolves. The formation of the soluble complex ion disrupts the solubility equilibrium of AgBr by decreasing the concentration of free $Ag^+(aq)$ ions. As a result, the equilibrium shifts toward the products (ie, the dissolved ions) to replace the $Ag^+(aq)$ ions bound by the added ligands (see Figure 7.9).

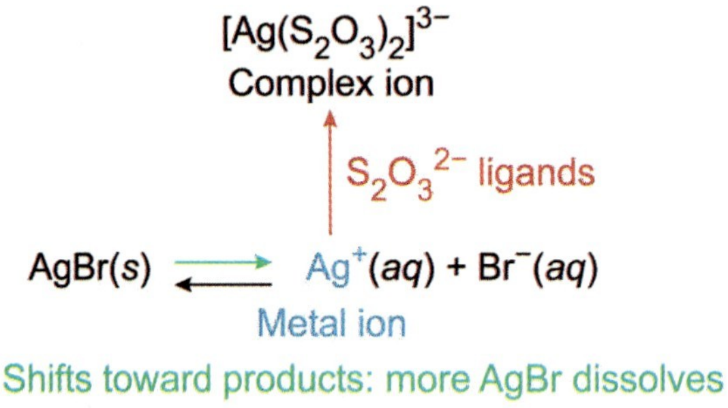

Figure 7.9 Effect of complex ion formation on the solubility equilibrium of an ionic compound.

Concept Check 7.9

Lead(II) thiosulfate (PbS_2O_3) has a low solubility in distilled water with an equilibrium of:

$$PbS_2O_3(s) \underset{}{\overset{K_{sp}}{\rightleftarrows}} Pb^{2+}(aq) + S_2O_3^{2-}(aq) \implies K_{sp} = 4.0 \times 10^{-7}$$

In contrast, the nitrate (NO_3^-) compounds $Pb(NO_3)_2$, $AgNO_3$, and HNO_3 (nitric acid) have high solubility in distilled water. If a dilute aqueous solution of one of these three nitrate compounds is added to the PbS_2O_3 equilibrium mixture, which addition would most significantly increase the solubility of the PbS_2O_3?

Solution

Note: The appendix contains the answer.

7.2.06 Gases in Solution (Henry's Law)

The solubility of a solute in a solvent depends on the intrinsic characteristics of the substances (ie, the ability and tendency of the two substances to form favorable interactions) and on the conditions (ie, the temperature and pressure) under which the substances interact. However, gas solutes dissolved in a liquid solvent respond very differently to changes in temperature and pressure than solid and liquid solutes dissolved in a liquid solvent.

As discussed in Concept 7.2.02, the effect of a change in temperature on solubility depends on whether the process of dissolving the solute (Concept 7.1.01) is exothermic or endothermic. For an exothermic process (ie, heat as a "product"), raising the temperature decreases the solubility of the solute according to Le Châtelier's principle. Conversely, for an endothermic process (ie, heat as a "reactant"), raising the temperature increases the solubility of the solute.

Dissolving solid and liquid solutes is usually an endothermic process because more energy is needed to overcome the attractions between the solute molecules than is released when forming new solute-solvent interactions (see Concept 7.1.01). As a result, increasing the temperature generally increases the solubility of solid and liquid solutes in a liquid solvent. Exceptions occur only in the uncommon cases in which breaking the solute-solute interactions releases energy.

In contrast, dissolving a gas solute into an inorganic liquid solvent (eg, water) is generally an exothermic process because gas molecules entering a liquid phase lose both entropy (ie, liquids are more ordered than gases) and kinetic energy (ie, liquid molecules move less freely and more slowly than gas molecules). As such, increasing the temperature *decreases* the solubility of gas solutes (the opposite of the effect for solid and liquid solutes) because the added thermal energy makes escaping the liquid phase easier for a gas particle. A comparison of the effect of temperature on solid and gas solutes is shown in Figure 7.10.

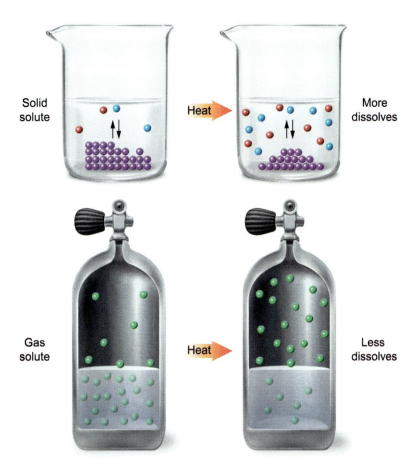

Figure 7.10 Typical effect of increasing temperature on the solubility of a solid solute and a gas solute.

Although pressure has a negligible effect on the solubilities of solid and liquid solutes in liquid solvents, the solubilities of gas solutes are significantly affected by pressure. Gas molecules above the surface of a solvent (eg, water) can dissolve into the solvent and form a solution that establishes an equilibrium involving the exchange of gas molecules across the gas-liquid interface.

For example, an unspecified gas X(g) dissolved into water to form a solution of X(aq) has a gas exchange equilibrium described by:

$$X(g) \overset{k_H}{\rightleftarrows} X(aq) \implies k_H = \frac{P_{X(g)}}{C_{X(aq)}}$$

The equilibrium constant k_H is equal to the ratio of the partial pressure of the gas $P_{X(g)}$ above the solution to the molar concentration $C_{X(aq)}$ (ie, the solubility) of the dissolved gas in the solution. Rearranging the k_H equation yields the mathematical expression for **Henry's law**, which states that the amount of a gas that will dissolve in a liquid at a given temperature is directly proportional to the partial pressure of that gas above the liquid:

$$C_{X(aq)} = k_H P_{X(g)}$$

An illustration of Henry's law is given in Figure 7.11.

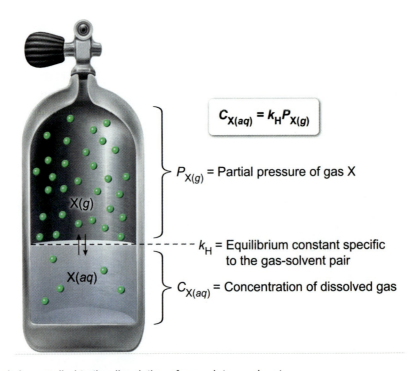

Figure 7.11 Henry's law applied to the dissolution of a gas into a solvent.

Lesson 8.1

Definitions of Acids and Bases

Introduction

In chemistry, acids and bases can be defined and categorized in different ways depending on the characteristics that need to be emphasized in a given context. Some important characteristics of acids and bases are summarized in Table 8.1.

Table 8.1 Common properties of acids and bases.

Properties of acids	Properties of bases
• Taste sour	• Taste bitter and chalky
• Sting or burn on contact with skin	• Feel slippery on contact with skin
• Turn blue litmus paper red	• Turn red litmus paper blue
• Decrease the pH of aqueous solutions	• Increase the pH of aqueous solutions
• Neutralize bases	• Neutralize acids

Over time, chemists developed different definitions to describe acids and bases according to their interactions within chemical systems. The three primary sets of acid-base definitions are the Arrhenius definitions, the Brønsted-Lowry definitions, and the Lewis definitions. These definitions are distinguished in the following ways:

- **Arrhenius definitions** are the most limited and define acids and bases in terms of interactions with water (ie, apply only to aqueous systems).
- **Brønsted-Lowry definitions** are wider in scope and define acids and bases in terms of H^+ ion transfers in both aqueous and nonaqueous systems.
- **Lewis definitions** are the broadest and define acids and bases by electron pair interactions within all types of systems.

As illustrated in Figure 8.1, every Arrhenius acid is both a Brønsted-Lowry acid and a Lewis acid, but a Brønsted-Lowry acid is not necessarily an Arrhenius acid. Likewise, a Lewis acid is not necessarily either a Brønsted-Lowry acid or an Arrhenius acid.

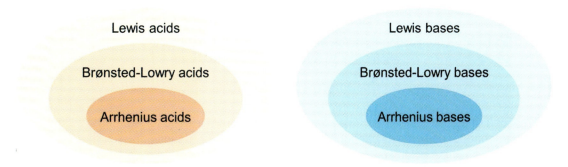

Figure 8.1 Relationships and inclusivity of different definitions of acids and bases.

In this lesson, each of the definitions of acids and bases is explored along with the contexts in which they are useful.

8.1.01 Arrhenius Acids and Bases

The Arrhenius definitions of acids and bases are the oldest and most limited. An **Arrhenius acid** is defined as any substance that dissociates in water to produce hydronium (H_3O^+) ions, and an **Arrhenius base** is defined as any substance that dissociates in water to produce hydroxide (OH^-) ions (Figure 8.2).

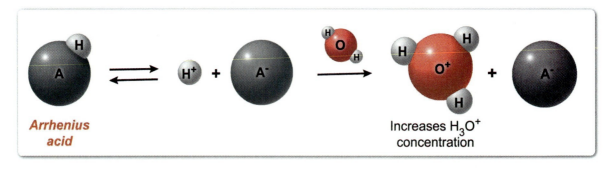

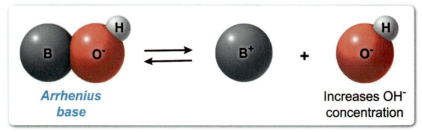

*Arrhenius definitions are limited to aqueous systems because they are based on the behavior of substances in water.

Figure 8.2 Arrhenius acids and bases.

When an Arrhenius acid reacts with an Arrhenius base, they neutralize one another. The product of this type of neutralization reaction is always an ionic salt and water, as shown in Figure 8.3.

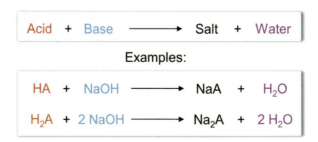

Arrhenius acids and bases react and neutralize each other to form ionic salts and water.

Figure 8.3 Neutralization reactions between Arrhenius acids and bases.

A major limitation of the Arrhenius definitions is that they can only be used for aqueous systems, and substances in nonaqueous systems cannot be classified using these definitions. Another major limitation is that they do not describe acids and bases in terms of their interactions with each other in chemical reactions. Thus, some substances that can act either as an acid or a base (depending on the situation) cannot be adequately classified by these definitions. In addition, only compounds that already contain an OH^- ion can be considered a base.

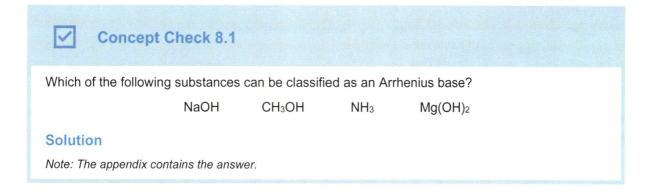

8.1.02 Brønsted-Lowry Acids and Bases

The Brønsted-Lowry theory of acids and bases defines acids and bases in terms of their interactions with each other in chemical reactions. Accordingly, a **Brønsted-Lowry acid** is a chemical species that donates a proton (ie, an H⁺ ion) in a chemical reaction, and a **Brønsted-Lowry base** is a chemical species that accepts a proton in a chemical reaction, as shown in Figure 8.4.

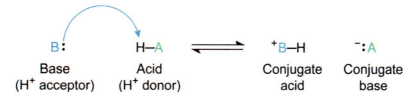

Figure 8.4 General representation of a Brønsted-Lowry acid-base reaction.

After a Brønsted-Lowry acid donates a proton, it becomes a conjugate base because this deprotonated species can later act as a base and accept a proton. In a similar way, after a Brønsted-Lowry base accepts a proton, it becomes a conjugate acid because this protonated species can later donate its newly obtained proton to another molecule. As such, two chemical species with structures that differ only by an acidic proton are referred to as a **conjugate acid-base pair**.

Some chemical species can act as either an acid or a base, depending on the reaction. These chemical species are called **amphoteric**. The amphoteric property of water, a particularly prominent example, is illustrated in Figure 8.5.

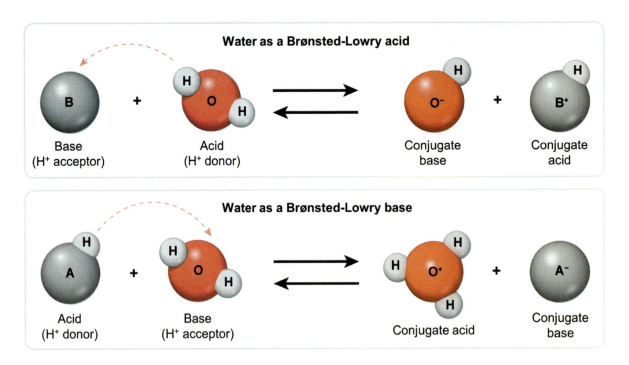

Figure 8.5 Water can behave as an acid or a base (an amphoteric species) in solution.

Unlike the Arrhenius definitions of acids and bases, which can only be used for specific aqueous systems, the Brønsted-Lowry theory defines acids and bases in terms of a chemical interaction (ie, proton exchange) and can therefore be used in a far greater variety of chemical contexts, including organic and gaseous systems. Another advantage of the Brønsted-Lowry definitions is the ability to describe the behavior of amphoteric substances based on their different chemical reactions.

Despite the practical usefulness of the Brønsted-Lowry definitions, they are not able to describe acid-base relationships in systems that do not involve proton transfers. For such systems, a more universal model of acids and bases is needed.

☑ Concept Check 8.2

The dihydrogen phosphate ion ($H_2PO_4^-$) is an amphoteric species. Write a reaction equation showing $H_2PO_4^-$.

1) acting as a Brønsted-Lowry acid in water and label the conjugate base.

2) acting as a Brønsted-Lowry base in water and label the conjugate acid.

Solution

Note: The appendix contains the answer.

8.1.03 Lewis Acids and Bases

Many substances exhibit acidic behavior in chemical systems but lack acidic protons in their structure. Likewise, many bases do not behave according to the Arrhenius or Brønsted-Lowry definitions. To overcome these limitations, the Lewis theory of acids and bases was developed.

According to the Lewis theory, the movement of electrons is the basis for acid-base behavior. As such, a **Lewis acid** is defined as an electron pair acceptor, and a **Lewis base** is defined as an electron pair donor (Figure 8.6).

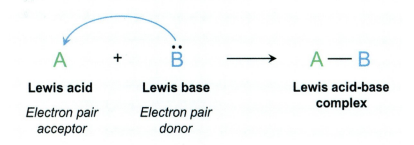

Figure 8.6 Lewis acids and bases.

Coordination chemistry (examined in more detail in Concept 2.1.07) makes extensive use of the Lewis model of acids and bases. In coordination chemistry, positive metal ions act as Lewis acids, accepting lone pairs of electrons from molecules called **ligands**, which act as Lewis bases. When ligands (Lewis bases) donate their electrons to metal ions (Lewis acids), they form coordinate covalent bonds.

Figure 8.7 shows ammonia molecules acting as Lewis bases, forming coordinate covalent bonds with an Ag^+ ion functioning as a Lewis acid.

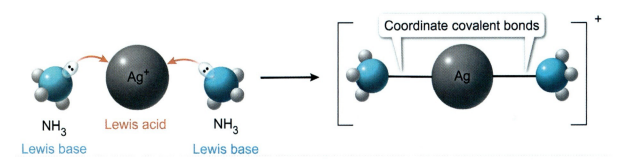

Figure 8.7 Formation of coordinate covalent bonds by a Lewis acid-base interaction between a silver ion and ammonia ligands.

Concept Check 8.3

Which substance acts as a Lewis acid in the following reaction?

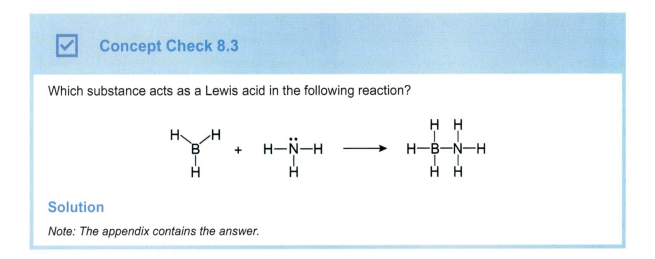

Solution

Note: The appendix contains the answer.

Lesson 8.2

Acids and Bases in Aqueous Solution

Introduction

Although acid-base chemistry is not limited to aqueous systems, the vast majority of acid-base chemistry in biological systems takes place in an aqueous environment. Therefore, understanding the behavior of acids and bases in an aqueous environment is crucial to understanding many biological processes, not to mention the numerous nonbiological processes that involve aqueous acid-base reactions.

This lesson provides a foundation for understanding aqueous acid-base systems by exploring the ionization behavior of water, as well as the pH (and pOH) scale used for measuring the degree of acidity (and basicity) in aqueous systems.

8.2.01 Ionization of Water

As explained in Concept 8.1.02, water is amphoteric, capable of acting either as an acid or as a base. This behavior is true not only for the interaction of water with other substances but also for the interactions of water molecules with each other. As such, water molecules can react with each other to form hydronium ions (H_3O^+) and hydroxide ions (OH^-) by a process referred to as the **autoionization of water**, illustrated in Figure 8.8.

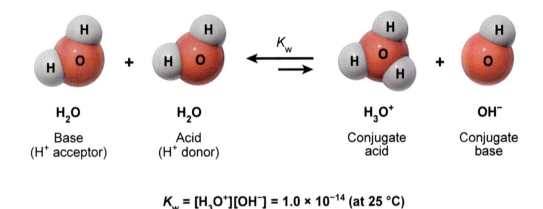

$$K_w = [H_3O^+][OH^-] = 1.0 \times 10^{-14} \text{ (at 25 °C)}$$

Water molecules interact and ionize into hydronium and hydroxide ions

- Water is amphoteric (can act as either an acid or a base)
- $[H_3O^+]$ is a proton shuttle (source of H^+): $[H^+] = [H_3O^+]$

Figure 8.8 The autoionization equilibrium of water.

Although the value of the **autoionization equilibrium constant of water** K_w varies with temperature, its value at 25 °C ($K_w = 1.0 \times 10^{-14}$) is a useful standard for comparing the molar concentrations of H_3O^+ and OH^- in aqueous solutions. The K_w equation (Figure 8.8) shows that the hydronium concentration $[H_3O^+]$ is inversely proportional to the hydroxide concentration $[OH^-]$. When an acid ionizes in water to produce H_3O^+ ions, the OH^- concentration drops proportionally (and vice versa), as shown in Figure 8.9.

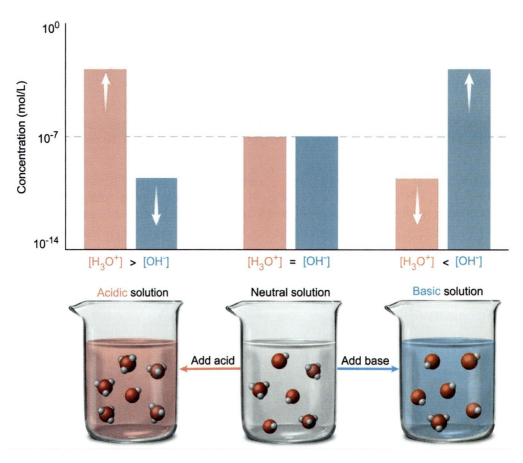

Figure 8.9 Hydronium-hydroxide balance in aqueous solutions.

In aqueous media, acidity and basicity are evaluated based on the hydronium-hydroxide balance of the K_w equilibrium. A neutral solution has equal concentrations of H_3O^+ and OH^- ions. When $[H_3O^+]$ exceeds $[OH^-]$, the solution is acidic, and when $[OH^-]$ is greater than $[H_3O^+]$, the solution is basic.

✓ Concept Check 8.4

The hydronium concentration in the mitochondrial matrix is approximately 2×10^{-8} M.

1. What is $[OH^-]$ for this solution? (Assume $K_w = 1.0 \times 10^{-14}$)
2. Is this solution acidic or basic?

Solution

Note: The appendix contains the answer.

8.2.02 The pH and pOH Scales

The values of the concentrations involved when measuring the acidity of a solution are often extremely small. Using the **pH scale** makes the task of expressing these values easier. The **pH** of a solution is expressed mathematically as the negative logarithm of the molar concentration of hydrogen ions (H^+) in solution:

$$pH = -\log[H^+]$$

where [H⁺] is in units of molarity (M or mol/L). The pH can also be thought of as the power (ie, exponent) of hydrogen ion concentration [H⁺], an idea expressed in equation form as:

$$[H^+] = 10^{-pH}$$

No free H⁺ ions actually exist in aqueous solutions because H₂O molecules bind to acidic hydrogens and form H3O⁺ ions before complete acid dissociation can occur. Nevertheless, the use of H⁺ ions is a convenient shorthand notation for describing acids, and many chemistry textbooks use H⁺ and H₃O⁺ interchangeably when referring to acids and pH.

A major advantage of the logarithmic scale for pH is that it reduces the broad range of hydronium concentrations to a much smaller range of numbers that is easier to work with and can more quickly give a sense of the degree of acidity in a solution. Because pH is a negative logarithmic function, lower pH values correspond to higher hydronium concentrations (and vice versa), as Figure 8.10 illustrates.

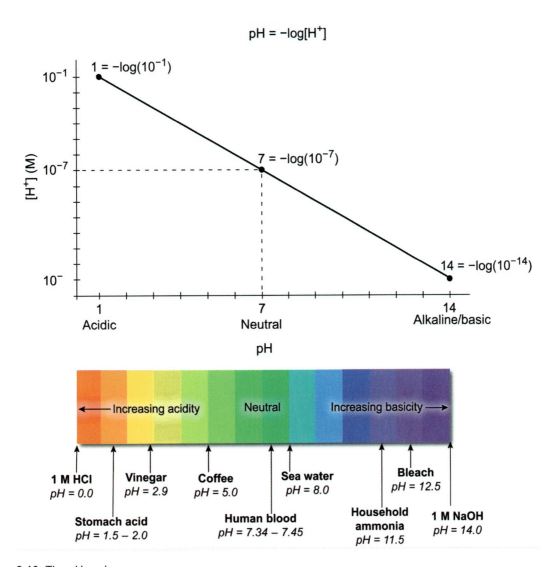

Figure 8.10 The pH scale.

In the same way that the pH scale expresses the degree of acidity of a solution, the pOH scale indicates the degree of basicity of a solution. The pOH is expressed mathematically as the negative logarithm of the molar concentration of OH⁻ in solution:

$$pOH = -\log[OH^-]$$

The relationship between pH and pOH can be derived from the K_w equilibrium expression (Figure 8.11):

$$K_w = [H_3O^+][OH^-] = 1.0 \times 10^{-14}$$

$$-\log([H^+][OH^-]) = -\log(1.0 \times 10^{-14})$$

$$-\log[H^+] - \log[OH^-] = 14$$

$$-\log(10^{-pH}) - \log(10^{-pOH}) = 14$$

$$\text{pH} + \text{pOH} = 14$$

Figure 8.11 Derivation of the relationship between pH and pOH from the K_w equilibrium equation.

✓ Concept Check 8.5

A sample of diluted household bleach has an OH⁻ concentration of 0.021 M. What is the pH of this solution?

Solution

Note: The appendix contains the answer.

Lesson 8.3

Strengths of Acids and Bases

Introduction

The strengths of acids and bases can be measured according to the degree to which they ionize (ie, react or dissociate to form ions) in aqueous solution. The equilibrium constant from the ionization equilibrium of an acid or a base gives an effective measure of the extent of its ionization and provides a useful means of comparing the strengths of different acidic or basic species.

This lesson explores how acids and bases are categorized in terms of strength and examines the relationship between the strength of an acid or a base and its conjugate. This lesson also examines equilibria in acid-base systems and the calculations associated with those equilibria.

8.3.01 Relative Strengths of Acids and Bases

Acids and bases can be broadly categorized as either strong or weak. **Strong acids** are substances that completely ionize in aqueous solution to form H_3O^+ ions. For example, a 1.0 M solution of any monoprotic strong acid (ie, an acid with only one easily released H^+ ion) has an H_3O^+ concentration of 1.0 M (ie, all the acid molecules form an H_3O^+ ion). In contrast, only a fraction of **weak acid** molecules ionizes in aqueous solution, as shown in Figure 8.12.

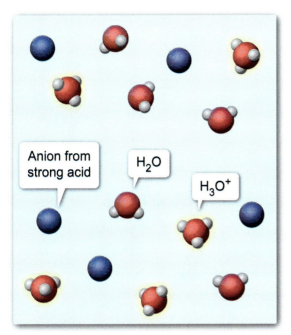
Strong acids ionize completely.
$HA + H_2O \longrightarrow A^- + H_3O^+$

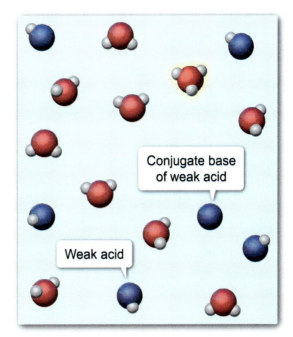
Weak acids only ionize to a limited extent.
$HA + H_2O \rightleftharpoons A^- + H_3O^+$

Figure 8.12 Ionization of strong acids versus weak acids.

Like strong acids, **strong bases** completely ionize in aqueous solution but with the ionization of bases producing OH⁻ ions. As such, a 1.0 M solution of any strong base (ie, a compound with the general formula $M(OH)_n$, where M^{n+} is a metal cation) fully ionizes and contains $(1.0 \times n)$ M OH⁻ ions. Conversely, **weak bases**, like weak acids, ionize to only a small degree, as illustrated in Figure 8.13.

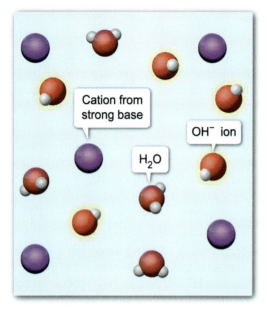
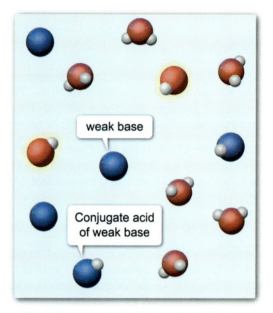

Strong bases ionize completely

$$MOH \longrightarrow M^+ + OH^-$$
OR
$$M(OH)_2 \longrightarrow M^{2+} + 2OH^-$$

Weak bases only ionize to a limited extent

$$B + H_2O \rightleftharpoons HB^+ + OH^-$$
OR
$$B^- + H_2O \rightleftharpoons HB + OH^-$$

Figure 8.13 Ionization of strong bases versus weak bases.

A relatively small number of acids and bases are strong, and recognizing (and possibly memorizing) them may be helpful (Table 8.2).

Table 8.2 Names and formulas for common strong acids and bases.

Strong acids

Name	Formulas
Perchloric acid	$HClO_4$
Hydroiodic acid	HI
Hydrobromic acid	HBr
Sulfuric acid	H_2SO_4
Hydrochloric acid	HCl
Nitric acid	HNO_3
Chloric Acid	$HClO_3$
Hydronium ion	H_3O^+

Strong bases

Name	Formulas
Lithium hydroxide	LiOH
Sodium hydroxide	NaOH
Potassium hydroxide	KOH
Rubidium hydroxide	RbOH
Cesium hydroxide	CsOH
Calcium hydroxide*	$Ca(OH)_2$
Strontium hydroxide*	$Sr(OH)_2$
Barium hydroxide	$Ba(OH)_2$

*Although $Ca(OH)_2$ and $Sr(OH)_2$ are not very soluble in water, they are still considered strong bases because the amount that does dissolve dissociates completely.

As outlined in Figure 8.14, the strength of an acid is inversely proportional to the strength of its conjugate base (ie, the stronger an acid, the weaker its conjugate base and vice versa). The same is true for bases and their conjugate acids.

Acid			Conjugate base
$HClO_4$			ClO_4^-
H_2SO_4			HSO_4^-
HI	100% acid ionization in water	No base ionization in water	I^-
HBr			Br^-
HCl			Cl^-
HNO_3			NO_3^-
H_3O^+			H_2O (water)
HSO_4^-			SO_4^{2-}
H_3PO_4			$H_2PO_4^-$
HF			F^-
HNO_2			NO_2^-
CH_3CO_2H			$CH_3CO_2^-$
H_2CO_3			HCO_3^-
H_2S			HS^-
NH_4^+			NH_3
HCN			CN^-
HCO_3^-			CO_3^{2-}
H_2O (water)			OH^-
HS^-	No acid ionization in water	100% base ionization in water	S^{2-}
C_2H_5OH			$C_2H_5O^-$
NH_3			NH_2^-

(Increasing acid strength ↑ on left; Increasing base strength ↓ on right)

Figure 8.14 Relative strengths of acids and their conjugate bases.

Understood from a Brønsted-Lowry perspective, the relationship between the strength of acids and their conjugate bases can be explained by comparing the two species' attraction to H^+ ions (protons).

A stronger acid (HA) more easily donates an H^+, meaning that the conjugate base (A^-) is a weaker base that has little attraction for the donated proton. Conversely, a stronger base has a very strong attraction for protons; consequently, when a strong base gains an H^+, the resulting conjugate acid is weakly acidic because it is much less likely to donate the acquired proton.

8.3.02 Weak Acid Equilibria

Weak acids, which ionize to only a limited extent, reach dynamic equilibrium in aqueous solution according to the following chemical equation:

$$HA(aq) + H_2O(l) \overset{K_a}{\rightleftharpoons} H_3O^+(aq) + A^-(aq)$$

The equilibrium constant for this reaction of an acid (HA) with water is known as its **acid ionization constant** K_a, which is expressed as:

$$K_a = \frac{[H_3O^+][OH^-]}{[HA]}$$

Although acids are classified as either strong or weak, acid strength is better understood as a continuum. Within this continuum, the strength of different acids can be compared using their respective K_a values, which indicate the degree to which the acid ionizes.

Strong acids have large K_a values ($K_a > 1$), indicating that the equilibrium lies heavily to the right side of the equation (ie, nearly complete ionization). In contrast, weak acids have small K_a values ($K_a < 1$), indicating that the equilibrium lies to the left side (ie, little ionization occurs), as summarized in Figure 8.15.

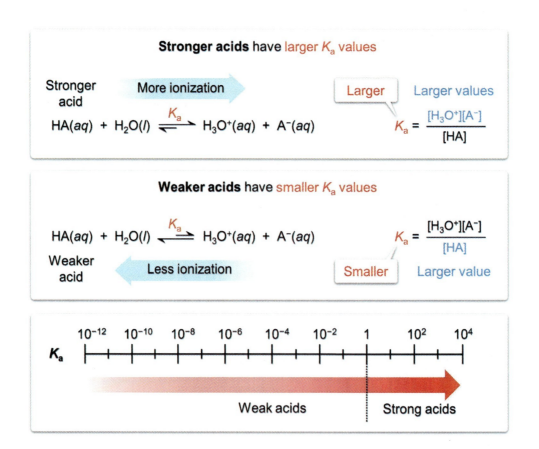

Figure 8.15 Correlation between K_a and acid strength.

The strength of an acid can also be assessed based on its percent ionization. Whereas the percent ionization of a strong acid is assumed to be 100%, the percent ionization of a weak acid can be calculated using:

$$\text{Percent ionization} = \frac{[H_3O^+]_{\text{equilibrium}}}{[HA]_{\text{initial}}} \times 100\%$$

As expected, higher K_a values typically correspond to larger percent ionizations. Furthermore, when an aqueous weak-acid system is diluted with water, the system responds by shifting to the right to reestablish equilibrium, resulting in a greater percent ionization.

> **Concept Check 8.6**
>
> What is the percent ionization of a 0.10 M aqueous solution of acetic acid ($HC_2H_3O_2$) with K_a = 1.76 × 10^{-5}?
>
> **Solution**
>
> *Note: The appendix contains the answer.*

Using a logarithmic scale for K_a (eg, like the pH scale does for [H_3O^+]) condenses the broad range of K_a values into a much smaller range of numbers that can be used to compare acid strength more easily. Accordingly, the pK_a of an acid is defined mathematically as:

$$pK_a = -\log K_a$$

Just as smaller pH values correspond to larger H_3O^+ concentrations, smaller pK_a values correspond to larger K_a values, as shown in Figure 8.16.

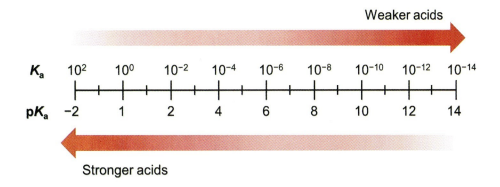

Figure 8.16 The relationship between acid strength, K_a, and pK_a.

> ## ✓ Concept Check 8.7
>
> Consider the following table of weak acid solutions and their corresponding hydronium ion concentrations at equilibrium.
>
> | Weak acid solution | $[H_3O^+]_{eq}$ |
> |---|---|
> | 0.1 M HClO | 6×10^{-5} M |
> | 0.1 M HCN | 8×10^{-6} M |
> | 0.1 M HNO$_2$ | 6×10^{-3} M |
> | 0.1 M HC$_7$H$_5$O$_2$ | 3×10^{-3} M |
>
> Which acid in this table would have the largest pK_a?
>
> ### Solution
> *Note: The appendix contains the answer.*

8.3.03 Polyprotic Acids

Polyprotic acids are chemical species that can undergo more than one ionization in an aqueous solution to produce H_3O^+ ions. Polyprotic acids have a different K_a value for each successive ionization, as illustrated in Figure 8.17.

$$H_3A(aq) + H_2O(l) \xrightleftharpoons[]{K_{a1}} H_2A^-(aq) + H_3O^+(aq) \qquad K_{a1} = \frac{[H_3O^+][H_2A^-]}{[H_3A]}$$

$$H_2A^-(aq) + H_2O(l) \xrightleftharpoons[]{K_{a2}} HA^{2-}(aq) + H_3O^+(aq) \qquad K_{a2} = \frac{[H_3O^+][HA^{2-}]}{[H_2A^-]}$$

$$HA^{2-}(aq) + H_2O(l) \xrightleftharpoons[]{K_{a3}} A^{3-}(aq) + H_3O^+(aq) \qquad K_{a3} = \frac{[H_3O^+][A^{3-}]}{[HA^{2-}]}$$

In a triprotic acid: $K_{a1} > K_{a2} > K_{a3}$

Figure 8.17 Successive acid ionizations of a triprotic acid.

From a Brønsted-Lowry perspective, polyprotic acids have more than one acidic proton (H^+ ion) that they can donate to another species, and the K_a values of a polyprotic acid decrease with each subsequent donation of an H^+. This trend is due in part to charge stability. With each successive H^+ removed, the negative charge of the conjugate base species increases, resulting in a less stable electrostatic state (ie, more charge repulsion). Thus, each donation of an acidic proton makes another proton donation much more difficult.

> **Concept Check 8.8**
>
> Concentrated phosphoric acid (H_3PO_4) is diluted in pure water to form a 0.1 M solution. Excluding H_2O, H_3O^+, and OH^-, list the chemical species present in order from most abundant to least abundant.
>
> **Solution**
>
> *Note: The appendix contains the answer.*

Compared to the first ionization of a polyprotic acid, subsequent ionizations produce a negligible amount of H_3O^+. As such, only the first ionization is considered when calculating the pH of a polyprotic acid solution in pure water. Nevertheless, all the acidic protons are available to ionize, and, as a result, the number of acidic protons determines the number of equivalents of strong base needed for complete neutralization of a polyprotic acid (Figure 8.18).

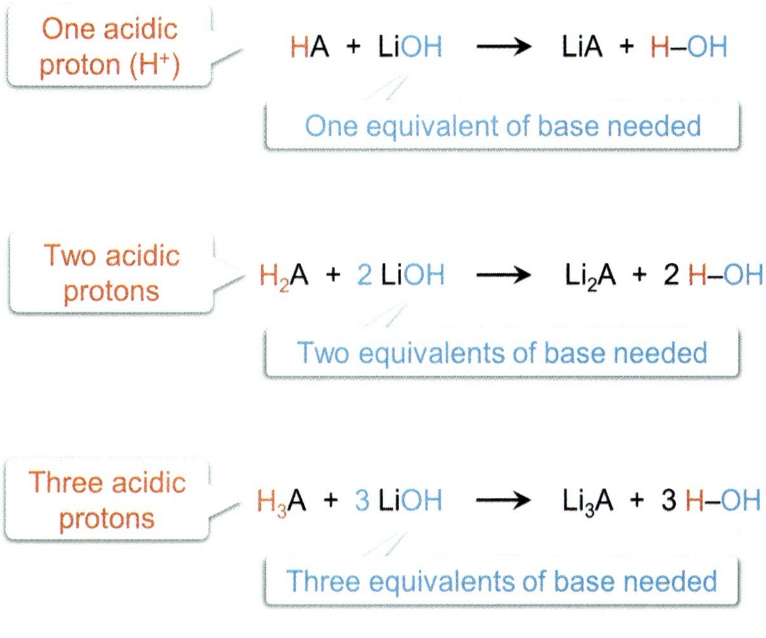

One acidic proton (H^+) — HA + LiOH ⟶ LiA + H–OH
One equivalent of base needed

Two acidic protons — H_2A + 2 LiOH ⟶ Li_2A + 2 H–OH
Two equivalents of base needed

Three acidic protons — H_3A + 3 LiOH ⟶ Li_3A + 3 H–OH
Three equivalents of base needed

Figure 8.18 Correlation between the equivalents of base needed to fully neutralize an acidic species and the number of acidic protons in the acid.

8.3.04 Weak Base Equilibria

Weak bases, like weak acids, partially ionize in aqueous solution and reach dynamic equilibrium as follows:

$$B(aq) + H_2O(l) \rightleftarrows HB^+(aq) + OH^-(aq)$$

The equilibrium constant for a weak base (B) reacting with water is known as its **base ionization constant** K_b, which is expressed as:

$$K_b = \frac{[HB^+][OH^-]}{[B]}$$

A larger value of K_b indicates a greater degree of base ionization and thus a stronger base. To simplify comparisons of the strengths of different bases, a logarithmic pK_b scale is often used, where **pK_b** is defined mathematically as:

$$pK_b = -\log K_b$$

As with all negative logarithmic scales, a larger value of pK_b corresponds to a smaller K_b (and vice versa), as illustrated in Figure 8.18.

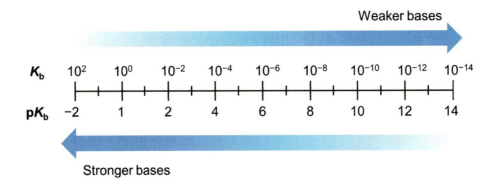

Figure 8.19 The relationship between base strength, K_b, and pK_b.

 Concept Check 8.9

For hydroxylamine (HONH$_2$), $K_b = 1.1 \times 10^{-8}$. Determine the pH of a 1.0 M HONH$_2$ solution.

Solution

Note: The appendix contains the answer.

When two equilibria are added together, the product of the corresponding equilibrium constants is equal to the equilibrium constant for the new net reaction. As shown in Figure 8.19, combining the aqueous ionization equilibrium of a weak acid with the aqueous ionization equilibrium of its conjugate base always results in the autoionization equilibrium of water.

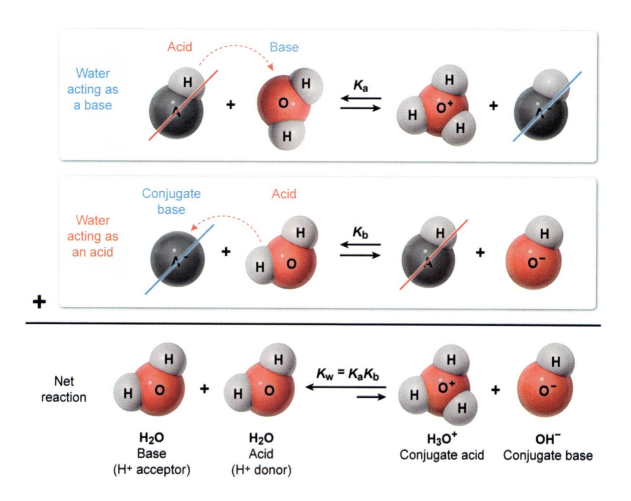

Figure 8.20 The net result of the aqueous ionization equilibrium of a weak acid and that of its conjugate base is always the autoionization equilibrium of water.

Therefore, the K_a of a weak acid is related to the K_b of its conjugate base by:

$$K_w = K_a K_b$$

When using the power scales, the mathematical relationship between pK_a and pK_b is derived from this equation, as shown in Figure 8.20:

$$K_a K_b = K_w$$
$$-\log(K_a K_b) = -\log(K_w)$$
$$-\log(K_a) - \log(K_b) = -\log(K_w)$$
$$pK_a + pK_b = pK_w$$
$$pK_a + pK_b = 14$$

Figure 8.21 Derivation of the mathematical relationship between pK_a and pK_b.

Concept Check 8.10

Sodium hypochlorite (NaClO) is the bleaching agent present in commercial bleach. Compare the hypochlorite ion to the bases in the following table by ranking them in order of increasing base strength. (Note: For HClO, $K_a = 3.0 \times 10^{-8}$.)

Weak base	pK_b
$(CH_3)_3N$	4.19
C_5H_5N	8.77
$HONH_2$	7.96
N_2H_4	5.89

Solution

Note: The appendix contains the answer.

Lesson 8.4
Buffers

Introduction

Buffers are solutions that make use of the properties of weak acids and their conjugate bases to regulate pH and keep [H_3O^+] within optimal ranges for both biological and nonbiological systems. This lesson examines the behavior and characteristics of buffers, how they function, and the calculations associated with them.

8.4.01 Characteristics and Behavior of Buffer Systems

A **buffer** is a solution containing significant concentrations of both components of a **conjugate acid-base pair** (ie, a weak acid with its conjugate base or a weak base with its conjugate acid). Buffer solutions can be made using any of the following combinations:

- A weak acid (HA) with a salt containing its conjugate base (A^-)
- A weak base (B) with a salt containing its conjugate acid (HB^+)
- A lesser amount of strong acid with a greater amount of weak base
- A lesser amount of strong base with a greater amount of weak acid

The presence of both components of a conjugate acid-base pair in solution enables a buffer system to resist substantial changes to its pH by neutralizing any strong acid or strong base that enters the system, as illustrated in Figure 8.22.

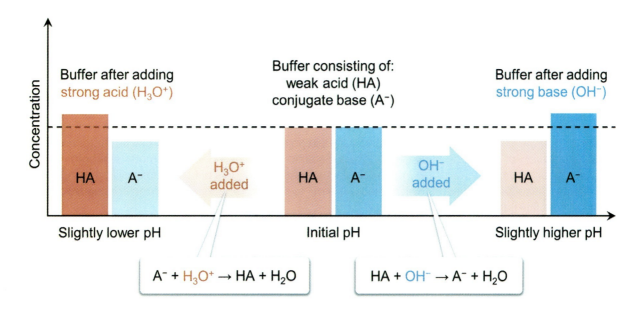

Figure 8.22 Effects of adding a strong acid or strong base to a buffer system.

Every buffer has a limit to the amount of strong acid or base that can be added to it before the pH begins to rapidly change. This limit, known as the **buffer capacity**, is determined by the amount of weak acid and conjugate base present in the buffer (ie, more buffer molecules can neutralize more strong acid or strong base), as shown in Figure 8.23.

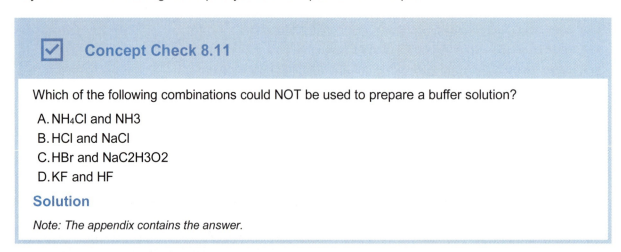

Figure 8.23 Higher versus lower buffer capacity.

A buffer solution can only neutralize added strong acid or strong base in an amount proportional to the concentrations of its acidic and basic components. If enough strong base molecules are added to neutralize all the weak acid molecules in the buffer, the buffer is "broken" (ie, its capacity is reached) and any added base exceeding this capacity causes a rapid increase in pH. Likewise, the buffer can also be broken by adding enough strong acid molecules to neutralize all the weak base molecules in the buffer. Any added acid exceeding this capacity causes a rapid decrease in pH.

> ☑ **Concept Check 8.11**
>
> Which of the following combinations could NOT be used to prepare a buffer solution?
>
> A. NH_4Cl and NH_3
> B. HCl and NaCl
> C. HBr and $NaC_2H_3O_2$
> D. KF and HF
>
> **Solution**
>
> *Note: The appendix contains the answer.*

8.4.02 Henderson-Hasselbalch Equation

In Lesson 8.3, the behavior of weak acids and bases in solution was explored, along with methods for calculating the pH of each type of solution. When a solution contains significant amounts of both components of a conjugate acid-base pair, calculations of the solution's pH can be simplified using the Henderson-Hasselbalch equation, which relates the pH to the pK_a, and the ratio of the concentrations of the weak acid [HA] and its conjugate base [A⁻]:

$$pH = pK_a + \log\left(\frac{[A^-]}{[HA]}\right)$$

From this equation, it can be shown that a buffer solution containing equal amounts of weak acid and conjugate base has a pH equal to the pK_a of the weak acid (Figure 8.24).

$$pH = pK_a + \log\frac{[A^-]}{[HA]}$$

When $[A^-] = [HA]$
$pH = pK_a + \log(1) = pK_a + 0$
$pH = pK_a$

Figure 8.24 pH and pK_a are equal when both species of a conjugate acid-base pair are present in equal amounts.

When $[A^-] > [HA]$, $\log\left(\frac{[A^-]}{[HA]}\right)$ is positive and pH is greater than pK_a. Conversely, when $[A^-] < [HA]$, $\log\left(\frac{[A^-]}{[HA]}\right)$ is negative and pH is less than pK_a. As expected, a higher $[HA]$ results in a lower pH, whereas a higher $[A^-]$ results in a higher pH, as illustrated in Figure 8.25.

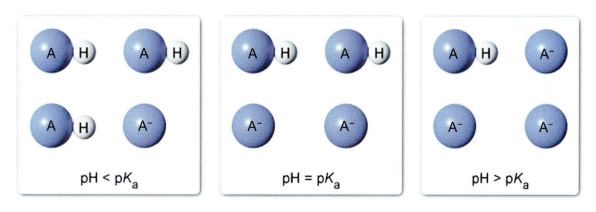

Figure 8.25 pH versus pK_a under different concentrations of weak acid (HA) and conjugate base (A^-).

In any case where the Henderson-Hasselbalch equation is used, both species of the conjugate acid-base pair (eg, HA and A^-) are present in the same solution and therefore exist in the same solution volume (V_{soln}). This means that the molarity ratio $\left(\frac{[A^-]}{[HA]}\right)$ of the conjugate acid-base pair is equal to its mole ratio $\left(\frac{\text{mol } A^-}{\text{mol } HA}\right)$, as shown in Figure 8.26. Recognizing this fact can help simplify buffer-related calculations where grams or moles are involved.

Given: $[HA] = \dfrac{\text{mol HA}}{V_{soln}}$ and $[A^-] = \dfrac{\text{mol } A^-}{V_{soln}}$

Substitution: $pH = pK_a + \log\left(\dfrac{[A^-]}{[HA]}\right)$ becomes $pH = pK_a + \log\left(\dfrac{\frac{\text{mol } A^-}{V_{soln}}}{\frac{\text{mol HA}}{V_{soln}}}\right)$

Yields: $pH = pK_a + \log\left(\dfrac{\text{mol } A^-}{\text{mol HA}}\right)$

Figure 8.26 The molarity ratio is equivalent to the mole ratio in the Henderson-Hasselbalch equation.

An alternative form of the Henderson-Hasselbalch equation derived from the K_b equilibrium expression is:

$$\text{pOH} = pK_b + \log\left(\frac{[HB^+]}{[B]}\right)$$

where [B] is the concentration of weak base and [HB$^+$] is the concentration of weak conjugate acid. In cases where K_b is given or when solving for [OH$^-$], using this form of the Henderson-Hasselbalch equation may be preferable. However, either form may be used for any situation involving a buffer.

> ☑ **Concept Check 8.12**
>
> What is the pH of a buffer solution made by adding 16.4 g NaC$_2$H$_3$O$_2$ to 1.0 L of 2.0 M HC$_2$H$_3$O$_2$? (Note: For HC$_2$H$_3$O$_2$, pK_a = 4.74)
>
> **Solution**
>
> *Note: The appendix contains the answer.*

8.4.03 Formulation of Buffers and Buffering Range

As shown in Concept 8.4.02, when a buffer solution contains equal concentrations of weak acid and conjugate base, pH equals pK_a. Adjusting the conjugate base-to-acid ratio $\left(\frac{A^-}{HA}\right)$ in a buffer allows the buffer's pH to shift to a specific desired value.

According to the Henderson-Hasselbalch equation, shifting the pH of a buffer solution by 1 unit requires the conjugate base-to-acid ratio to change by a factor of 10, as shown in Figure 8.27.

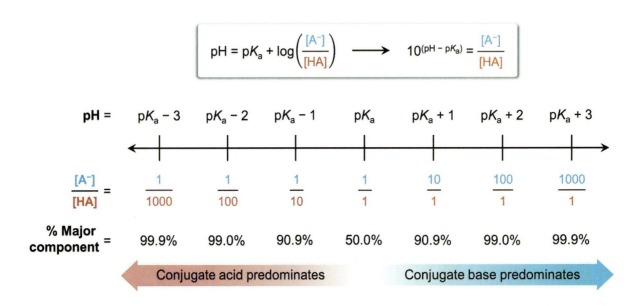

Figure 8.27 The effect of the conjugate base-to-acid ratio on the pH of a buffer.

> **Concept Check 8.13**
>
> How many grams of sodium acetate ($NaC_2H_3O_2$) should be added to 500 mL of 1.0 M acetic acid ($HC_2H_3O_2$) to make a buffer with a pH of 5.74? (Note: For $HC_2H_3O_2$, $pK_a = 4.74$)
>
> **Solution**
>
> *Note: The appendix contains the answer.*

Since a limit exists of how concentrated the components of a buffer can be, once the ratio of weak acid to conjugate base (or vice-versa) exceeds a 10:1 ratio, a buffer begins to lose its ability to moderate changes in pH. As such, the **buffering range** (ie, the pH range over which a buffer can effectively resist large changes in pH) is generally defined as the range of pH values within 1 unit of the pK_a (ie, $pK_a \pm 1$) of the buffer's weak acid component, as illustrated in Figure 8.28.

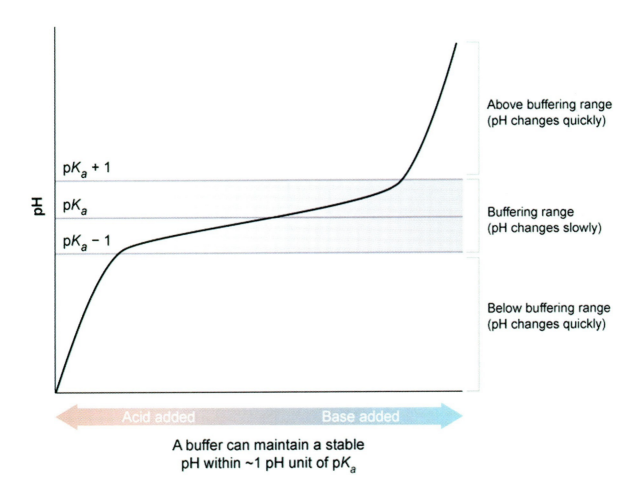

Figure 8.28 Buffering range of a buffer solution.

The buffering range is a key consideration when making a buffer solution. The most effective buffer has both a high buffer capacity and a pH value in the middle of its buffering range (ie, a pH near its pK_a).

When making a buffer of a specific pH, the conjugate acid-base pair that should be chosen depends on the pK_a of the weak acid. For example, if a buffer with a pH of 8.0 is needed, the pK_a of the weak acid component should be between 7.0 and 9.0, with the ideal choice having a pK_a as close to 8.0 as possible.

Concept Check 8.14

Based on the information in the table shown, what conjugate acid-base pair would be the best choice for making a buffer solution with a pH of 6.5?

Weak acid	pK_a
$H_2PO_4^-$	7.21
H_2CO_3	6.36
$H_2AsO_4^-$	7.00
$HC_4H_4O_4^-$	5.64

Solution

Note: The appendix contains the answer.

Chapter 8: Acid-Base Chemistry

Lesson 8.5

Hydrolysis of Salts

Introduction

As explained in Concept 8.1.01, combining an Arrhenius acid and base causes an acid-base neutralization reaction to occur in which a salt and water are formed. Although salts can form in other ways, every salt can be thought of as the product of a hypothetical acid-base neutralization. From this perspective, the constituent ions of a salt are the conjugate acid and conjugate base formed during the neutralization. Depending on the strengths of these conjugate species, one or more of the ions making up the salt may have acidic or basic properties when dissolved in water.

Salt hydrolysis is the process by which an ion from a salt participates in an acid or a base ionization reaction with water to produce H^+ or OH^- ions. The solution resulting from salt hydrolysis is either acidic (pH < 7) or basic (pH > 7) depending on the properties of the salt ions. If a salt does not hydrolyze, the solution is neutral (pH = 7). This lesson explores how to determine if a salt has acidic or basic properties and examines the pH calculations associated with salt hydrolysis.

8.5.01 Qualitative Analysis of the Acid-Base Properties of Salts

Each ion that makes up a salt can potentially hydrolyze in water by acting either as an acid or a base. As shown in Figure 8.29, cations that hydrolyze are generally acidic and cause the formation of H_3O^+, whereas anions that hydrolyze tend to be basic and cause the formation of OH^- unless they are weak amphiprotic anions (eg, $H_2PO_4^-$, HSO_4^-).

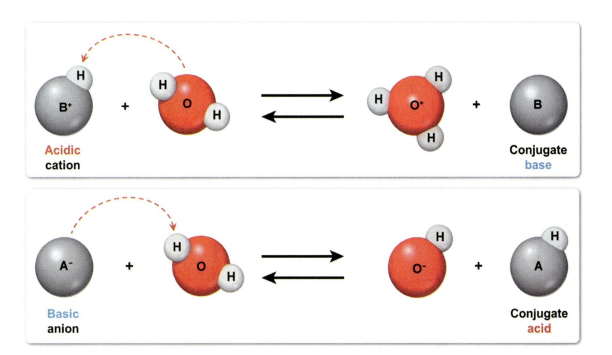

Figure 8.29 Hydrolysis of ions.

Many ions do not hydrolyze in water and are neutral. Analyzing the conjugate species of a given ion can determine if that ion is a strong enough acid or base to hydrolyze in an aqueous solution.

Determining Acid-Base Properties of Ions Using Conjugates

The two members of a conjugate acid-base pair are inversely proportional in strength. As such, the conjugate of a weak acid or base is often strong enough to exhibit acid-base properties and hydrolyze in water.

Conversely, the conjugate of a strong acid or base is usually too weak to hydrolyze in water and has no effect on pH. These principles can be used to analyze the acid-base properties of ions and their ability to affect the pH of a solution, as illustrated in Figure 8.30.

	Ion → Conjugate		Conclusion
NH_4^+ ?	Remove H^+ →	NH_3 Weak base	NH_4^+ is a weak acid
Cl^- ?	Add H^+ →	HCl Strong acid	Cl^- is neutral
BrO^- ?	Add H^+ →	HBrO Weak acid	BrO^- is a weak base

Figure 8.30 Analysis of the acid-base properties of ions.

Acid-Base Properties of Metal Cations

Some metal cations produce weakly acidic solutions when dissolved in water. Although metal cations themselves do not have H^+ ions to donate in solution, metal cations can function as Lewis acids and form complex ions in which water molecules surround the cation as ligands (ie, a hydrated complex). In some cases, these complex ions are weakly acidic and prone to release an H^+ ion from one of the ligands, as illustrated in Figure 8.31.

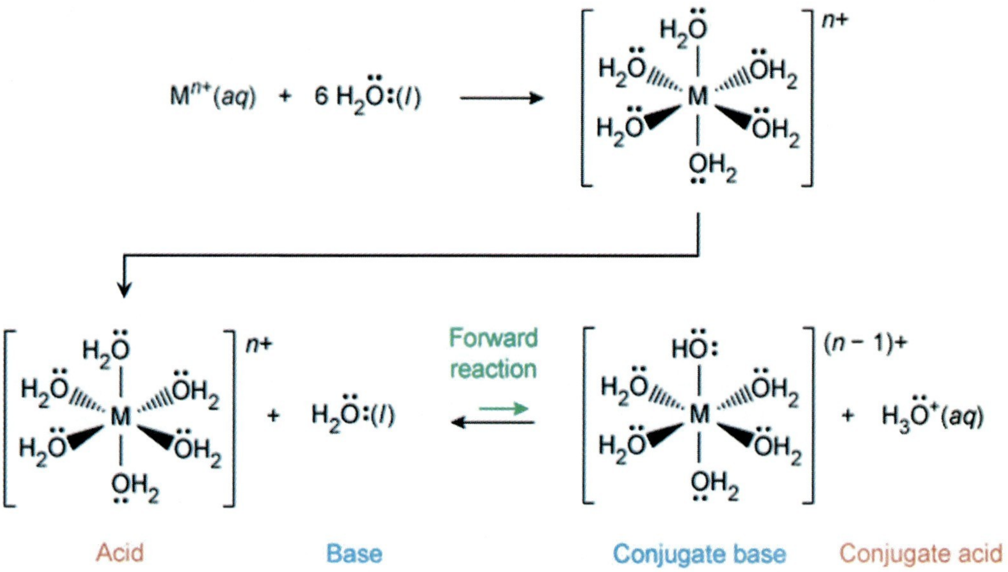

$$M^{n+}(aq) + 6\,H_2\ddot{O}{:}(l) \longrightarrow [M(H_2O)_6]^{n+}$$

$$[M(H_2O)_6]^{n+} + H_2\ddot{O}{:}(l) \underset{}{\overset{\text{Forward reaction}}{\rightleftharpoons}} [M(OH)(H_2O)_5]^{(n-1)+} + H_3\ddot{O}^+(aq)$$

Acid   Base   Conjugate base   Conjugate acid

Figure 8.31 The acidic behavior of metal ions in aqueous solution. M^{n+} is an unspecified metal cation with a charge of $+n$.

Not all metal cations form acidic hydrated complexes. The acidity of a hydrated complex depends primarily on the charge ($+n$) of the metal cation and its ionic radius. Smaller, highly charged metal cations (eg, Al^{3+}, Fe^{3+}) are more acidic than larger cations with less charge (eg, K^+, Ca^{2+}).

Metal cations with a greater charge and smaller radius attract ligands more strongly and pull them closer. This in turn distorts the electron density around the ligands, making it easier to release an H^+ ion. In general, metal cations from Group 1 (eg, Na^+, K^+) do not exhibit acidic behavior in aqueous solutions. However, most other metal cations are at least slightly acidic, reacting with water to generate H_3O^+ ions.

Acid-Base Properties of Amphiprotic Ions

For amphiprotic ions (ie, ions capable of accepting or donating an H^+ ion), such as the conjugate bases of polyprotic acids (eg, HCO_3^-), both the acidic and basic behavior of the ion must be considered to determine the net effect on the pH of a solution. Amphiprotic ions have both a K_a and a K_b value, and which ionization constant is greater determines whether the ion is predominantly acidic or basic. Figure 8.32 gives an example of this type of analysis.

$H_2PO_4^-$ function	Reaction	Ionization constant
As an acid	$H_2PO_4^-(aq) + H_2O(l) \rightleftharpoons HPO_4^{2-}(aq) + H_3O^+(aq)$	$K_a = 6.2 \times 10^{-8}$
As a base	$H_2PO_4^-(aq) + H_2O(l) \rightleftharpoons H_3PO_4(aq) + OH^-(aq)$	$K_b = 1.3 \times 10^{-12}$

$H_2PO_4^-$ is predominately an acidic ion because its K_a is greater than its K_b.

Figure 8.32 Analysis of the acid-base properties of $H_2PO_4^-$.

Overall Acid-Base Properties of a Salt

When qualitatively analyzing the acid-base properties of a salt, one effective method is to consider the "parent" Arrhenius acid and base whose neutralization reaction forms the salt (ie, the "daughter" compound). If the acid and base parents are both strong, the daughter salt formed is neutral. If one parent is stronger than the other, the stronger parent determines the acid-base behavior of the daughter salt formed, as illustrated in Figure 8.33.

		Anion from:	
		Strong acid	Weak acid
Cation from:	Strong base	Neutral salt (pH = 7)	Basic salt (pH > 7)
	Weak base	Acidic salt (pH < 7)	Conditional (pH = ?)*

*The pH produced by these salts depends upon the relative strength of the weak acid and weak base.

Figure 8.33 pH of salts formed from acid-base neutralization reactions.

The Arrhenius acid-base parents of a salt can be determined by adding OH⁻ to the cation to give the parent base and adding H⁺ to the anion to give the parent acid, as illustrated in Figure 8.34 using NH₄Cl salt as an example:

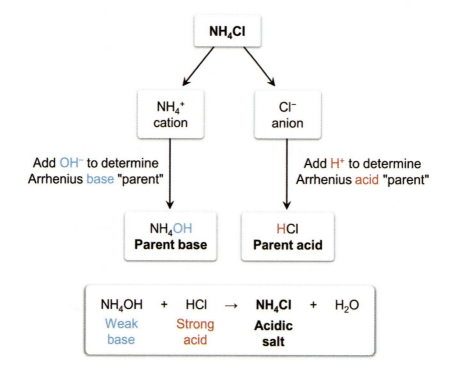

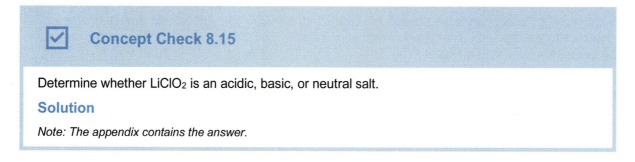

Figure 8.34 The classification of NH₄Cl as determined by parent acid-base species.

In cases where the parent acid and parent base of a particular salt are both weak, the acid-base behavior of the salt can be qualitatively determined by comparing the K_a of the parent acid to the K_b of the parent base. If K_a is larger than K_b, the salt is acidic overall; if K_b is larger than K_a, the salt is basic overall.

☑ Concept Check 8.15

Determine whether LiClO₂ is an acidic, basic, or neutral salt.

Solution

Note: The appendix contains the answer.

8.5.02 pH Calculations for Hydrolysis of Salts

In Lesson 8.3, calculations relating to weak acid and weak base equilibria were examined. These calculations also apply to salt solutions in which the salt has one or more ions with weak acid-base properties. Such calculations typically fall into one of two categories: the pH of the solution or the pK_a or pK_b of the ionic species.

pH of Salt Solutions

The complex process of calculating the pH of a salt solution can be simplified by breaking down the process into the following steps:

1. Determine if the salt has an ion that hydrolyzes in water and, if so, whether the ion is acidic or basic.

2. If the salt has an ion with acid-base properties, write the equation for the hydrolysis equilibrium of the ion based on whether the ion is acidic or basic.

3. If the K_a or K_b of the ion's conjugate species is given, use that constant and the autoionization constant of water (ie, $K_w = 1.0 \times 10^{-14}$ at 25 °C) to calculate the unknown ionization constant of the acidic or basic ion (ie, $K_w = K_a \cdot K_b$).

4. Analyze the hydrolysis equilibrium using an ICE chart.

5. Use the ion's K_a or K_b expression, along with the data from the ICE chart, to calculate $[H_3O^+]$ or $[OH^-]$, respectively.

6. Calculate the pH using pH = $-\log[H_3O^+]$. If $[OH^-]$ is determined in Step 5, first calculate $[H_3O^+]$ using $[OH^-]$ and K_w (ie, $K_w = [H_3O^+][OH^-]$).

For example, these steps can be applied to determine the pH of a 0.10 M $NaNO_2$ solution at 25 °C, as illustrated in Figure 8.35.

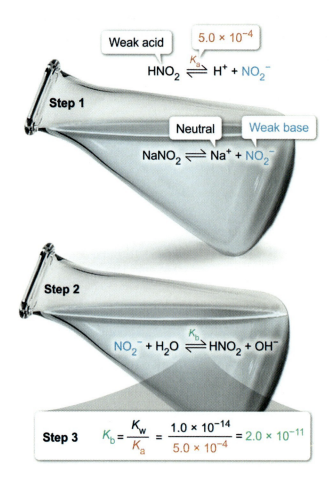

Figure 8.35 Stepwise method for calculating the pH of a basic salt solution (0.10 M NaNO$_2$).

Calculations of pH that involve either amphiprotic ions or salt species with both acidic and basic ions are extremely complex and beyond the scope of general chemistry. As such, they are not covered here.

pK_a or pK_b of Ions in Solution

To calculate the pK_a or pK_b of a non-amphiprotic ionic species, the pH of the solution and the concentration of the salt must be known. Analysis of the constituent ions of the salt is unnecessary in this case because the pH of the solution indicates if an acidic or basic ion is present. With these considerations in mind, the following steps can be used to carry out this type of calculation:

1. If the solution is acidic (ie, pH < 7), write an equation for the hydrolysis equilibrium of the acidic cation. If the solution is basic (ie, pH > 7), write an equation for the hydrolysis equilibrium of the basic anion.

2. Write the expression for the ionization constant (K_a or K_b) associated with the hydrolysis equilibrium from Step 1.

3. Use the solution pH to calculate [H_3O^+] if solving for pK_a or [OH^-] if solving for pK_b. The key relationships for this step are [H_3O^+] = 10^{-pH}, pH + pOH = 14, and [OH^-] = 10^{-pOH}.

4. (Optional) Use an ICE chart to determine the equilibrium concentrations of the other species in the equilibrium.

5. Solve for K_a or K_b using the calculated equilibrium concentrations.

6. Solve for pK_a or pK_b using the calculated K_a or K_b.

For example, these steps can be followed to evaluate the pK_a of a 0.25 M $Cu(NO_3)_2$ solution that has a pH of 4.30 at 25 °C, as illustrated in Figure 8.36.

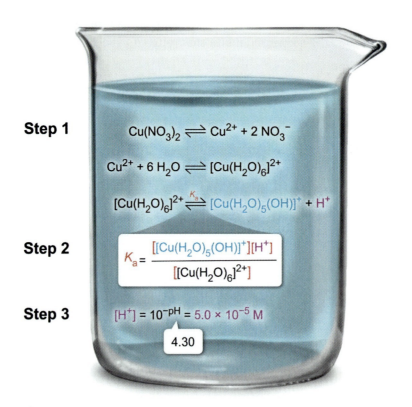

Step 1

$Cu(NO_3)_2 \rightleftharpoons Cu^{2+} + 2\,NO_3^-$

$Cu^{2+} + 6\,H_2O \rightleftharpoons [Cu(H_2O)_6]^{2+}$

$[Cu(H_2O)_6]^{2+} \overset{K_a}{\rightleftharpoons} [Cu(H_2O)_5(OH)]^+ + H^+$

Step 2

$$K_a = \frac{[[Cu(H_2O)_5(OH)]^+][H^+]}{[[Cu(H_2O)_6]^{2+}]}$$

Step 3

$[H^+] = 10^{-pH} = 5.0 \times 10^{-5}\,M$

4.30

Step 4

	$[Cu(H_2O)_6]^{2+}$ $\rightleftharpoons$	$[Cu(H_2O)_5(OH)]^+$	+	H^+
Initial	0.25 M	0		0
Change	-5.0×10^{-5} M	$+5.0 \times 10^{-5}$ M		$+5.0 \times 10^{-5}$ M
Equilibrium	0.25 M	5.0×10^{-5} M		5.0×10^{-5} M

Step 5

$$K_a = \frac{[[Cu(H_2O)_5(OH)]^+][H^+]}{[[Cu(H_2O)_6]^{2+}]} = 1.0 \times 10^{-8}$$

with $[Cu(H_2O)_5(OH)]^+ = 5.0 \times 10^{-5}$ M, $[H^+] = 5.0 \times 10^{-5}$ M, $[Cu(H_2O)_6]^{2+} = 0.25$ M

Step 6 $pK_a = -\log K_a = -\log(1.0 \times 10^{-8}) = 8.00$

Figure 8.36 Stepwise method for calculating the pK_a of a metal ion (Cu^{2+}).

Concept Check 8.16

What is the pH of a 0.10 M solution of NH_4Cl? (Note: For NH_3, $K_b = 1.8 \times 10^{-5}$)

Solution

Note: The appendix contains the answer.

Lesson 8.6

Acid-Base Titrations

Introduction

A titration is an experimental procedure that dispenses measured amounts of a reactant solution into another solution containing an unknown concentration of a substance of interest (the analyte). By applying the concepts of molarity and reaction stoichiometry (Lesson 1.6 and Lesson 2.5), a titration can determine the precise quantity of an analyte in an unknown sample.

Acid-base titrations are a common and useful type of titration, variations of which not only help determine the concentration or amount of an acidic or basic analyte but also provide additional useful information, such as ionization constants (K_a or K_b) and the number of acidic protons in an acidic substance. This lesson focuses specifically on acid-base titrations, including experimental methods, indicators and how to choose them, interpreting titration curves, and calculations associated with titrations.

8.6.01 Titration Method

Titration is method of chemical analysis that involves adding a carefully measured amount of a solution of known concentration (ie, the **titrant**) to a solution with an unknown concentration of a sample being analyzed (ie, the **analyte**, or **titrand**). The titrant is typically dispensed from a graduated tube called a buret.

The analyte solution contains a substance of interest with which the titrant quantitatively reacts (ie, the reaction goes to completion according to known stoichiometry). In many cases, an indicator is added to the analyte that changes color to signal the end of the titration.

An example of the experimental setup for a typical acid-base titration is illustrated in Figure 8.37.

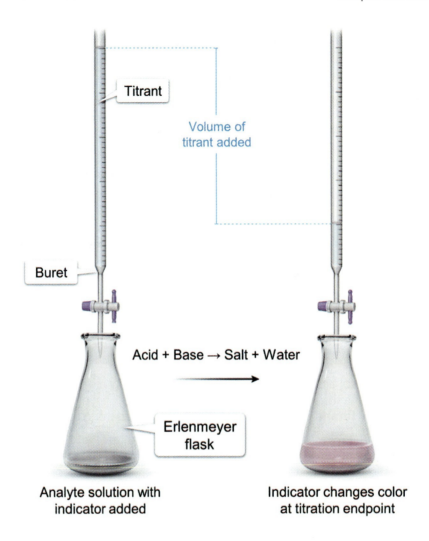

Figure 8.37 Experimental setup for a typical acid-base titration.

The equivalence point of a titration is reached when stoichiometrically equivalent amounts of titrant and analyte species have been combined and all the analyte species has reacted. The volume of titrant required to reach the equivalence point is referred to as the **equivalence volume**. A slight excess of titrant beyond the equivalence volume is often needed to trigger the color change of an indicator. This color change marks the **end point** of a titration.

For acid-base titrations, a pH meter can be used in place of an indicator. Graphing pH versus the volume of titrant added at various points during the titration generates a titration curve, from which the equivalence point and other useful information can be determined. Analyzing titration curves is discussed in more detail in Concept 8.6.03.

Chapter 8: Acid-Base Chemistry

> ☑ **Concept Check 8.17**
>
> A lab technician performs a titration to analyze a sample for quality control. The analyte is a drug with weak acid properties, which is titrated with NaOH. After twice the expected amount of NaOH titrant is added, no color change has been observed. Which of the following is the most likely reason for this phenomenon?
>
> A. The technician forgot to add the drug sample when making the analyte solution.
>
> B. The analyte solution has carbonic acid (H_2CO_3) present due to dissolved CO_2.
>
> C. The technician forgot to add an indicator.
>
> D. The analyte sample contains a basic contaminant that reacts with the drug being tested.
>
> **Solution**
>
> Note: The appendix contains the answer.

8.6.02 Indicators

As previously noted, indicators are substances that change colors to signal the end of a titration. An indicator typically does not react with the titrant until all (or nearly all) the analyte species has reacted. Indicators for acid-base titrations are usually weak acids (or weak base) whose protonated and deprotonated forms exhibit notably different colors, as illustrated in Figure 8.38.

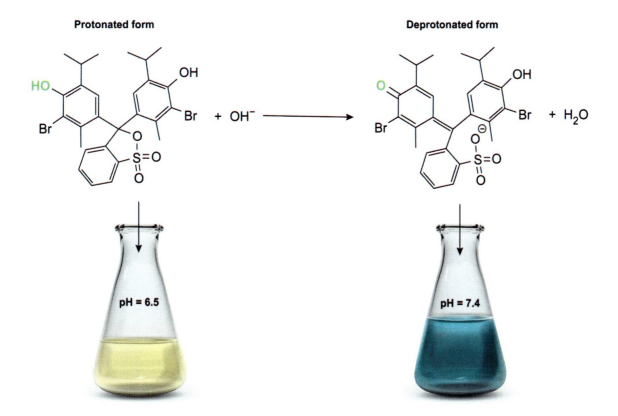

Figure 8.38 Protonated versus deprotonated forms of the indicator bromothymol blue.

An acid-base indicator chosen for a particular titration must be a weaker acid (or base) than the analyte species being tested to prevent the indicator reacting with the titrant before the analyte. Ideally, indicators should also be vibrantly colored so that only a few drops are needed for a notable color change at the end point.

This visual change ensures that the indicator has an extremely low concentration, and thus a negligible effect on pH. A chart of some common indicators showing their color changes and pH ranges is shown in Figure 8.39.

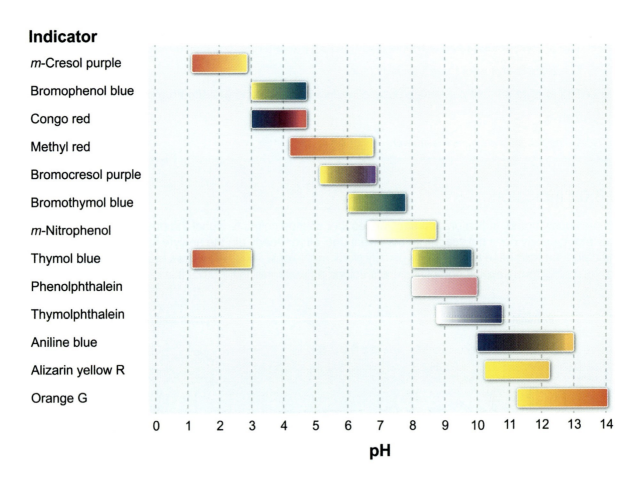

Figure 8.39 Chart of the pH ranges and color changes of common indicators.

In acid-base titrations, a rapid change in pH is observed just before and after the equivalence point is reached. Therefore, choosing a good indicator for a particular acid-base titration requires knowing the pH at the equivalence point.

Ideally, the pK_a of the indicator should be as close as possible to the pH at equivalence, ensuring that the color change of the indicator occurs over a narrow volume range within which the equivalence point is to be found, as shown in Figure 8.40.

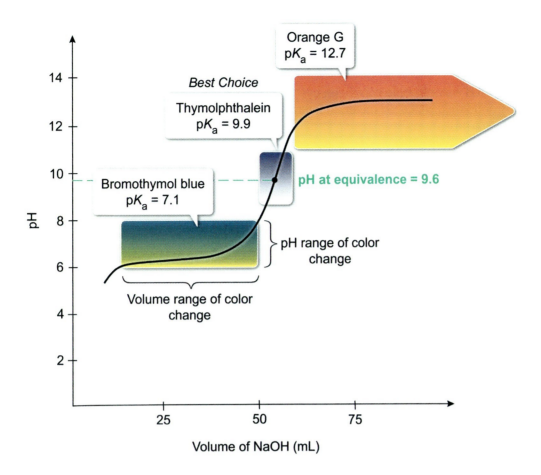

Figure 8.40 Comparison of indicators for use in a titration.

In addition to their use as additives in titrations, indicators can also be embedded and dried in a paper medium to make test strips (ie, pH paper) used for a quick analysis of solution pH. Some test strips, like litmus paper, only indicate whether a sample is acidic (red) or basic (blue). In contrast, universal indicator test strips contain a mixture of indicators and display a range of colors across the pH scale, and are used to make more precise pH estimates.

Chapter 8: Acid-Base Chemistry

✓ Concept Check 8.18

A contaminated water sample containing cyanide ions (CN^-) is titrated with HCl. The initial titration is performed with a pH meter to determine the pH at equivalence, generating the following titration curve.

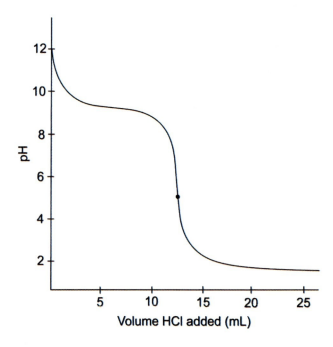

Which indicator from the table shown is best suited for subsequent titrations of similar samples?

Indicator	pK_a
Methyl orange	3.5
Bromophenol blue	4.1
Methyl red	5.1
Bromothymol blue	7.1
Phenolphthalein	9.0

Solution

Note: The appendix contains the answer.

8.6.03 Interpreting Titration Curves

Titration curves (ie, graphs of pH versus volume of titrant added) are used to analyze the data from acid-base titrations. Five common shapes of titration curves reflect the five main types of useful acid-base titrations:

- Strong acid titrated with strong base
- Weak acid titrated with strong base
- Strong base titrated with strong acid
- Weak base titrated with strong acid
- Polyprotic acid titrated with strong base

In each case, a strong species is used as the titrant to drive the reaction to completion.

Weak acid-weak base titrations are generally not performed because both species establish an equilibrium and the equivalence points are not well defined. Thus, their titration curves are not discussed here.

Strong Acid-Strong Base Titration Curves

Complete neutralization of a strong acid with a strong base results in a neutral solution (ie, pH = 7). As such, the pH at the equivalence point of any strong acid-strong base titration is always 7. As Figure 8.41 illustrates, if the analyte is a strong acid, the titration curve shows a very low starting pH that increases as the strong base titrant is added. The opposite is seen if the analyte is a strong base and a strong base titrant is added.

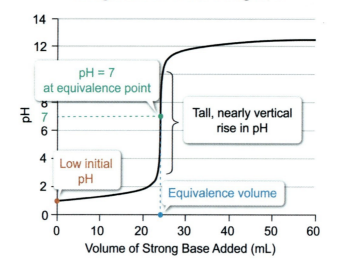

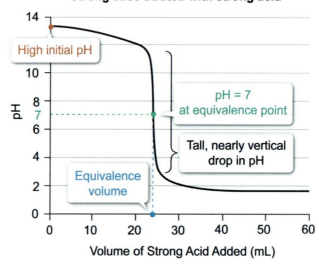

Figure 8.41 Comparison of the titration curve of a strong acid titrated with a strong base versus that of a strong base titrated with a strong acid.

A key feature of strong acid-strong base titration curves is the long, nearly vertical segment of the curve near the equivalence point. Before and after this sharp change in pH, the curve levels off and becomes relatively horizontal.

Titration Curve for a Weak Acid Titrated with a Strong Base

The neutralization of a weak acid with a strong base produces a conjugate base. Therefore, any titration of this sort always has a pH greater than 7 at the equivalence point. This type of curve also has a higher initial acidic pH (eg, usually between 2 and 6) than the initial pH observed for the titration of a strong acid with a strong base.

As the neutralization begins, an initial short, sharp increase in pH is observed that quickly levels off as the concentration of the conjugate base increases. The resulting mixture of the weak acid and its conjugate base produces a transient buffer mixture, which appears on the titration curve as a slightly upward-sloping buffer region. Subsequently, a steep increase in pH is seen as the amount of added base nears the equivalence point, and the pH then levels off as excess base accumulates. These characteristic features are illustrated in Figure 8.42.

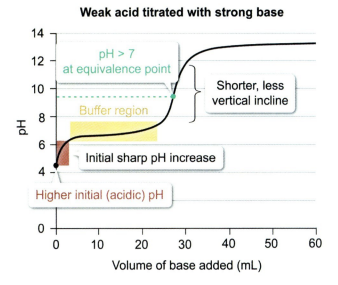

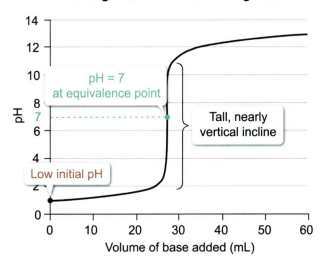

Figure 8.42 Comparison of the titration curve for a weak acid titrated with a strong base to that of a strong acid titrated with a strong base.

At the half-equivalence point (ie, the midpoint of the buffering region), half of the weak acid has been converted to its conjugate base, resulting in equal concentrations of each. According to the Henderson-Hasselbalch equation (Concept 8.4.02), the solution pH is equal to the pK_a of the weak acid when equal concentrations of weak acid and conjugate base are in solution. Therefore, pH = pK_a at the half-equivalence point.

This correlation of the pK_a to the half-equivalence point functions as a baseline on the titration curve and causes the change in pH (ie, the steeply inclined segment) around the equivalence point to have different heights, depending on the strength of the acid titrated. Stronger acids (ie, lower pK_a values) have larger changes in pH with longer, more steeply inclined segments through the equivalence point, as shown in Figure 8.43.

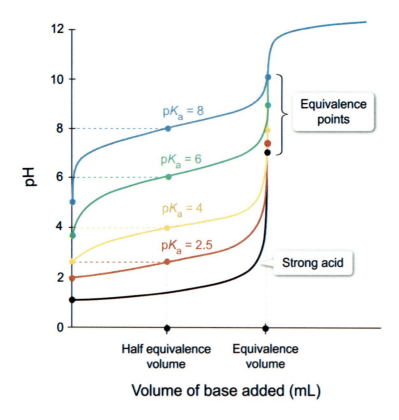

Figure 8.43 Effect of pK_a on the shape of the titration curves for weak acids titrated with strong bases.

Titration Curve for a Weak Base Titrated with a Strong Acid

When a weak base is titrated with a strong acid, the neutralization reaction between these two species produces a weak conjugate acid. At the equivalence point, the weak acid is the only species present that affects the pH, which causes the pH at the equivalence point to be notably acidic (ie, less than 7). The shape of the curve for a weak base titrated with a strong acid, detailed in Figure 8.44, resembles an upside-down version of the titration curve for a weak acid titrated with a strong base.

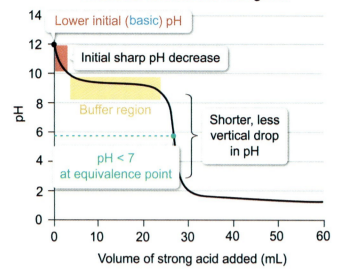

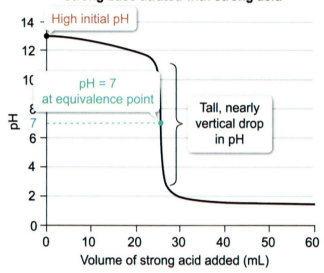

Figure 8.44 Comparison of the titration curve for a weak base titrated with a strong acid to that of a strong base titrated with a strong acid.

For the titration curve of a weak base with a strong acid, the pH at the half-equivalence volume is equal to the pK_a of the *conjugate acid* of the weak base being analyzed, from which the pK_b of the base may be determined (ie, pK_b = 14 − pK_a). As with the weak acid-strong base curve, the height of the nearly vertical part of the weak base-strong acid titration curve is dependent on the strength of the weak base being analyzed, with stronger bases having taller, steeper drops in pH as acid is added near the equivalence point (Figure 8.45).

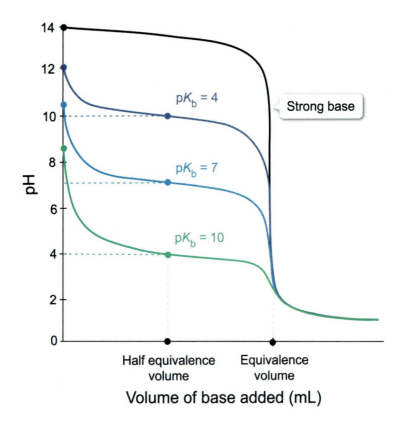

Figure 8.45 Effect of pK_b on the shape of the titration curves for weak bases titrated with strong acids.

Chapter 8: Acid-Base Chemistry

✓ Concept Check 8.19

An aqueous solution of HF (pK_a = 3.17) is titrated with a strong base. Which of the following is the most likely curve for this titration?

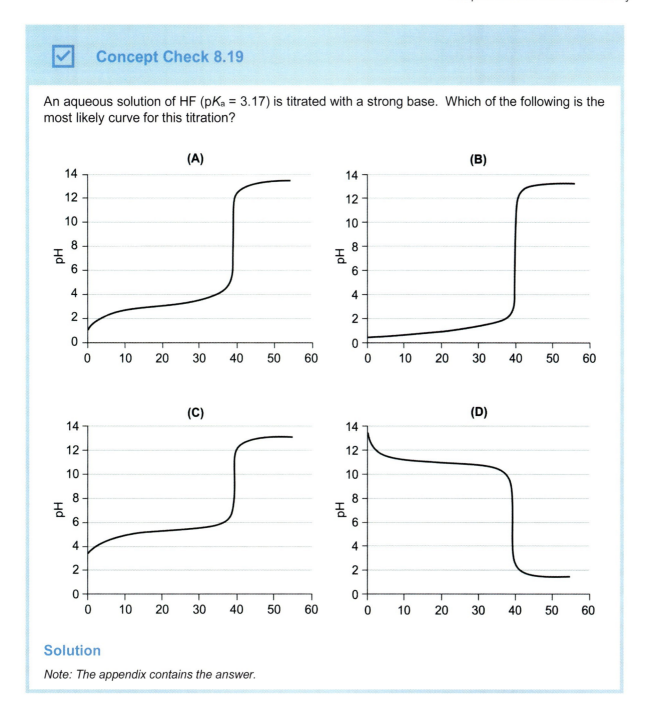

Solution
Note: The appendix contains the answer.

Titration Curve for a Polyprotic Acid Titrated with a Strong Base

When a polyprotic acid is titrated with a strong base, the initial shape of the titration curve is similar to that of a weak acid titrated with a strong base. However, the complete neutralization of the first acidic proton produces a conjugate species with at least one additional acidic proton. Thus, not only does the increase in pH at the first equivalence point tend to be smaller (an acid is still present), but the curve pattern repeats for each additional acidic proton until the acid is fully neutralized, as illustrated in Figure 8.46.

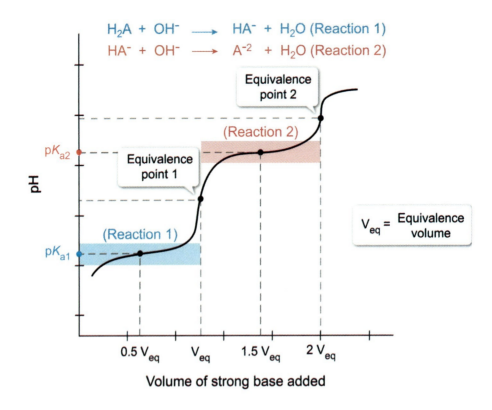

Figure 8.46 Titration curve of a weak diprotic acid titrated with a strong base.

The pH at the first half-equivalence point is equal to the pK_a for the first ionization (ie, the loss of the first and most acidic H$^+$ ion) of the polyprotic acid (ie, pK_{a1}). Likewise, the pH at the second half-equivalence point is equal to the pK_a for the second ionization (ie, the loss of the second H$^+$ ion) of the polyprotic acid (ie, pK_{a2}), and so on, for any additional half-equivalence points.

The number of equivalence points on a titration curve is equal to the number of distinct acidic protons in the acid structure. Because each equivalence point marks the complete removal of an acidic proton, equivalence points are useful markers for identifying the species present at each point in the titration (Figure 8.47).

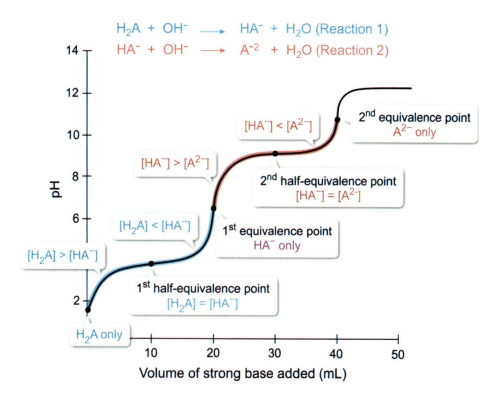

Figure 8.47 Major analyte species present at different points of a polyprotic acid (H₂A) titration with a strong base (H₂O, H₃O⁺, and OH⁻ excluded).

Concept Check 8.20

A solution of the amino acid histidine at a pH of 1 has three acidic protons. As the pH increases, histidine loses its acidic protons as follows:

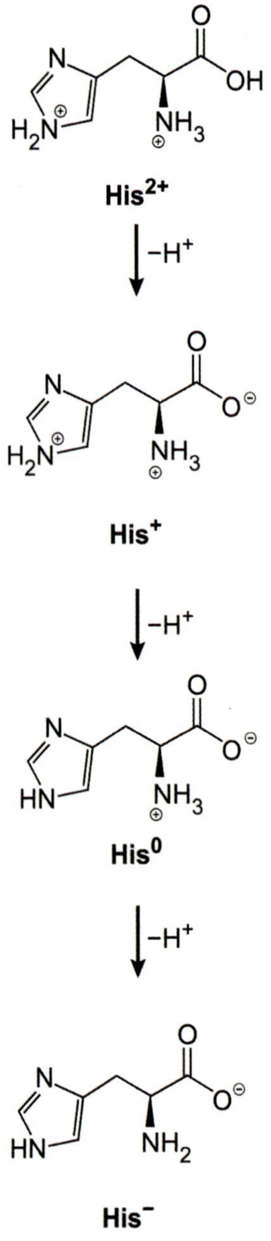

His^{2+}

$\downarrow$ $-H^+$

His$^+$

$\downarrow$ $-H^+$

His^0

$\downarrow$ $-H^+$

His$^-$

The titration of His^{2+} with a strong base yields the following titration curve. Which forms of histidine are present during the part of the titration indicated by the highlighted region of the curve?

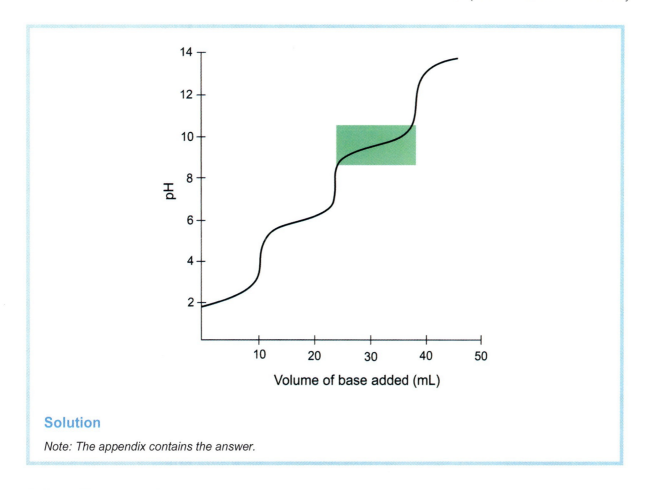

Solution

Note: The appendix contains the answer.

8.6.04 Titration Calculations

Titrations are often performed to determine either the amount or concentration of an analyte in an unknown sample. The amount of titrant species dispensed at the equivalence point is calculated, and the stoichiometric ratio from the balanced chemical equation is used to convert the amount of titrant species to the amount of analyte species.

For acid-base titrations, the titrant and analyte (ie, acid and base) have a 1:1 stoichiometric ratio for the first deprotonation, which is fully complete at the first equivalence point (Figure 8.48). The first equivalence point is the *only* equivalence point in titrations that does not involve polyprotic acids.

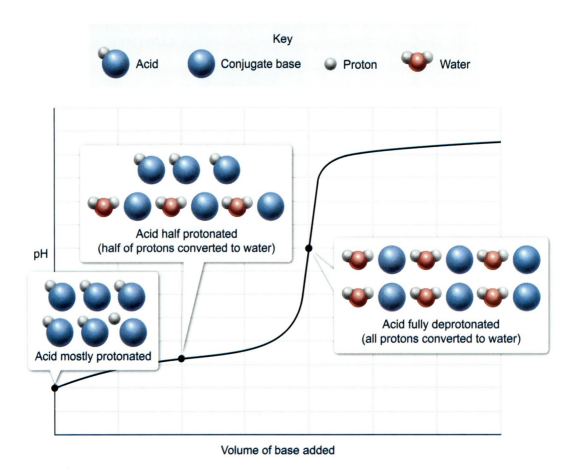

Figure 8.48 At the first equivalence point, the most acidic proton is completely deprotonated in all molecules.

The relationship between the moles of acid (n_{acid}) and the moles of base (n_{base}) that have been combined at the first equivalence point is:

$$n_{acid} = n_{base}$$

This relationship holds true for all acid-base titrations, regardless of whether the acid is the titrant and the base is the analyte or vice versa. Because moles of solute are equal to the molarity M multiplied by the solution volume V in liters (ie, $n = M \times V$), the previous equation can be restated in terms of molar concentration:

$$M_{acid} \cdot V_{acid} = M_{base} \cdot V_{base}$$

where M_{acid} and V_{acid} are the molarity and volume of the acid and M_{base} and V_{base} are the molarity and volume of the base combined at the first equivalence point.

Chapter 8: Acid-Base Chemistry

☑ Concept Check 8.21

A 10.0 mL sample of water contaminated with cyanide ions (CN^-) is titrated with 0.010 M HCl, and the following titration curve is generated from the data.

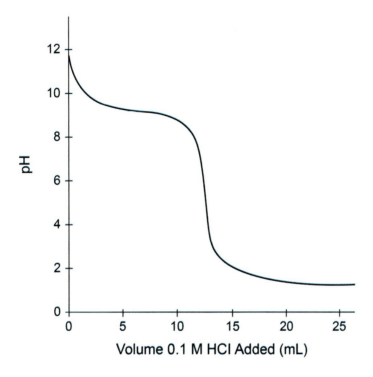

What is the concentration of CN^- in the contaminated water?

Solution

Note: The appendix contains the answer.

Chapter 8: Acid-Base Chemistry

Lesson 9.1

Review of Redox Reactions

9.1.01 Important Principles Associated with Redox Reactions

In contrast to Chapter 2, which first introduces oxidation-reduction (redox) reactions as one of several reaction types that are classified according to the chemical process involved, this chapter focuses on the application of redox reactions in useful analyses (eg, redox titrations) and devices (eg, batteries, electrolytic cells). As such, this lesson provides a brief summary of the important concepts and general principles associated with redox reactions:

- **Oxidation and reduction** processes
- **Oxidizing and reducing agents**
- **Oxidation numbers** to track the transfer of electrons during a redox reaction
- **Balancing redox reactions**

In subsequent lessons, these concepts and principles are presented in applied scenarios rather than as isolated reactions as in Chapter 2.

Oxidation and Reduction

As discussed in Concept 2.4.02, redox reactions involve a transfer of electrons (e^-) in which one species is oxidized (ie, loses electrons) as another species is simultaneously reduced (ie, gains electrons). During this process, the species containing the oxidized atom functions as the **reducing agent** because it *causes reduction* in another species while undergoing oxidation itself. In contrast, the species with the reduced atom acts as the **oxidizing agent** because it causes the oxidation of another species while being reduced in the process. These processes are summarized in Figure 9.1.

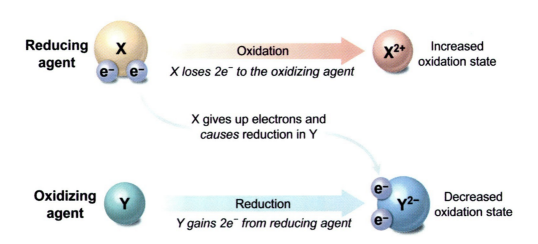

- Reducing agents *cause reduction* but get oxidized in the process
- Oxidizing agents *cause oxidation* but get reduced as a result

Figure 9.1 Oxidation and reduction caused by the oxidizing and reducing agents.

Oxidation Numbers

To track the transfer of electrons between atoms and determine which species is oxidized or reduced during a redox reaction, the oxidation numbers of each atom before and after the reaction must be determined using the oxidation number rules outlined in Concept 2.4.03. If the oxidation number of an atom decreases (becomes less positive), the atom has been reduced whereas a number that increases indicates the element has been oxidized. This is illustrated in Figure 9.2 using a reaction between bismuth and hydrochloric acid.

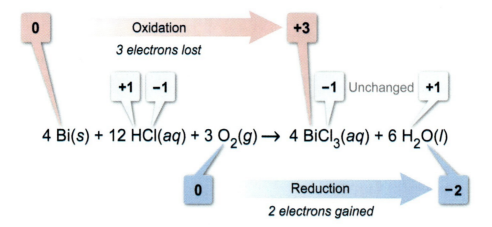

Figure 9.2 Oxidation numbers in the reaction of bismuth with hydrochloric acid in the presence of oxygen.

Balancing Redox Reactions

Redox reactions are fully balanced when the same number of each type of atom and the same net charge are present on both sides of the reaction equation. An overall (net) redox reaction can be expressed as the sum of the balanced oxidation and reduction half-reactions, as shown in Figure 9.3.

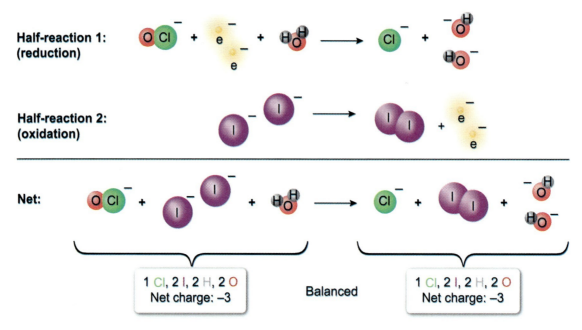

Figure 9.3 Balanced redox reaction between aqueous bleach (OCl⁻) and iodide (I⁻) under basic conditions.

Note that in some reactions, multiples of one or both half-reactions may be needed to achieve an electron balance that cancels the electrons in the overall summation. For redox reactions taking place in acidic or basic conditions, the balancing process often requires the addition of H^+ or OH^- ions and H_2O (from the solution medium) to the half-reactions to balance the atoms and charges (as outlined in Concept 2.4.06).

Lesson 9.2
Redox Titrations

9.2.01 Titrations Involving Redox Reactions

Redox titrations are similar to acid-base titrations (Lesson 8.6) in that both techniques slowly add a titrant solution with a known concentration to another solution containing an unknown concentration of an analyte (ie, a species to be measured). However, the key difference is that acid-base titrations track the movement of *protons* whereas redox titrations track the movement of *electrons* to determine the unknown concentration of the analyte.

During a redox titration, the titrant (either an oxidizing or reducing agent) is slowly added to the analyte solution, causing the analyte to be either oxidized or reduced, respectively. This process progressively changes the solution's electric potential (ie, the voltage, defined as potential energy per unit charge). The voltage changes sharply when the titration passes through the equivalence point, which occurs when all the analyte has been oxidized or reduced, depending on the titrant.

One simple method for detecting the equivalence point involves the use of a redox indicator that changes color near the equivalence point voltage. Note that this is analogous to an acid-base indicator that changes color near the equivalence point pH of an acid-base titration (see Concept 8.6.02). For example, a common type of redox titration involves titrating a solution containing elemental iodine with thiosulfate ($S_2O_3^{2-}$), a reducing agent, and detecting the equivalence point using a starch indicator.

$$I_2 + 2\ S_2O_3^{2-} \rightarrow S_4O_6^{2-} + 2\ I^-$$

Elemental iodine (I_2) forms a dark blue complex with starch; however, when the $S_2O_3^{2-}$ titrant reduces all the iodine ions (I^-) (ie, the oxidation state decreases from 0 to −1), the blue color disappears. This color change indicates that the equivalence point has been reached (Figure 9.4).

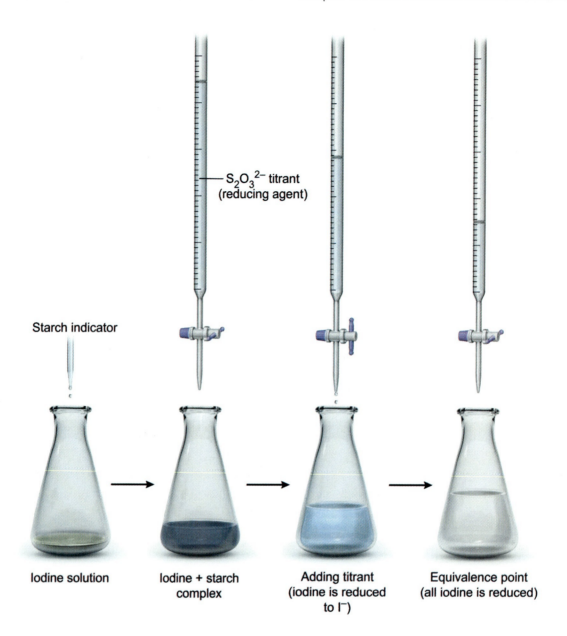

Figure 9.4 Detecting the equivalence point of a redox titration using a starch indicator.

Alternatively, the equivalence point of a redox titration can be determined by its titration curve, which is a plot of the solution's potential (measured using an electrode) as a function of the volume of titrant added. An example of a redox titration curve is shown in Figure 9.5.

Chapter 9: Redox Reactions and Electrochemistry

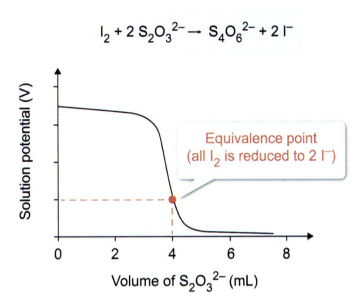

Figure 9.5 Redox titration curve for the titration of 0.1 L of an elemental iodine solution with 0.10 M thiosulfate.

On a redox titration curve, the equivalence point is found at the midpoint of the nearly vertical segment (ie, a symmetric titration curve) when the titrant and analyte react in a 1:1 molar ratio (as with acid-base titrations). However, if the titrant and analyte do *not* react in a 1:1 molar ratio, the equivalence point lies elsewhere along the steep gradient (ie, an asymmetric titration curve), as seen in Figure 9.5.

According to Figure 9.5, 4 mL (0.004 L) of $S_2O_3^{2-}$ are needed to reach the equivalence point of the titration. Therefore, because 2 moles of $S_2O_3^{2-}$ are needed to reduce 1 mole of I_2 (ie, a 2:1 molar ratio), the unknown molar concentration of I_2 in the 0.1 L analyte solution can be calculated as:

$$0.004 \; \cancel{L \, S_2O_3^{2-}} \times \frac{0.1 \; \cancel{mol \, S_2O_3^{2-}}}{1 \; \cancel{L}} \times \frac{1 \; mol \; I_2}{2 \; \cancel{mol \, S_2O_3^{2-}}} \times \frac{1}{0.1 \; L} = 0.002 \; M \; I_2$$

Chapter 9: Redox Reactions and Electrochemistry

Lesson 9.3
Electrochemical Cells

Introduction

An **electrochemical cell** is an apparatus that converts chemical energy into electrical energy (and vice versa) by harnessing an oxidation-reduction (redox) reaction that takes place within the cell. Fundamentally, an electrochemical cell consists of two electrodes—an anode and a cathode—immersed in a conductive electrolyte solution. The cell operates in one of two ways:

- **Passive operation** yields an electric current as electrons are transferred from one electrode to the other during a spontaneous redox reaction in the cell.
- **Active operation** involves supplying an external electric current through the electrodes to achieve a nonspontaneous redox reaction within the cell.

In this lesson, the key components of the cell apparatus and their functions are discussed in detail. Aspects related to the cell potential and its relationship to spontaneity are also explored.

9.3.01 Cell Components and Their Function

A typical electrochemical cell is divided into two separate compartments known as **half-cells**, where either oxidation or reduction occurs. Each half-cell is composed of a conductive implement (usually a wire or strip of metal) known as an **electrode**, which is immersed in an electrolyte solution. The oxidation and reduction half-cells are connected by a wire running from one electrode to the other and by a porous structure known as a **salt bridge** that connects the two electrolyte solutions. The major components of an electrochemical cell are shown in Figure 9.6.

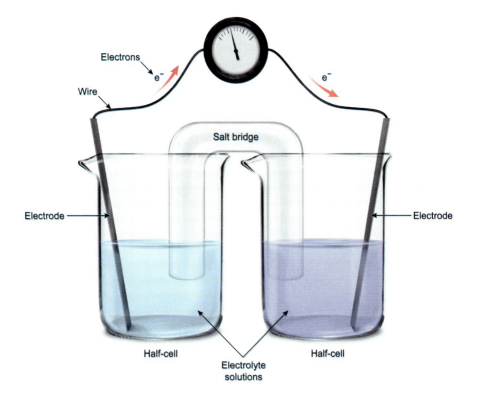

Figure 9.6 Major components of an electrochemical cell.

Electrodes

An electrode is made of a solid conducting material that facilitates the transfer of electrons from one half-cell to the other. Electrodes are classified as one of two types:

- **Active electrode**. This type of electrode is a metal that actively participates in the redox reaction occurring in the cell. The electrode either gains or loses mass depending on whether it is reduced or oxidized. For example, if a magnesium electrode is oxidized, Mg atoms are converted to Mg^{2+} ions, which enter the electrolyte solution. In this scenario, the electrode is consumed and its mass decreases. Conversely, if the ions in solution are reduced, they deposit as metal atoms on the electrode surface and cause its mass to increase.
- **Inert electrode**. This type of electrode is made of a conductive material that does not participate or interfere with the redox reaction occurring in the cell. Instead, the reactants in the electrolyte solution are oxidized or reduced at the surface of the inert electrode without altering the electrode. Thus, the main purpose of an inert electrode is to act as a medium for electrons. Common inert electrodes include graphite and platinum.

Wire

A piece of wire made of a conducting material (eg, silver, copper) is attached to both electrodes, creating a pathway for the electrons generated during the redox reaction to flow from one electrode to the other. Typically, a device called a voltmeter is inserted along the wire between the electrodes to measure the cell voltage. Alternatively, an external power source can be inserted along the wire to provide an electric current for nonspontaneous reactions.

Electrolyte Solution

The electrolyte solution surrounding the electrodes is typically an aqueous solution of dissolved ionic compounds whose ions act as charge-carrying electrolytes. Furthermore, electrolyte solutions also serve as reservoirs of ions produced or consumed as the cell operates. If an active electrode is used, it is common for the solution to contain ions of the same metal type as the electrode (eg, a copper electrode submerged in a solution containing Cu^{2+} ions).

Salt Bridge

A salt bridge is typically a U-shaped glass tube (or porous filter paper) filled (or soaked) with an ionic salt solution, such as $NaNO_3(aq)$. As the cell operates, ions migrate from the salt bridge to maintain a net neutral ionic charge within each half-cell. Negative ions migrate from the salt bridge to the half-cell containing the oxidizing electrode, and positive ions migrate to the half-cell containing the reducing electrode. An example of ion migration from a salt bridge is shown in Figure 9.7.

Chapter 9: Redox Reactions and Electrochemistry

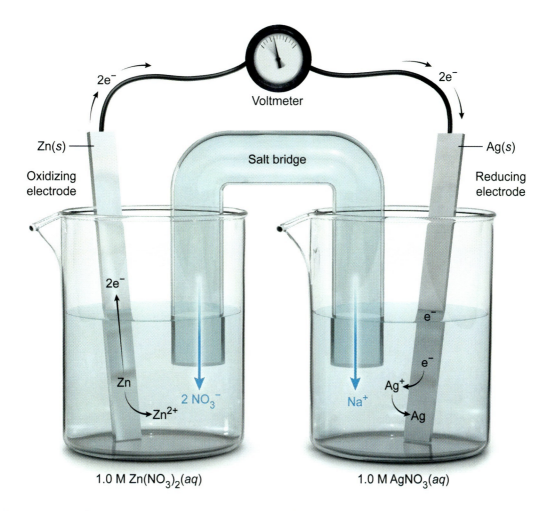

Figure 9.7 Migration of ions from a salt bridge containing NaNO$_3$(aq) during the operation of an electrochemical cell.

Without the salt bridge, a buildup of positive charges would occur in the oxidation half-cell and a buildup of negative charges would occur in the reduction half-cell. This buildup of charges would result in a charge imbalance, which would prevent the flow of ions and halt the redox reaction.

> ✓ **Concept Check 9.1**
>
> Consider the electrochemical cell operating as shown.
>
>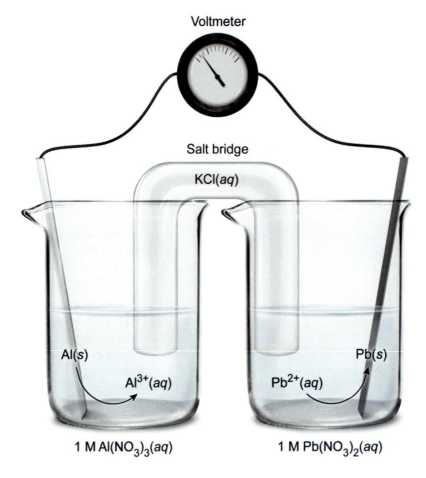
>
> In which direction, toward the Al(s) or Pb(s) electrode, will the ions in the salt bridge flow?
>
> **Solution**
>
> *Note: The appendix contains the answer.*

9.3.02 Redox Processes at the Electrodes

In every electrochemical cell, the electrode where oxidation takes place is called the anode and the electrode where reduction takes place is called the cathode. Because electrons are generated at the anode by the redox reaction, the anode is the more *negatively* charged electrode and the cathode is the more *positively* charged electrode. Therefore, electrons always flow from the anode to the cathode (ie, from negative to positive) through the connecting wire, as depicted in the electrochemical cell seen on the left in Figure 9.8.

Chapter 9: Redox Reactions and Electrochemistry

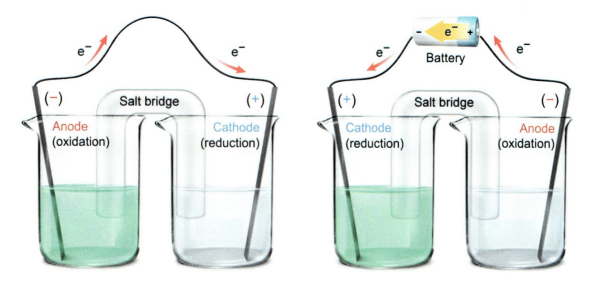

Figure 9.8 Electrons flow from the anode to the cathode.

If an external power source (eg, a battery) is inserted along the wire, the redox reaction is reversed and the roles of the electrodes are switched (ie, the anode becomes the cathode, and the cathode becomes the anode), as illustrated in the modified electrochemical cell seen on the right in Figure 9.8. Although the role of each electrode changes, the electrons still flow from the anode to the cathode.

Recall from Concept 2.4.06 that a redox reaction can be broken into two half-reactions (ie, one for oxidation and one for reduction). Consequently, a half-reaction can be written for each electrode in the electrochemical cell. For example, consider an electrochemical cell in which a Cr(s) anode is submerged in a solution containing $Cr^{3+}(aq)$ ions and a Cu(s) cathode is submerged in a solution containing $Cu^{2+}(aq)$ ions. The half-reaction occurring at each electrode can be written as:

Anode (oxidation) half-reaction: $Cr(s) \rightarrow Cr^{3+}(aq) + 3\ e^-$

Cathode (reduction) half-reaction: $Cu^{2+}(aq) + 2\ e^- \rightarrow Cu(s)$

At the Cr(s) anode, electrons (e^-) are generated as chromium atoms are oxidized to $Cr^{3+}(aq)$ ions. The electrons flow through the wire to the Cu(s) cathode, where $Cu^{2+}(aq)$ ions are reduced to Cu(s) atoms. This process is illustrated in Figure 9.9.

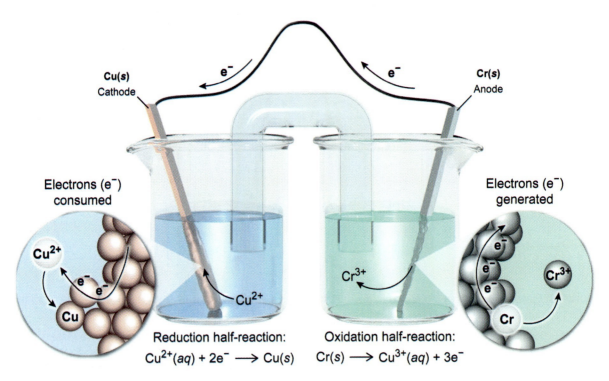

Figure 9.9 Reduction and oxidation half-reactions occurring in an electrochemical cell with a Cu(s) cathode and a Cr(s) anode.

As noted in Concept 9.3.01, an active electrode either gains or loses mass depending on whether it is reduced or oxidized. In this example, the Cr(s) electrode is oxidized and some of the chromium atoms are converted to Cr^{3+} ions, which enter the electrolyte solution. Thus, the Cr(s) anode is partially consumed (ie, appears "eaten away") and its mass decreases. In contrast, some of the $Cu^{2+}(aq)$ ions surrounding the Cu(s) cathode are reduced and deposited as Cu(s) atoms on the active electrode surface and the cathode's mass increases (ie, the electrode appears to "grow").

9.3.03 Standard Reduction Potentials and Cell Potential

The tendency for a species participating in a redox reaction to be either reduced or oxidized is predicted by its **standard reduction potential** $E°_{red}$, which is the potential (voltage) measured during a *reduction* reaction occurring in a cell under standard state conditions.

The value of $E°_{red}$ is measured relative to a reference electrode known as the **standard hydrogen electrode (SHE)**, which by convention is assigned a potential of exactly zero. As such, the SHE can be used to measure the potential of different types of electrodes. The standard reduction potentials of some selected common half-reactions are provided in Table 9.1.

Table 9.1 Standard reduction potentials for selected common half-reactions at 25° C.

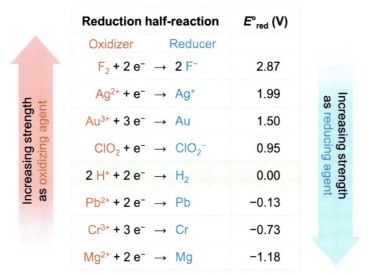

In Table 9.1, $E°_{red}$ becomes increasingly positive for values above the SHE potential and increasingly negative for values below it. The more positive (ie, less negative) $E°_{red}$ is, the greater the tendency for the species on the *left* of the half-reaction arrow to be *reduced* (ie, a stronger oxidizing agent). Conversely, the less positive (ie, more negative) $E°_{red}$ is, the greater the tendency for the species on the *right* of the half-reaction arrow to be *oxidized* (ie, a stronger reducing agent). As such, the table is arranged from strongest oxidizing agent (weakest reducing agent) at the top to weakest oxidizing agent (strongest reducing agent) at the bottom.

In an electrochemical cell, the $E°_{red}$ values of the two electrodes can be compared to determine the anode and the cathode. When no external power source is applied, the electrode with the more positive $E°_{red}$ acts as the cathode (where reduction occurs) and the electrode with the less positive $E°_{red}$ acts as the anode (where oxidation occurs). As such, the $E°_{red}$ values predict the electrode roles for the spontaneous redox reaction between the electrodes. Consequently, if an external power source is applied, the redox reaction is reversed (ie, the nonspontaneous direction) and the electrode roles are switched.

For example, suppose an electrochemical cell without an external power source consists of Cu(s) and Zn(s) electrodes, each submerged in an electrolyte solution of their respective metal cations (ie, Cu^{2+}, Zn^{2+}). Note that $E°_{red} = 0.34$ V for $Cu^{2+}(aq)$ and $E°_{red} = -0.76$ V for $Zn^{2+}(aq)$. Because $E°_{red}$ is more positive for $Cu^{2+}(aq)$ than for $Zn^{2+}(aq)$, reduction occurs at the Cu(s) electrode (ie, the cathode) and oxidation occurs at the Zn(s) electrode (ie, the anode) when the cell operates spontaneously, as summarized in Figure 9.10.

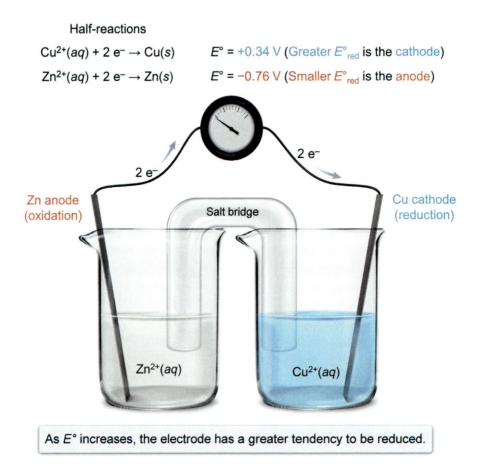

Figure 9.10 Cathode and anode determination using standard reduction potentials $E°_{red}$.

Typically, tabulated potentials are given for only reduction half-reactions. Because reduction and oxidation are reverse processes, **standard oxidation potentials** $E°_{ox}$ can be determined by simply reversing the reduction half-reaction and switching the sign of $E°_{red}$. Accordingly, the oxidation half-reaction and $E°_{ox}$ of the Zn(s) anode are:

$$Zn(s) \rightarrow Zn^{2+}(aq) + 2\ e^- \qquad E°_{ox} = +0.76\ V$$

Because reduction and oxidation occur simultaneously, the net standard potential of the redox reaction occurring in the cell, referred to as the **standard cell potential** $E°_{cell}$, is equal to the *sum* of $E°_{red}$ and $E°_{ox}$:

$$E°_{cell} = E°_{red} + E°_{ox}$$
$$E°_{cell} = 0.34\ V + 0.76\ V = 1.10\ V$$

Alternatively, $E°_{cell}$ can be determined by the difference between the standard reduction potentials of the cathode $E°_{cathode}$ and anode $E°_{anode}$:

$$E°_{cell} = E°_{cathode} - E°_{anode}$$
$$E°_{cell} = 0.34\ V - (-0.76\ V) = 1.10\ V$$

Note that if a multiple of either (or both) half-reactions is needed to balance the electrons in the net reaction (see Concept 2.4.06), the multiple *does not* need to be applied to the standard potentials.

Concept Check 9.2

$$Al^{3+}(aq) + 3\ e^- \rightarrow Al(s) \quad E°_{red} = -1.66\ V$$

$$Sn^{2+}(aq) + 2\ e^- \rightarrow Sn(s) \quad E°_{red} = -0.14\ V$$

The half-reactions and standard reduction potentials of an electrochemical cell are shown. Based on this information, what is $E°_{cell}$? (Assume no external power source is used.)

Solution

Note: The appendix contains the answer.

9.3.04 Gibbs Free Energy of an Electrochemical Cell

In an electrochemical cell, a spontaneous redox reaction converts chemical energy to electrical energy to produce an electric current, or a nonspontaneous redox reaction enabled by an applied external current converts electrical energy into chemical energy. The change in Gibbs free energy $\Delta G°$ resulting from these energy conversions represents the maximum amount of work that can be performed during the reaction. The amount of work required to move one unit of charge through an electrochemical cell is the **electromotive force (EMF)** and is inferred from $E°_{cell}$. As such, $\Delta G°$ and $E°_{cell}$ are directly related by:

$$\Delta G° = -nFE°_{cell}$$

or equivalently,

$$E°_{cell} = -\frac{\Delta G°}{nF}$$

where *n* is the number of electrons transferred per reaction and *F* is Faraday's constant (see Concept 9.5.01 for discussion of Faraday's constant).

Recall from Concept 3.7.02 that a negative $\Delta G°$ indicates a spontaneous reaction and a positive $\Delta G°$ indicates a nonspontaneous reaction (ie, cannot proceed without adding energy). Therefore, a spontaneous redox reaction has a positive $E°_{cell}$ and a nonspontaneous redox reaction has a negative $E°_{cell}$ (Figure 9.11).

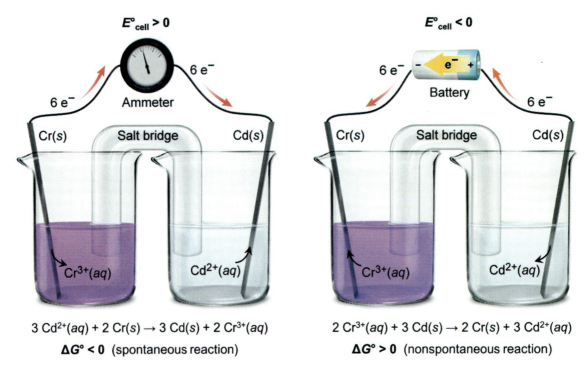

Figure 9.11 Gibbs free energy ($\Delta G°$) and cell potential ($E°_{cell}$) for spontaneous and nonspontaneous redox reactions in electrochemical cells.

✓ Concept Check 9.3

The overall redox reaction occurring in an electrochemical cell is shown here.

$$2\ Fe^{3+}(aq) + 2\ I^-(aq) \rightarrow 2\ Fe^{2+}(s) + I_2(s) \qquad E°_{cell} = 0.23\ V$$

1) What is $\Delta G°$ for the reaction?
2) Is the reaction spontaneous or nonspontaneous?

Solution

Note: The appendix contains the answer.

Lesson 9.4
Types of Electrochemical Cells

Introduction

Electrochemical cells perform conversions between chemical and electrical forms of energy. As such, an electrochemical cell can generate electrical energy (ie, current) from a spontaneous redox reaction or use electrical energy from an external source to drive a nonspontaneous redox reaction.

$$\text{Reactants} \underset{\text{nonspontaneous}}{\overset{\text{spontaneous}}{\rightleftarrows}} \text{Products} + \text{Electrical energy}$$

The type of energy conversion a cell performs depends on the setup of the cell. In this lesson, three fundamental types of electrochemical cells are discussed: galvanic cells, concentration cells, and electrolytic cells.

9.4.01 Galvanic Cells

A **galvanic cell** (also called a voltaic cell) is a type of electrochemical cell that uses a spontaneous ($\Delta G°$ < 0, $E°_{cell}$ > 0) redox reaction to generate an electric current. In a typical galvanic cell, two distinct electrodes (often two different metals) are placed in separate vessels (ie, half-cells) containing an aqueous electrolyte solution. Typically, the electrolyte solution surrounding each electrode contains ions of the same metal type as the electrode. A general galvanic cell is presented in Figure 9.12.

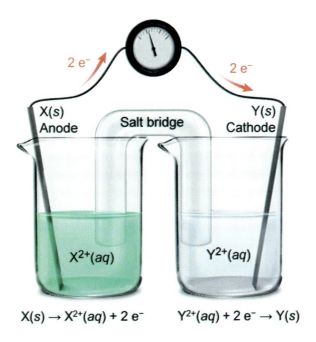

Figure 9.12 The general setup of a galvanic (voltaic) cell.

When the electrodes are connected by a conducting wire, the species with the lower standard reduction potential $E°$ (see Concept 9.3.03) is spontaneously oxidized at the anode, and the species with the higher $E°$ is spontaneously reduced at the cathode. As the reaction proceeds, an electric current is generated as the electrons continually migrate from the anode to the cathode.

9.4.02 Concentration Cells

A **concentration cell** is a unique type of galvanic cell. Both types of cells are similar in that they are composed of two-half cells connected by a conductive wire, which allows for a spontaneous ($\Delta G° < 0$, $E°_{cell} > 0$) redox reaction to occur. However, in a concentration cell the anode and cathode half-cells are composed of the *same* type of electrode material and ionic solution, and the *concentration* of the ionic solutions differs between the half-cells. As a result, the same half-reaction occurs in opposite directions at both electrodes. The setup of a general concentration cell is shown in Figure 9.13.

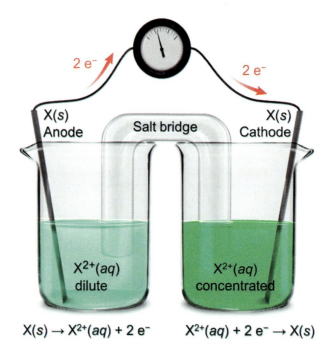

Figure 9.13 The general setup of a concentration cell.

Concentration cells produce a flow of electrons caused by the drive to equalize the ion concentrations in the two half-cells. In the less-concentrated half-cell, cations are generated by oxidizing metal atoms at the anode, and in the more concentrated half-cell, cations are consumed by reduction at the cathode until the concentrations in each half-cell are equal. As such, electrons flow from the anode to the cathode until both half-cells have the same cation concentration, at which point the reaction stops and $E°_{cell} = 0$. Therefore, a concentration cell generates current spontaneously as a function of the concentration gradient.

9.4.03 Electrolytic Cells

An **electrolytic cell** is a type of electrochemical cell that uses an electric current from an external power source (such as a battery) to enable a nonspontaneous ($\Delta G° > 0$, $E°_{cell} < 0$) redox reaction to occur within the cell. The configuration of an electrolytic cell is nearly identical to that of a galvanic cell (Concept 9.4.01), with the key difference being the connection of a power source between the two electrodes, as shown in Figure 9.14.

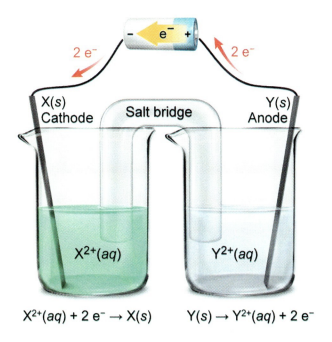

Thermodynamically unfavorable chemical reaction ($\Delta G > 0$) driven by current supplied from battery

Figure 9.14 The general setup of an electrolytic cell.

Using a loosely applied analogy, an electrolytic cell can be conceptualized as a galvanic cell operating in reverse. In both types of cells, reduction takes place at the cathode and oxidation takes place at the anode; however, the role of a given electrode switches in each cell type. The electrode that functions as the anode in a galvanic cell functions as the cathode in an electrolytic cell, and vice versa. As such, the direction of electron flow in an electrolytic cell is opposite the direction of the electron flow in a galvanic cell.

Electrolytic cells are useful to perform electrolysis (ie, an electrochemical decomposition reaction) or electroplating (ie, electrochemically depositing metal layers onto surfaces). For electrolysis, an electrolytic cell can sometimes be constructed using a single cell instead of two half-cells. The application of electrolytic cells for electrolysis is discussed in Lesson 9.5.

Lesson 9.5

Applications of Electrochemical Cells

Introduction

The versatility of electrochemical cells has led to their widespread adoption in numerous industries, ranging from aerospace and defense to medical and renewable energy sectors. Electrochemical cells harness the power of redox reactions to either convert chemical energy into electrical energy or vice versa, depending on their application. This lesson gives a detailed description of two common applications of electrochemical cells: electrolysis and rechargeable batteries.

9.5.01 Electrolysis

Electrolysis is a process in which an electric current is used to drive the nonspontaneous, electrochemical decomposition of a compound. This process is often used in industry for extracting metals or depositing metals onto a solid surface (ie, electroplating). An electrolytic cell is used to carry out electrolysis and is typically constructed using two inert electrodes (eg, platinum, graphite), both of which are immersed in the same liquid or electrolyte solution.

For example, consider an electrolytic cell constructed using two inert electrodes connected to a battery and placed in a vessel containing a $Cr(NO_3)_3(aq)$ solution (Figure 9.15).

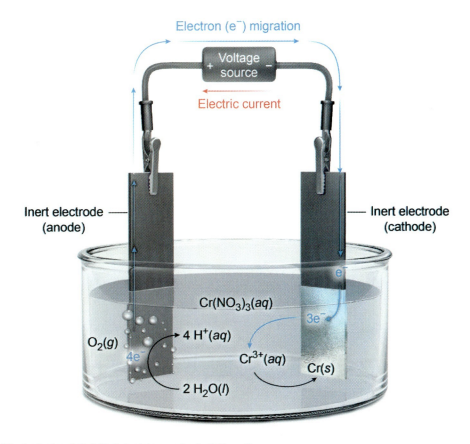

Figure 9.15 Electrolysis of $Cr(NO_3)_3(aq)$ in an electrolytic cell.

As current flows through the cell, water is oxidized to form $O_2(g)$ at the anode while $Cr^{3+}(aq)$ ions are reduced to form Cd(s) at the cathode (ie, Cd(s) is plated onto the surface of the cathode), as described by the following half-reactions:

Reaction at anode: $\quad 2\,H_2O(l) \rightarrow O_2(g) + 4\,H^+(aq) + 4\,e^-$

Reaction at cathode: $\quad Cr^{3+}(aq) + 3\,e^- \rightarrow Cr(s)$

The relationship between the mass of a chemical species produced at an electrode surface and the electric current within the cell are quantitatively described by Faraday's laws of electrolysis. These laws state that the mass of chemical species produced is directly proportional to the molar mass of the species and the quantity of electric charge (carried by the electrons) transferred through the electrode.

For example, suppose 3.82 A of electric current passes through the electrolytic cell in Figure 9.15 for 3,400 seconds, resulting in a certain mass of Cr(s) being deposited on the cathode surface. To determine the mass of Cr(s) deposited, the number of moles of electrons transferred through the cell must first be calculated.

Note that electric current is measured in amperes, where 1 ampere (A) is equal to 1 coulomb (C) of electric charge flowing through a conductor per second (ie, 1 A = 1 C/s). As such, multiplying the time by the amount of current yields the amount of electric charge flowing through the cell:

$$3{,}400\text{ s} \times \frac{3.82\text{ C}}{\text{s}} = 13{,}000\text{ C}$$

The electric charge carried by 1 mole of electrons (e^-) is equal to 96,485 C, which is known as Faraday's constant. As such, dividing the amount of charge transferred through the cell by Faraday's constant gives the moles of electrons transferred:

$$13{,}000\text{ C} \times \frac{1\text{ mol }e^-}{96{,}485\text{ C}} = 0.13\text{ mol }e^-$$

Based on the stoichiometric coefficients of the cathode half-reaction, 3 moles of electrons are transferred for each 1 mole of Cr(s) deposited. Therefore, the total number of moles of Cr(s) produced is evaluated as:

$$0.13\text{ mol }e^- \times \frac{1\text{ mol Cr}(s)}{3\text{ mol }e^-} = 0.043\text{ mol Cr}(s)$$

Multiplying the moles of Cr(s) by its molar mass yields the mass of Cr(s) deposited at the cathode surface:

$$0.043\text{ mol Cr}(s) \times \frac{112.4\text{ g Cr}(s)}{1\text{ mol Cr}(s)} \approx 5.0\text{ g Cr}(s)\text{ deposited}$$

Alternatively, the mass m of a chemical species generated at an electrode during time t can be evaluated by the expression:

$$m = \frac{ItM}{zF}$$

where I is the average electric current passing through the cell, z is the number of electrons transferred per reaction (obtained from the balanced half-reaction), M is the molar mass of the species produced, and F is Faraday's constant. Substituting the values into the equation yields the same numerical result:

$$m = \frac{\left(3.82\,\frac{C}{s}\right)(3{,}400\,s)\left(112.4\,\frac{g}{mol}\right)}{(3\,e^-)\left(96{,}485\,\frac{C}{mol\,e^-}\right)} \approx 5.0\,g\,Cd(s)$$

9.5.02 Batteries

A rechargeable battery is a portable power source consisting of one or more electrochemical cells that convert chemical energy into electrical energy. When a rechargeable battery is discharged, a spontaneous redox reaction occurs, producing an electric current; consequently, a discharging battery operates as a galvanic cell. After the battery has been fully discharged, electric current from an external source is used to drive the reverse, nonspontaneous reaction, which recharges the battery. As such, a charging battery operates as an electrolytic cell. Two common types of rechargeable batteries are discussed in the remainder of this lesson.

Lead-Acid Batteries

A lead-acid (or lead storage) battery is a type of rechargeable battery commonly used as an electrical power source for automobiles. Typically, a lead-acid battery consists of several electrochemical cells connected in series (eg, a 12 V lead-acid battery contains six 2 V cells). Each cell, when fully charged, is comprised of a grid packed with Pb(s), acting as the anode, and a grid packed with $PbO_2(s)$, acting as the cathode. Both electrodes are immersed in an aqueous solution of 4 M $H_2SO_4(aq)$, acting as the electrolyte, as illustrated in Figure 9.16.

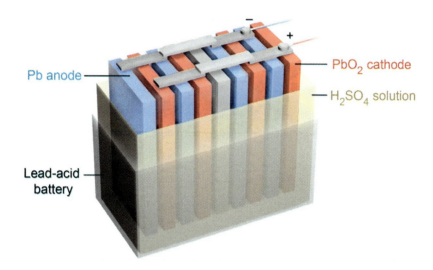

Figure 9.16 The electrochemical components of a lead-acid battery.

When a lead-acid battery is discharging, the oxidation half-reaction occurring at the Pb(s) anode (ie, the negative terminal of the battery) is:

$$Pb(s) + HSO_4^-(aq) \rightarrow PbSO_4(s) + H^+(aq) + 2\,e^- \qquad E°_{anode} = -0.356\,V$$

The reduction half-reaction occurring at the $PbO_2(s)$ cathode (ie, the positive terminal of the battery) is:

$$PbO_2(s) + HSO_4^-(aq) + 3\,H^+(aq) + 2\,e^- \rightarrow PbSO_4(s) + 2\,H_2O(l) \qquad E°_{cathode} = 1.685\,V$$

Therefore, the overall discharging reaction and $E°_{cell}$ for each cell in a lead-acid battery are:

$$Pb(s) + PbO_2(s) + 2\ HSO_4^-(aq) + 2\ H^+(aq) \rightarrow 2\ PbSO_4(s) + 2\ H_2O(l)$$

$$E°_{cell} = E°_{cathode} - E°_{anode} = 2.041\ V$$

Because the spontaneous reaction produces insoluble $PbSO_4(s)$ and water, both electrodes eventually become coated with $PbSO_4(s)$ and the acid solution becomes increasingly dilute, rendering the battery inoperable. However, the battery can be recharged by applying current from an external source, which decomposes $PbSO_4(s)$ and reconcentrates the acid solution (ie, the reverse of the discharge reaction).

Nickel-Cadmium Batteries

Another type of rechargeable battery is the nickel-cadmium (Ni-Cd, or NiCad) battery, which is commonly used in small, battery-operated devices (eg, power tools, portable medical equipment, flashlights). Similar to a lead-acid battery, a Ni-Cd battery is constructed using three layered components: a NiO(OH)(s) cathode, followed by a porous separator containing KOH(aq), and then a Cd(s) anode. All three components are rolled together and placed inside a cylindrical case, as demonstrated in Figure 9.17.

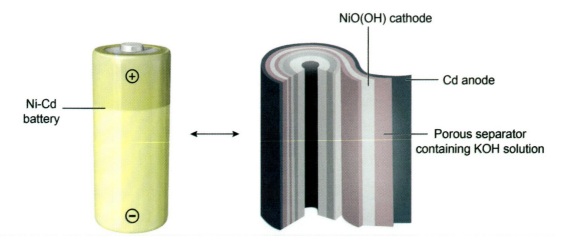

Figure 9.17 The electrochemical components of a nickel-cadmium battery.

During discharge of the Ni-Cd battery, the oxidation half-reaction occurring at the Cd(s) anode is:

$$Cd(s) + 2\ OH^-(aq) \rightarrow Cd(OH)_2(s) + 2\ e^- \qquad E°_{anode} = -0.86\ V$$

The reduction half-reaction occurring at the NiO(OH)(s) cathode is:

$$2\ NiO(OH)(s) + 2\ H_2O(l) + 2\ e^- \rightarrow 2\ Ni(OH)_2(s) + 2\ OH^-(aq) \qquad E°_{cathode} = 0.49\ V$$

Therefore, the overall discharging reaction and $E°_{cell}$ for a Ni-Cd battery are:

$$Cd(s) + 2\ NiO(OH)(s) + 2\ H_2O(l) \rightarrow Cd(OH)_2(s) + 2\ Ni(OH)_2(s)$$

$$E°_{cell} = E°_{cathode} - E°_{anode} = 1.35\ V$$

Over time, the electrodes become coated with their respective insoluble hydroxide compounds, which causes the cell voltage to drop. As discussed with lead-acid batteries, recharging the Ni-Cd battery with an external power source reverses the discharge reaction and removes the buildup of the insoluble compound, allowing the battery to be discharged again.

END-OF-UNIT MCAT PRACTICE

Congratulations on completing **Unit 4: Solutions and Electrochemistry**.

Now you are ready to dive into MCAT-level practice tests. At UWorld, we believe students will be fully prepared to ace the MCAT when they practice with high-quality questions in a realistic testing environment.

The UWorld Qbank will test you on questions that are fully representative of the AAMC MCAT syllabus. In addition, our MCAT-like questions are accompanied by in-depth explanations with exceptional visual aids that will help you better retain difficult MCAT concepts.

TO START YOUR MCAT PRACTICE, PROCEED AS FOLLOWS:

1) Sign up to purchase the UWorld MCAT Qbank
   IMPORTANT: You already have access if you purchased a bundled subscription.
2) Log in to your UWorld MCAT account
3) Access the MCAT Qbank section
4) Select this unit in the Qbank
5) Create a custom practice test

Appendix
Concept Check Solutions

You will find detailed, illustrated, step-by-step solutions for each concept check in the digital version of this book.

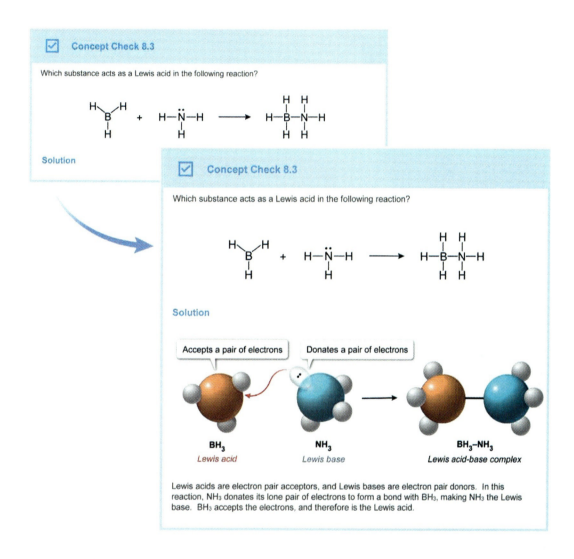

In this section of the print book, you will only find short answers to the concept checks included in each chapter. Please go online for an interactive and enhanced learning experience with visual aids.

Unit 1. Atomic Theory and Chemical Composition

Chapter 1. Structure and Properties of Atoms

Lesson 1.1

    1.1   22 neutrons and 18 electrons

    1.2   (191 amu)(0.373) + (193 amu)(0.627) = 192.2 amu

Lesson 1.2

    1.3 $^{231}_{91}Pa$

    1.4 ^{23}Al

    1.5   30 s (half-life); 663 Bq (activity)

Lesson 1.3

    1.6   [Xe]$6s^2 5d^{10}$

Lesson 1.5

    1.7   Cl

    1.8   Mg → Mg^{2+} requires the loss of both valence electrons (the entire outer shell is lost)

    1.9   N (highest) > As > In > Sr (lowest)

    1.10   barium

    1.11   K (least) < Se < S (greatest)

Lesson 1.6

    1.12   3.2 mL

    1.13   96.09 amu

    1.14   64.79%

    1.15 NO_2 (empirical formula); N_2O_4 (molecular formula)

    1.16   4.1 × 10^{21} $CaCl_2$ molecules

    1.17   32 g oxygen atoms

    1.18   0.6 M Cl^-

Unit 2. Interactions of Chemical Substances

Chapter 2. Chemical Bonding, Reactions, and Stoichiometry

Lesson 2.1

    2.1 Al_2O_3

    2.2

2.3 Structure 1 has too many total electrons and violates the octet rule for the oxygen.

Structure 2 has a net charge (the given H_3PO_4 formula is neutral).

2.4 +1 (central S atom); −1 (outer S atom); +1 (N atom); −1 (O atom with only one bond)

2.5

[Lewis structures showing two resonance forms of a thiosulfate-like species with H-O-S bonds]

Lesson 2.2

2.6 O–H bond

2.7 S–Br (lowest) < H–Cl < K–Br < K–Cl (highest)

2.8 trigonal planar

2.9 T-shaped molecular geometry

2.10 Four atoms are sp^3 hybridized (the three C atoms and one O atom, *not* in the C=O bond)

2.11 XeF_2 is nonpolar

Lesson 2.3

2.12 neopentane (weakest) < isopentane < n-pentane < n-hexane (strongest)

2.13 H–Cl

2.14 Mixture A: dipole–induced dipole interactions; Mixture B: ion-dipole interactions;

Mixture C: dipole-dipole interactions

2.15 Interactions II and IV

2.16 IV (weakest) < V < I < II < III (strongest)

2.17 Dodecane requires more energy to increase its vapor pressure because the molecule is much larger and can form a greater number of London dispersion forces.

Lesson 2.4

2.18 single replacement reaction

2.19 Carbon is oxidized and CH_4 is the reducing agent.

Oxygen is reduced and O_2 is the oxidizing agent.

2.20 Zn gets oxidized and Mn gets reduced

2.21 $FeCl_2$ and H_2S

2.22 $2\ C_2H_6(g) + 7\ O_2(g) \rightarrow 4\ CO_2(g) + 6\ H_2O(g)$

2.23 $2\ Al(s) + OH^-(aq) + NO_2^-(aq) + H_2O(l) \rightarrow 2\ AlO_2^-(aq) + NH_3(g)$

Lesson 2.5

2.24 95 g Cu(s)

2.25 331 g Ni_2O_3(s)

2.26 0.72 L of 0.050 M NaOH(aq)

2.27 Al(s) is the limiting reactant; 1.50 mol of excess $O_2(g)$

2.28 52%

Unit 3. Thermodynamics, Kinetics, and Gas Laws

Chapter 3. Thermodynamics

Lesson 3.1

3.1 1) open system; 2) isolated system; 3) closed system; 4) open system

3.2 Temperature indicates the average kinetic energy of molecules due to thermal energy but does not quantify the amount of total heat in a sample. The tub of water has a greater mass and contains a larger quantity of thermal energy overall.

3.3 295 K

Lesson 3.2

3.4 The system loses 119 J of energy to the surroundings.

Lesson 3.3

3.5 1) The entropy of the ice cubes increases.

2) The entropy of the glass of water decreases.

3) The entropy of the universe increases.

3.6 1) increased entropy; 2) increased entropy; 3) decreased entropy

3.7 1) −582.2 J/K; 2) 266.7 J/K

Lesson 3.4

3.8 Deposition releases the most energy because it involves going from the state of matter with the highest energy (gas) to the state of matter with the lowest energy (solid).

3.9 Salting a road introduces impurities into the structure of ice, which disrupts the crystal structure of the ice and lowers the melting point.

3.10 1) A pressure cooker dramatically reduces cooking time.

2) The boiling liquid produces steam which fills the chamber and increases the pressure above the liquid. The increased pressure raises the boiling point of the liquid because a higher temperature is needed for its vapor pressure to equal the higher chamber pressure.

3.11 1) 240 K at 1 atm

2) 160 K and 0.8 atm

3) The solid phase of this substance is denser than its liquid phase because the slope of the melting curve is positive (ie, increasing pressure favors the solid phase) and higher pressure always favors the denser phase of matter.

4) The normal melting and boiling points of the substance in this phase diagram are much lower than those of water. Therefore, the unknown substance is expected to have much weaker intermolecular forces than water because stronger intermolecular forces generally result in higher melting and boiling points.

3.12 −1154 kJ

Lesson 3.5

    3.13   152 kJ

    3.14   296.8 kJ

    3.15   155.7 kJ

    3.16   −57 kJ/mol

Lesson 3.6

    3.17   −45.2 kJ/mol

Lesson 3.7

    3.18   −72.6 kJ/mol

    3.19 $\Delta H°$ and $\Delta S°$ are both positive.

Chapter 4. Kinetics

Lesson 4.1

    4.1   Step 2 is the slowest, rate-determining step for the overall reaction.

Lesson 4.2

    4.2 1.4×10^{-5} M/s

    4.3   0.0060 mol $N_2O_5(g)$

    4.4   The reaction is third order overall ($m + n = 3$), and the rate law is: rate = $k[HgCl_2][C_2O_4^{2-}]^2$

    4.5   The net effect of the changes results in an increased reaction rate relative to the initial rate.

Chapter 5. Chemical Equilibrium

Lesson 5.2

    5.1 $K_{eq} = \dfrac{[CO_2(g)]^2}{[CO(g)]^2}$

Lesson 5.3

    5.2 $COCl_2(g)$

    5.3 9.6×10^{-4} M

    5.4   The reaction will must shift toward the reactants.

    5.5   The reaction is not spontaneous in the forward direction ($\Delta G > 0$).

    5.6 ΔG is negative for the forward reaction.

Lesson 5.4

    5.7   Numbers I, II, and III

    5.8   The partial pressure of $HF(g)$ will decrease.

    5.9   The reaction should be performed at low temperatures to increase the amount of $H_2(g)$.

Chapter 6. Gas Laws

Lesson 6.1

6.1 5.39 atm

6.2 64 °C

6.3 196 mL → 2.0×10^2 mL

6.4 1.0 atm

Lesson 6.2

6.5 0.192 mol He(g)

6.6 3.7 mol O_2(g)

Lesson 6.3

6.7 1.6 atm

6.8 68.3 L N_2(g)

6.9 CH_3F(g) will have the greatest deviation from ideal behavior under the stated conditions, and the volume of the CH_3F(g) sample will be is less than that predicted by the ideal gas law.

Unit 4. Solutions and Electrochemistry

Chapter 7. Solutions

Lesson 7.1

7.1 Numbers I and II

7.2 0.020 M; 22% by mass

7.3 0.30 Osmol/L; 7.4 atm

Lesson 7.2

7.4 PbI_2 is the precipitate.

7.5 A 3 M solution of $Mg(ClO_4)_2$(aq) is not possible at 20 °C because $[Mg^{2+}][ClO_4^-]^2 > K_{sp}$.

7.6 $K_{sp} = [Ag^+][Br^-] = 4.9 \times 10^{-13}$; AgCl is more soluble than AgBr.

7.7 Sample 2 (least precipitate) < Sample 1 < Sample 3 (most precipitate)

7.8 The amount of PbC_2O_4 that will dissolve can be increased by decreasing the solution pH (ie, making the solution more acidic).

7.9 Adding $AgNO_3$(aq) would cause a significant increase in the solubility of PbS_2O_3(aq) due to the formation of a complex ion by the Ag^+(aq) and $S_2O_3^{2-}$(aq) species.

Chapter 8. Acid-Base Chemistry

Lesson 8.1

8.1 NaOH and $Mg(OH)_2$ are both Arrhenius bases.

8.2 As an acid: $H_2PO_4^- + H_2O \rightleftarrows HPO_4^{2-} + H_3O^+$

As a base: $H_2PO_4^- + H_2O \rightleftarrows H_3PO_4 + OH^-$

8.3 BH_3 is the Lewis acid.

Lesson 8.2

8.4 The solution is basic with $[OH^-] = 5 \times 10^{-7}$ M.

8.5 pH = 12.32

Lesson 8.3

8.6 1.3% ionization

8.7 HCN has the smallest $[H_3O^+]$, giving it the smallest K_a and the largest pK_a.

8.8 H_3PO_4 (most abundant) > $H_2PO_4^-$ > HPO_4^{2-} > PO_4^{3-} (least abundant)

8.9 pH = 10.0

8.10 C_5H_5N (weakest base) < $HONH_2$ < ClO^- < H_2NNH_2 < $(CH_3)_3N$ (strongest base)

Lesson 8.4

8.11 HCl and NaCl

8.12 pH = 3.74

8.13 410 g of $NaC_2H_3O_2$

8.14 H_2CO_3 and HCO_3^-

Lesson 8.5

8.15 $LiClO_2$ is a basic salt.

8.16 pH = 5.12

Lesson 8.6

8.17 Not adding an indicator would result in no color change occurring regardless of the amount of NaOH titrant added (Choice C).

8.18 methyl red

8.19 The pK_a of the acid from titration Curve A (3.2) best matches the pK_a of HF (3.17).

8.20 His^0 and His$^-$

8.21 0.013 M

Chapter 9. Redox Reactions and Electrochemistry

Lesson 9.3

9.1 The $Cl^-(aq)$ ions flow toward the $Al(s)$ electrode, and the $K^+(aq)$ ions flow toward the $Pb(s)$ electrode.

9.2 $E°_{cell} = +1.52$ V

9.3 The reaction is spontaneous with $\Delta G° = -44{,}000$ J/mol.

Index

A
absolute zero, 143
absorption, 16–17, 19, 147
absorption line spectrum, 17
acid-base properties, 324–25, 327
acidic protons, 299–300, 312–13, 333, 346, 348, 350
acid ionization constant, 310–12, 315–16, 325–27, 329, 333
acids
  polyprotic, 312–13, 325, 339, 345–47, 349
  strong, 307–8, 310–11, 317–18, 324, 339–44
  weak, 307–8, 310–11, 313–15, 317–22, 324, 327, 335, 339–42, 345
activation energy, 200–201, 207–10, 221–23, 237
activity, 12–14, 231, 380
adsorption, 225
alkali metals, 34, 117
alkaline-earth metals, 34, 117
alpha decay, 9–10
amphiprotic, 325, 329
amphoteric, 299, 303
analyte, 333, 335–36, 339, 349–50, 355, 357
anions, 25, 35, 40, 42, 64, 77–78, 85, 112, 119, 323, 326
anode, 359, 362–66, 370–71, 374–76
antineutrinos, 10–11
Arrhenius
  acid, 297–98
  base, 298–99, 384
Arrhenius equation, 221–22
atomic number, 4, 6, 9–11, 23, 40
atomic radii, 38–40, 45, 66–67, 69
atomic weight, 7
Aufbau principle, 23–25
autoionization of water, 303–4, 306, 314–15, 327
average kinetic energy, 142–43, 147, 165, 220, 251, 253, 265, 382
average rate, 213, 215–17
Avogadro's law, 138, 260–61, 265
Avogadro's number, 52–53, 126–27, 129

B
base ionization constant, 314–15, 320, 325–27, 329, 331, 333
bases
  strong, 308, 313, 317–18, 339–43, 345–48
  weak, 293, 308, 313–14, 316–17, 320, 335, 339, 342–44
battery, 276, 353, 363, 371, 373, 375–76
  lead-acid, 375–76
  nickel-cadmium, 376
beta decay, 9–12
bimolecular processes, 205
Bohr model, 15–19
boiling point, 106–8, 110, 157, 169–73, 175, 278, 382
Boltzmann's constant, 156
bond angles, 87, 89, 91
bond dipole, 79, 84, 96
bond dissociation energy, 70–71, 190
bond enthalpies, 190, 192–94
bond length, 66, 69–71, 79, 84
bond order, 69–71, 79
bonds
  double, 67–68, 71, 79, 88
  single, 71, 74, 79–80, 88
  triple, 67–68, 71, 74, 79–80, 88, 93
Boyle's law, 138, 252–53, 256, 258, 265
Brønsted-Lowry acid, 297, 299
Brønsted-Lowry base, 299
buffer, 317–18, 320–21

C
calorimeter, 196–97
catalysts, 111, 138, 201, 205, 222–25, 237–38
cathode, 359, 362–66, 370–71, 374–76
cations, 26, 34, 40, 44–45, 64, 72–74, 112, 119, 323–24, 326, 370
chalcogens, 34
Charles's law, 138, 252, 255–56, 258, 265
chelate, 73
chemical formula, 50–51, 53, 63, 74–75, 111, 117, 121
chemical reactions, 111, 113, 118, 120–21, 125–26, 128–31, 158, 183–84, 186, 195–96, 205, 207, 298–300

colligative properties, 278
collisions
  bimolecular, 206
  productive, 220, 222
  single molecular, 206
  termolecular, 206
combination reactions, 111–12, 114, 118
combined gas law, 138, 252, 256, 258
combustion reactions, 113
common-ion effect, 290–92
complex ions, 73, 294, 324, 384
compressibility, 269–70
concentration cells, 276, 369–70
condensation, 162–63, 169
conduction, 144–45
conjugate acid, 299–300, 309, 317, 323, 343
conjugate acid-base pair, 299, 317–19, 321–22, 324
conjugate base, 299–300, 309, 314–15, 317–21, 323, 325, 340–41
conjugation, 80
conservation of energy, 151
convection, 144, 146
conversion factors, 49, 53–55, 126–29, 185, 217, 267, 281
coordinate covalent bonds, 72–73, 114, 301
coordination number, 72–73
core electrons, 24, 37–39, 61–62
Coulomb's law, 9
covalent bonds, 63, 65–68, 70, 72, 74, 76–77, 83, 92, 99
critical pressure, 174
critical temperature, 174

D

Dalton's law, 138, 261–62
decomposition, 119, 186, 203, 205, 213–16, 227, 233–34
decomposition reactions, 112, 119, 128, 224
definite proportions, 51, 125–27
density, 6, 48–50, 128–30, 146, 175, 177, 266, 281, 286
denticity, 73
deposition, 162, 164, 382
desorption, 225
diamagnetic, 27
diatomic molecules, 32–33, 35, 66, 96
diffusion, 278, 282
diluting, 245

dipole-dipole interactions, 99, 101, 103, 108, 381
dipoles, 86, 94, 99, 101–2, 381
disproportionation reaction, 115
dissolution, 279–80, 288, 290–91, 296
double replacement reactions, 112–14, 119–20, 287
dynamic equilibrium, 106, 138, 227–29, 232–33, 237, 243, 310, 313

E

E_a. See activation energy
effective nuclear charge, 37–42, 45
electrochemical cells, 276, 359–73, 375
electrodes, 276, 356, 359–60, 362–65, 369–71, 374–76, 385
electrolysis, 276, 371, 373–74
electrolytes, 65, 67, 359–60, 364–65, 369, 373, 375
electrolytic cells, 276, 353, 369, 371, 373–75
electromagnetic spectrum, 16
electromotive force (EMF), 367
electron affinity, 42–44
electron capture, 10–11
electron configuration, 6, 23–26, 34, 61–62
electron density, 38, 116, 325
electron domains, 87–89, 91, 93
electronegativity, 41, 63, 65, 83–85, 94, 117
electronegativity difference, 84–86
electron geometry, 87–89, 91
electronic configuration, 29, 32, 37
electron shielding, 37, 43
electroplating, 371, 373
electrostatic force, 45
elemental state, 4–5, 23, 25, 40, 66, 112
elementary reaction steps, 208, 219, 223, 232
EMF. See electromotive force
emission, 9–11, 16–18, 147
emission line spectrum, 17, 19
empirical formula, 51, 380
endergonic, 200, 203
endothermic, 42, 184–85, 191, 199, 202–3, 248–49, 279, 290, 295
enthalpy, 140, 154, 159, 183–92, 194–95, 197, 199, 202, 279
enthalpy of dissolution, 279
entropy, 140, 155–59, 161, 199, 202–3, 280, 295, 382
enzymes, 225

equilibrium, 164–65, 169, 172–74, 231–36, 238, 240–41, 243–49, 285, 288, 290–91, 293–94, 296, 307, 310–12, 314
equilibrium constant, 138, 227, 231–38, 240–45, 248–49, 285, 307, 310, 314
equivalence point, 334, 336, 339–43, 346, 349, 355–57
evaporation, 106, 245
excited state, 12, 18
exergonic, 200–201, 203
exothermic, 42, 44, 184–85, 197, 202–3, 248–49, 279, 290, 295

F
Faraday's constant, 367, 374–75
first ionization energy, 44–45
first law of thermodynamics, 151, 153, 196
formal charges, 75–77, 80, 116
formation constant, 294
formula unit, 53, 55, 64–65, 73, 289
freezing point, 166–67, 278
fusion, 164–66, 180

G
galvanic cells, 276, 369–71, 375
gamma decay, 12
gamma emission, 9, 12
Gay-Lussac's law, 252, 254, 256, 258, 265
Gibbs free energy, 138, 199, 202, 238, 276, 367
ground state, 18
Guy-Lussac's law, 138, 253

H
half-equivalence point, 341
half-life, 12
half-reactions, 123, 354, 363, 366–67, 370, 374–75
halogens, 35, 66, 117
heat capacity, 178, 195–97
heating curve, 177–80
heat of fusion, 165
heat of vaporization, 169
heats of formation, 188–90, 194
heat transfer, 142, 144–48, 153–54, 157, 179–80, 183–84, 195–96
Heisenberg uncertainty principle, 19–20
Henderson-Hasselbalch equation, 318–20, 341
Henry's law, 295–96
Hess's law, 186–88, 191
heterogeneous equilibria, 229

homogeneous equilibria, 229
Hund's rule, 26–27
hybrid orbitals, 92–93
hydrogen bonding, 103–4, 109–10
hydrophilic, 109–10
hydrophobic, 109–10
hypervalent species, 63

I
ideal gas, 138, 253, 265–71, 384
indicators, 289, 333–38, 385
   acid-base, 336, 355
   redox, 355
inert gases, 247
instantaneous rate, 213
intermediates, 205, 223
intermolecular forces, 99–102, 104–8, 110, 161–64, 166, 170–71, 278
internal energy, 140, 151, 153–54, 159, 183
International System of Units, 47
ion-dipole interactions, 102, 105, 381
ionic bond, 63–64, 73, 83, 85, 116
ionic character, 85–86
ionic compounds, 65, 73, 77, 119, 287, 290–91, 294
ionic radii, 40, 65, 325
ionization, 307, 310–13, 385
ionization energy, 44–45
ionize, 6, 307–8, 310, 313
isoelectronic, 40
isotopes, 6–8, 12, 50

K
K_a. See acid ionization constant
K_b. See base ionization constant
K_{eq}. See equilibrium constant
kinetic control, 209–10
kinetic energy, 143, 145, 147, 163–64, 166–67, 169, 207, 209, 221, 295
kinetic molecular theory of gases, 251–52, 256, 265
K_{sp}. See solubility product constant
K_w. See autoionization of water

L
latent heat, 165, 169, 178
law of conservation of mass, 120, 122
law of definite proportions, 51, 125
law of mass action, 138, 231–33
Le Châtelier's principle, 138, 243, 248, 285, 290, 292, 294–95
Lewis acids, 72, 297, 300–301, 324, 384

Lewis bases, 72, 301
Lewis structure, 74–77, 79, 81, 88–89, 91, 93, 116
Lewis symbols, 61–62
ligands, 72–73, 294, 301, 324–25
limiting reactant, 130–31, 382
line spectrum, 17
London dispersion forces, 99–101, 108, 110, 381

M

mass action ratio, 231, 233, 235, 243
mass number, 4–6, 9–10
mass ratio, 48, 50–51, 266, 286
mass spectrum, 6–8
Maxwell-Boltzmann distribution, 207
melting curve, 174–75, 382
melting point, 164–68, 174–75, 181, 382
metalloids, 30–32
metals, 30–32, 34, 40, 45, 64, 85, 117, 359–60, 369
microstates, 156, 158–59
molar concentration, 54–55, 213, 217, 231–32, 283, 288, 303–5, 350. *See also* molarity
molarity, 49, 54–55, 128–29, 217, 281–83, 305, 333, 350
molar mass, 52–53, 126–29, 266, 282, 286, 374–75
mole, 52–55, 125–28, 130–31, 178, 184–85, 195, 260–63, 266, 350, 374
   fraction, 262
   ratios, 53–55, 126–27, 130, 235, 319
molecular formula, 51–52, 121, 380
molecular geometry, 87, 89–91, 94
molecular weight, 50, 108, 171
monatomic ions, 64, 77, 117

N

natural abundance, 6–8
neutralization, 298, 313, 323, 325–26, 339–40, 342, 345
neutrinos, 10–11
neutrons, 4, 6–7, 9–12, 15, 380
noble gas configuration, 62
noble gases, 25, 35, 43, 45, 61–62
nonmetals, 30–32, 34, 40, 64–65, 85
nonpolar covalent bonds, 83–84, 94–97, 116, 381
nonpolar molecules, 96, 100–101, 109
nuclear charge, 37–38
nuclear decay, 9, 11

nucleus, 4, 6, 9, 11–12, 15, 20–23, 25, 37–43, 45, 69, 84

O

octet rule, 61–68, 74, 79–80, 381
orbitals, 20–22, 26–27, 34–35, 66, 68–70, 80, 92–93, 99
osmolarity, 283
osmosis, 278, 282
osmotic pressure, 278, 282–83
oxidation, 114–16, 119, 123, 353, 359, 361–63, 365–66, 371. *See also* oxidation half-reaction
oxidation half-reaction, 123, 364, 366, 375–76
oxidation numbers, 115–18, 353–54
oxidation-reduction reactions. *See* redox reactions
oxidation states, 116–19, 355
oxidizing agent, 114–15, 353, 381
oxyanions, 78–79

P

paramagnetic, 27
partial charges, 84, 96, 102, 109, 279
partial pressures, 138, 159, 199, 231, 238, 242, 244, 246–47, 261–62, 296
path functions, 140–41, 154
Pauli exclusion principle, 22
Pauling scale, 41
period, 29–30, 37–43, 45, 63, 74, 233
periodic table, 6–7, 23–25, 29–32, 34, 37–39, 43, 50, 61–62, 74, 76, 80
periodic trends
   atomic radii, 39
   effective nuclear charge, 38
   electron affinity, 44
   electronegativity, 41
   ionic radii, 40
phase changes, 157, 161–62, 164, 166, 170, 177–78, 199
phase diagrams, 172–77, 382
photons, 15–18, 147
Planck's constant, 16, 19
polarity, 83–85, 96
   bonds, 83–85, 94, 96, 101
   molecular, 83, 87, 94–96, 100–102, 105
polyatomic ions, 77–78, 117, 121
positron, 10–11
precipitation, 112–14, 118–19, 220, 285, 287–88, 291–92, 384

probability distributions, 20–21, 99. *See also* orbitals
pure substances, 3

Q
quantum numbers, 20–22
  angular momentum, 21
  electron spin, 21–22, 27
  magnetic, 21
  principal, 21, 23–25, 61

R
radiation, 9, 144, 147
radioactive decay, 9, 11, 13–14
Raoult's law, 171
rate constant, 220–22, 232
rate-determining step, 207–9, 217, 219, 383
rate law, 213, 217–20, 383
reaction energy diagram, 199, 201, 207–8, 211
reaction order, 217, 219, 232
reaction quotient, 138, 233, 236–38, 241, 243
reaction rate, 201, 207, 213–14, 218, 220–22
redox reactions, 113–16, 118–19, 122–23, 276, 353–54, 359–66, 368–71, 373
redox titrations, 276, 353, 355–56
reducing agent, 114–15, 353, 355, 365, 381
reduction, 14, 114–16, 119, 123, 353, 359, 361–66, 370–71. *See also* reduction half-reactions
reduction half-reactions, 123, 354, 366, 375–76
representative elements, 32–33, 61–64
resonance structures, 79–81, 84, 94, 96

S
salt bridge, 359–61
second law of thermodynamics, 155, 157, 199
shielding constant, 37
single replacement reactions, 112, 114, 119, 381
solid phase, 166, 168, 174–75, 177, 382
solubility, 278, 285–89, 291–96, 384
solubility limit, 286, 288
solubility product constant, 288–90, 294, 384
solubility table, 286
solute, 54–55, 129, 171, 195, 277–83, 285–86, 290, 295, 350

solvent, 277–78, 285
spectroscopy, 147
spontaneity, 199, 238, 240, 359
standard cell potential, 366
standard reduction potentials, 364, 366–67
standard state conditions, 159, 240, 364
state functions, 139–41, 154, 157, 159, 186–87, 201
stoichiometry, 53, 61–134, 183, 214–15, 217, 219, 281, 288, 291, 374, 380
STP (standard temperature and pressure), 138, 266–68
sublimation, 162, 173
subshell, 15, 21–23, 25, 63
supercritical fluid, 174
surface catalysts, 225

T
temperature
  absolute, 251, 253–56, 266
  Kelvin, 48, 143–44, 147, 157, 199, 202, 238, 251, 253–55, 258, 283
theoretical yield, 130–31
thermal energy, 142, 153, 178, 199, 295, 382
thermal equilibrium, 139, 147–48, 196
thermodynamic control, 210
titration, 333–38, 340, 345–49, 355, 357
titration curves, 334, 339–46, 356–57, 385
transition metals, 32, 34, 63
transition state, 201, 207–8, 224–26
triple point, 173–74, 177

U
unimolecular processes, 205–6

V
valence electrons, 24, 29–30, 33–34, 37–38, 41–42, 45, 61–66, 74, 76, 85, 116
valence-shell electron-pair repulsion. *See* VSEPR theory
valence shells, 25, 34–35, 43, 62–63, 65–68, 74, 93
van der Waals equation, 270
van 't Hoff factor, 283
vaporization, 161–62, 169, 180
vapor pressure, 106–7, 169, 171–73, 175, 278, 381–82
vapor pressure curves, 173, 175–76
voltage, 355, 364
VSEPR theory (valence-shell electron-pair repulsion), 87

W
weighted averages, 7, 79
work, 139–41, 153–54, 157, 199, 305, 367

Z
Z_{eff}. *See* effective nuclear charge
zeroth law of thermodynamics, 147–49